IDENTITIES

Reciprocal

$$\sin\theta = \frac{1}{\csc\theta}$$

$$\cos\theta = \frac{1}{\sec\theta}$$

$$\tan\theta = \frac{1}{\cot\theta}$$

Ratio

$$\tan\theta = \frac{\sin\theta}{\cos\theta}$$

$$\cot\theta = \frac{\cos\theta}{\sin\theta}$$

Pythagorean

$$\cos^2\theta + \sin^2\theta = 1$$

$$\tan^2\theta + 1 = \sec^2\theta$$

$$1 + \cot^2\theta = \csc^2\theta$$

Sum and Difference

$$\sin(A + B) = \sin A\cos B + \sin B\cos A$$

$$\sin(A - B) = \sin A\cos B - \sin B\cos A$$

$$\cos(A + B) = \cos A\cos B - \sin A\sin B$$

$$\cos(A - B) = \cos A\cos B + \sin A\sin B$$

Multiple Angle

$$\sin 2A = 2\sin A\cos A$$

$$\cos 2A = \cos^2 A - \sin^2 A = 2\cos^2 A - 1 = 1 - 2\sin^2 A$$

$$\tan 2A = \frac{2\tan A}{1 - \tan^2 A}$$

$$\sin 3A = 3\sin A - 4\sin^3 A$$

$$\cos 3A = 4\cos^3 A - 3\cos A$$

Half Angle

$$\sin\frac{\theta}{2} = \pm\sqrt{\frac{1 - \cos\theta}{2}}$$

$$\cos\frac{\theta}{2} = \pm\sqrt{\frac{1 + \cos\theta}{2}}$$

$$\tan\frac{\theta}{2} = \frac{1 - \cos\theta}{\sin\theta} = \frac{\sin\theta}{1 + \cos\theta}$$

Other

$$\sin(-\theta) = -\sin\theta \qquad \sin\left(\theta + \frac{\pi}{2}\right) = \cos\theta \qquad \sin\left(\frac{\pi}{2} - \theta\right) = \cos\theta$$

$$\cos(-\theta) = \cos\theta \qquad \cos\left(\theta - \frac{\pi}{2}\right) = \sin\theta \qquad \cos\left(\frac{\pi}{2} - \theta\right) = \sin\theta$$

$$\tan(-\theta) = -\tan\theta \qquad \tan\left(\theta - \frac{\pi}{2}\right) = -\cot\theta \qquad \tan\left(\frac{\pi}{2} - \theta\right) = \cot\theta$$

Inverse Functions

$y = \mathrm{Sin}^{-1}x$ or $y = \mathrm{Arcsin}\,x$ means $x = \sin y$ for $-\frac{\pi}{2} \le y \le \frac{\pi}{2}$

$y = \mathrm{Cos}^{-1}x$ or $y = \mathrm{Arccos}\,x$ means $x = \cos y$ for $0 \le y \le \pi$

$y = \mathrm{Tan}^{-1}x$ or $y = \mathrm{Arctan}\,x$ means $x = \tan y$ for $-\frac{\pi}{2} \le y \le \frac{\pi}{2}$

TRIGONOMETRY
with applications

PATRICK J. BOYLE
Santa Rosa Junior College

TRIGONOMETRY
with applications

1817

HARPER & ROW, PUBLISHERS, New York
Cambridge, Philadelphia, San Francisco,
London, Mexico City, São Paulo, Sydney

Sponsoring Editor: Fred Henry
Project Editor: Eleanor Castellano
Designer: Dawn L. Stanley
Production Assistant: Jacqui Brownstein
Compositor: Progressive Typographers
Printer and Binder: R. R. Donnelley & Sons Company
Art Studio: Fine Line Illustrations

Trigonometry with Applications

Library of Congress Cataloging in Publication Data

Boyle, Patrick J., 1935–
Trigonometry with applications.

Includes index.
1. Trigonometry. I. Title.
QA531.B794 1983 516.2'4 82-15658
ISBN 0-06-040898-7

for JERRY and ROSE CIRNER
in appreciation for their
love and friendship

contents

six

complex numbers and polar coordinates, 234

appendix A

functions and graphs, 277

appendix B

trigonometric tables, 323

appendix C

answers to odd-numbered exercises, 333

index, 367

preface

Trigonometry with Applications covers all of the standard topics of a college trigonometry course offered in one semester or one quarter. It is assumed that the student has completed 1 to 2 years of high-school algebra and a course in geometry. The book outlines a course that should provide a solid foundation for students continuing their study of mathematics or for those applying trigonometry in a technical area.

Many features are included to provide the student with the means to make the most of classroom instruction. There are more than 300 completely worked-out examples. Each example is keyed directly to the Exercise Set so that there is a model for nearly every problem the student is asked to solve. There is an abundance of exercises: more than 1800, graded by difficulty into A, B, and C Exercises. The C Exercises provide the instructor with the means to extend an idea or a topic, or to challenge the better student, or to motivate the disinterested with unusual or recreational exercise formats. The exercises include more than 200 applied problems chosen from a variety of fields. This will provide strong appeal to the range of majors now found in a trigonometry course and will help students relate the work to their own discipline.

The trigonometric functions and their graphs are introduced as early as possible, and great flexibility is possible in the early sequence of chapters to achieve an order that is comfortable for the instructor. Complex numbers, DeMoivre's theorem, and polar coordinates are treated in a separate chapter that may be optional material in some curricula. Other optional material has, when possible, been left to the last section of a chapter, where it is easier to omit. Appendixes review algebra topics.

Answers to odd-numbered problems are included in Appendix C. Where approximate numerical results are necessary, they are given with eight digits of calculator accuracy and a rounded answer. Rules and conventions of rounding off are left as ground rules for each instructor to establish. The convention followed here attempts simply to support common sense. Each chapter concludes with a chapter review. All answers are given for the review so that it may be used as a self-test.

Much is done to contradict the stereotype of a dull, dry, difficult mathematics book, including applications, illustrations, historical asides, unusual problem formats, and other motivational commentary. The many applied problems are set up with realistic data. This is possible only because a suitable inexpensive means of performing such computation is available. The hand-held calculator and its operation are discussed so that a healthy attitude will be formed in addition to efficient and effective use. At least as much attention is given to using a calculator correctly as was formerly given to calculation with logarithms. Table look-up becomes button-pressing, and in both cases there is the danger of trigonometry getting lost in the swarm of numbers.

The use of the various features of a calculator is discussed as the need arises. Examples contain key schemes for the computation of results with a calculator, and answers are given with calculator accuracy where appropriate. Much work is done with exact results lest certain understandings are lost. The calculator model used should be a scientific model with a memory. A scientific model should have scientific notation, logarithmic and exponential functions, the trigonometric functions and their inverses. There will undoubtedly be a wide variety of brands and models within the average classroom. All examples assume an algebraic logic. Those students with RPN (reverse Polish notation) or arithmetic logic will have to make certain adjustments. Algebraic logic incorporates the order-of-operations convention from algebra. Arithmetic logic performs all operations serially as entered, and must have parentheses to properly order operations. A simple check can be carried out by pressing:

$$\boxed{2}\ \boxed{+}\ \boxed{3}\ \boxed{\times}\ \boxed{4}\ \boxed{=}$$

If the answer displayed is 14, the model is algebraic. The model is arithmetic if the answer is given as 20. Students with an arithmetic model will need to be aware that parentheses are necessary in many calculations shown in the text.

The author has many acknowledgments, but first I must thank my family for their help and encouragement because their support was essential. I would like to express my appreciation to Charles P. Benner, University of Houston; Professor C. Kenneth Bradshaw, San Jose State University; Dr. Claude V. Duplissey, University of Arkansas at Little Rock; Herbert Gindler, San Diego State University; Dr. Rosalie M. O'Mahony, College of San Mateo; Richard Plagge, Highline Community College; Warren Rund, Allan Hancock College; Terry Shell, Santa Rosa Junior College; Neil Stahl, University of Wisconsin Center-Fox Valley; James Streeter, Clackamas Community College; and Dr. J. Terry Wilson, San Jacinto College for their thoughtful reviews and constructive criticism. I wish also to thank the Harper & Row editorial staff for their patient help, especially project editor Eleanor Castellano; Gloria Langer for her careful check of all the answers and solutions; and certainly my students for enduring some of the early forms of the final product.

Patrick J. Boyle

the trigonometric ratios

one

1.1 TRIANGLES AND DISTANCE

The word **trigonometry** comes from the Greek words *trigonon* and *metron,* for triangle and measure. The study of trigonometry, however, is not limited to measuring triangles. A great deal of trigonometry deals with the concepts of distance and area. Thus, we begin by reviewing these notions, starting with a very basic relationship: the *Pythagorean theorem.*

PYTHAGOREAN THEOREM

If a, b, and c are the sides of a right triangle, then

$$\mathbf{a^2 + b^2 = c^2}$$

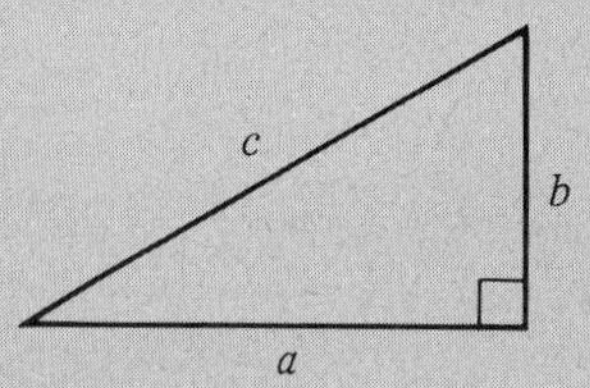

The sum of the squares of the sides of a right triangle is equal to the square of the hypotenuse (the side opposite the right angle).

EXAMPLE

1. If two sides of a triangle are $\sqrt{3}$ and $\sqrt{6}$, what is the length of the hypotenuse?

$$a^2 + b^2 = c^2$$

$$(\sqrt{3})^2 + (\sqrt{6})^2 = c^2$$

$$3 + 6 = c^2$$

$$9 = c^2$$

$$3 = c$$

The hypotenuse is 3 units.

The Pythagorean theorem states that given a right triangle, a certain relationship exists between the sides. It is also true that given the sides of the triangle, if the relationship holds, then the triangle is a right triangle. That is, if you know the sides of a triangle, it can be determined if the triangle is a right triangle.

CONVERSE OF THE PYTHAGOREAN THEOREM
If a, b, and c are the sides of a triangle, and if $a^2 + b^2 = c^2$, the triangle is a right triangle.

EXAMPLE

2. Determine if a triangle with sides 3, 4, and $\sqrt{7}$ is a right triangle. Notice that $3 < \sqrt{7} < 4$.

$$a^2 + b^2 = c^2$$
$$3^2 + (\sqrt{7})^2 = 4^2$$
$$9 + 7 = 16$$

The triangle is a right triangle.

The Pythagorean theorem is a workhorse result. It will be cited many times. We shall apply it immediately to the problem of distance. Suppose that we want to find the distance between two points P_1 and P_2 with coordinates (x_1, y_1) and (x_2, y_2). If the two points are connected with a line segment and horizontal and vertical segments are drawn, a right triangle is formed as shown in Figure 1.1. The lengths of the horizontal and vertical sides of the triangle are seen to be the differences of the coordinates of the two points. The third side can be found by the Pythagorean theorem.

$$d^2 = |x_2 - x_1|^2 + |y_2 - y_1|^2$$
$$d = \sqrt{(x_2 - x_1)^2 + (y_2 - y_1)^2}$$

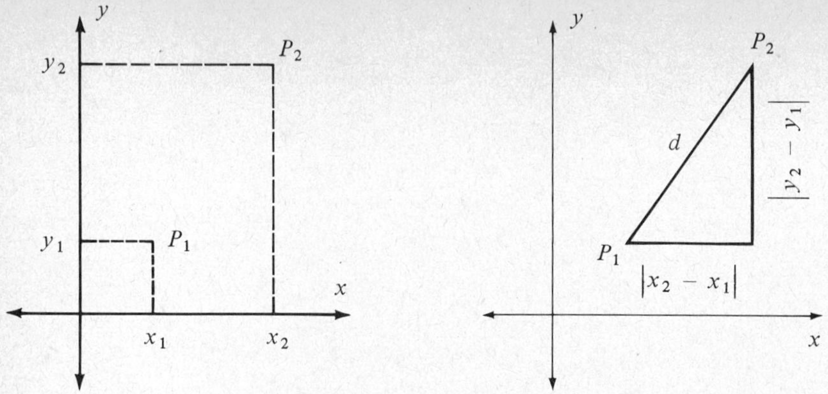

Figure 1.1

Notice the use of absolute values. It is necessary to use the absolute value because the lengths are positive.

DISTANCE FORMULA

The distance between (x_1, y_1) and (x_2, y_2) is given by

$$d = \sqrt{(x_2 - x_1)^2 + (y_2 - y_1)^2}$$

EXAMPLES

Find the distance between the given points.

3. $(-5, 3)$ and $(7, -2)$

Let $(x_1, y_1) = (-5, 3)$ and $(x_2, y_2) = (7, -2)$. Then

$$d = \sqrt{(x_2 - x_1)^2 + (y_2 - y_1)^2}$$

becomes

$$d = \sqrt{[7 - (-5)]^2 + (-2 - 3)^2}$$
$$= \sqrt{144 + 25} = \sqrt{169}$$
$$\mathbf{d = 13}$$

4. $(-3, 1)$ and $(-2, -3)$

$$d = \sqrt{[-2 - (-3)]^2 + (-3 - 1)^2}$$
$$= \sqrt{1 + 16}$$
$$= \sqrt{17}$$

$d \approx$ **4.123** (to the nearest thousandth)

Note that $\sqrt{17}$ is *approximately* 4.123. The symbol $\approx$ means "is approximately equal to." Use a calculator or tables to compute this approximation.

Angles and Triangles

A point divides a line into two parts. The end point together with either part is called a half-line or ray. An **angle** is a geometric figure formed by rotating a ray about its end point from an initial position to a new or terminal position. The two positions are called the **sides** of the angle, and the common point is called the **vertex** of the angle. The measure of the angle is the amount of rotation from the **initial side** to the **terminal side.** The measure is positive if the rotation is counterclockwise and negative if it is clockwise.

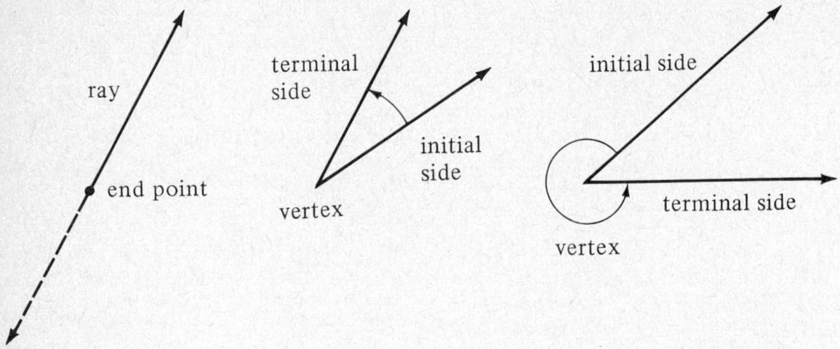

Greek Alphabet

A	α	Alpha
B	β	Beta
Γ	γ	Gamma
Δ	δ	Delta
E	ϵ	Epsilon
Z	ζ	Zeta
H	η	Eta
Θ	θ	Theta
I	ι	Iota
K	κ	Kappa
Λ	λ	Lambda
M	μ	Mu
N	ν	Nu
Ξ	ξ	Xi
O	o	Omicron
Π	π	Pi
P	ρ	Rho
Σ	σ	Sigma
T	τ	Tau
Υ	υ	Upsilon
Φ	ϕ	Phi
X	χ	Chi
Ψ	ψ	Psi
Ω	ω	Omega

Notice that the order of the letters does not normally make a difference. Some people, however, prefer the following convention:

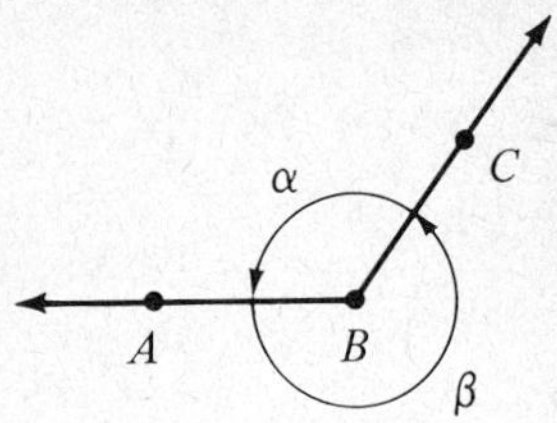

where $\alpha = \angle CBA$ and $\beta = \angle ABC$.

Angles are designated or named in a variety of ways. Greek letters $\alpha, \beta, \gamma, \ldots$ are often used; see the alphabet in the margin. If points are designated on the sides of the angle, then $\angle ABC$ and $\angle B$ are other names for θ. The middle letter in $\angle ABC$ is the vertex, and the other two are points on the sides of the angle.

In a triangle, the vertex is commonly used to name that angle, as $\angle A$, $\angle B$, and $\angle C$ in Figure 1.2, or the corresponding Greek letter is used: $\angle A = \alpha$, $\angle B = \beta$, and $\angle C = \gamma$. Three letters can be used as in $\angle A = \angle BAC$ and must be used if there is more than one angle at that vertex. For instance, in Figure 1.3, notice that $\angle S$ is ambiguous, because it could refer to $\theta = \angle PST$, $\lambda = \angle TSR$, or $\angle PSR$. Where many angles

Figure 1.2 DESIGNATION OF ANGLES.

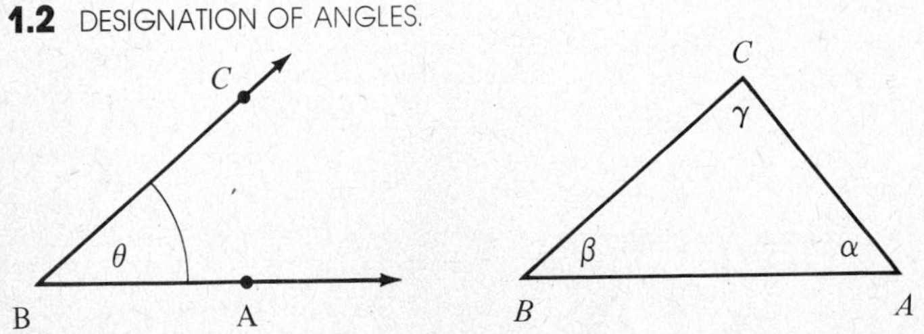

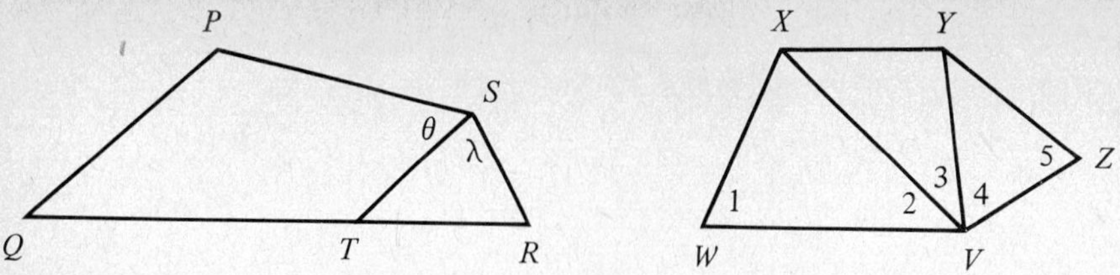

Figure 1.3

are to be named, they are usually numbered, as $\angle 1 = \angle W$ and $\angle 2 = \angle WVX$ in Figure 1.3.

EXAMPLES

Name each angle in at least one other way.

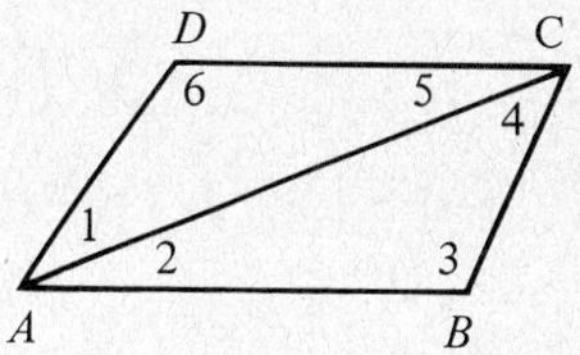

5. $\angle 1$

   $\angle 1 = \angle CAD$

   or $\angle DAC$

6. $\angle 3$

   $\angle 3 = \angle B$

   or $\angle ABC$

   or $\angle CBA$

7. $\angle ACB$

   $\angle ACB = \angle 4$

   or $\angle BCA$

8. $\angle CDA$

   $\angle CDA = \angle 6$

   or $\angle ADC$

The most common unit of measure for angles is the **degree.** Degree measure assigns 360 degrees (360°) to one complete rotation of a ray. One degree (1°) thus represents $\frac{1}{360}$ of a rotation, 180° is $180 \cdot \frac{1}{360} = \frac{1}{2}$ of a rotation, 90° represents $\frac{1}{4}$ of a rotation, and so forth. A degree is divided into 60 equal parts, called **minutes,** and a minute is divided into 60 **seconds.** An angle of 43°17′20″ is 43 degrees 17 minutes 20 seconds. An angle less than 90° is called an *acute* angle, and one greater than 90° but less than 180° is called *obtuse*.

Angles with the same initial and terminal sides are called **coterminal.** There are always infinitely many angles coterminal with a given angle, because a ray may be rotated more than one rotation in either direction. If an angle has a measure of x degrees, then for any integer k, $(x + 360k)$ degrees is always the measure of a coterminal angle. Notice

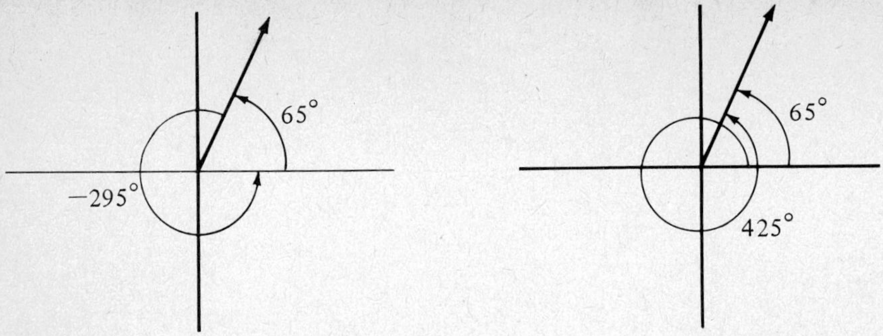

Figure 1.4

that if $x = 65$, as in Figure 1.4, then $k = 0$ yields 65°, $k = 1$ yields 425°, and $k = -1$ yields $-295°$. The smallest positive measure of an angle is generally used.

EXAMPLES

Determine the degree measure of each angle described.

9. $\frac{2}{9}$ of one complete rotation

$$\tfrac{2}{9} \cdot 360° = 80°$$

10. The sweep of the minute hand of a clock during a 24-min time interval. Remember that there are 60 min (time) in one rotation of the minute hand.

$$\tfrac{24}{60} = \tfrac{2}{5} \qquad \tfrac{2}{5} \cdot 360° = 144°$$

11. The smallest positive and negative angles coterminal with a 407° angle

$$(407 - 360)° = 47°$$
$$(47 - 360)° = -313°$$

Angle Theorems

A number of angle relationships are very useful and worth reviewing here. Two angles whose sum is 180° are called **supplementary. Complementary** angles have a sum of 90°. See Figure 1.5. The sum of the interior angles of a triangle is 180°, and an exterior angle of a triangle is equal to the sum of the two remote interior angles.

Two parallel lines L_1 and L_2 are shown in Figure 1.6. They are intersected by a third line T, called a transversal. A number of pairs of equal angles result. $\angle 1 = \angle 4$ and $\angle 2 = \angle 3$ are called **vertical angles.**

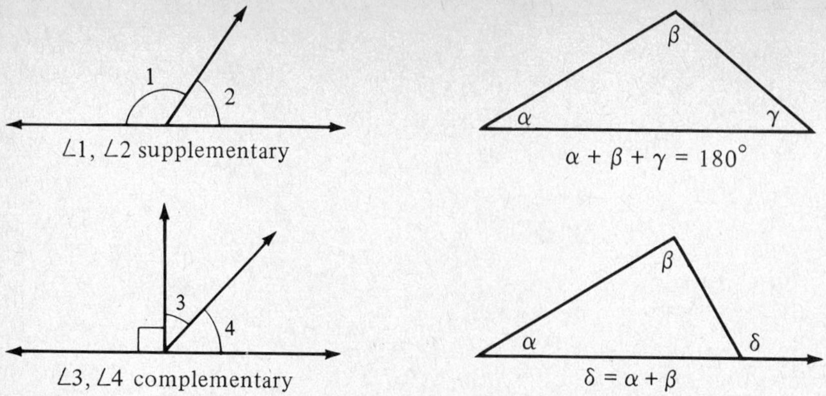

Figure 1.5 ANGLES AND TRIANGLES.

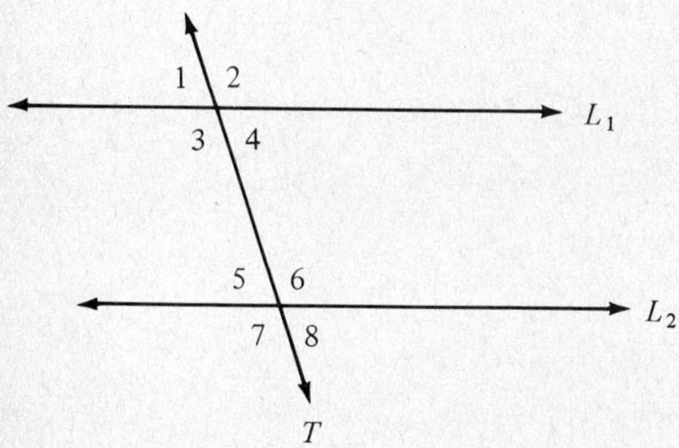

Figure 1.6 PARALLEL LINES CUT BY A TRANSVERSAL.

Line L_1 is parallel to line L_2, usually written as $L_1 \parallel L_2$, cut by transversal T.

$\angle 1 = \angle 5$, $\angle 2 = \angle 6$, $\angle 3 = \angle 7$, and $\angle 4 = \angle 8$ are pairs of **corresponding angles.** $\angle 1 = \angle 8$ and $\angle 2 = \angle 7$ are pairs of **alternate exterior angles. Alternate interior angles** are also equal: $\angle 3 = \angle 6$ and $\angle 4 = \angle 5$.

EXAMPLES

If $AB \parallel CD$, $\angle 1 = 80°$, and $\angle 2 = 35°$, find each of the following angles.

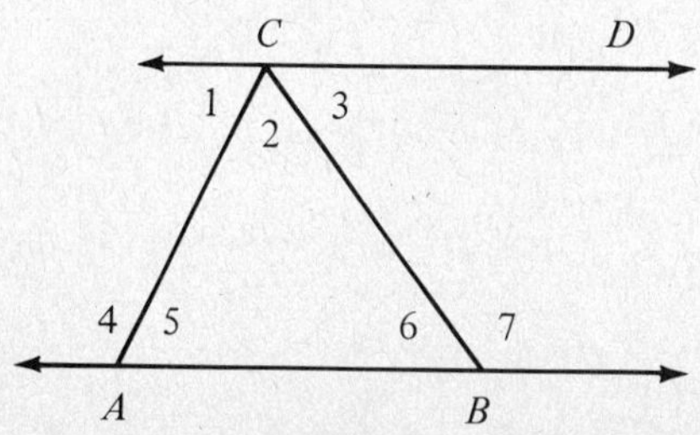

12. $\angle 3$

$(\angle 1 + \angle 2) + \angle 3 = 180°$ Supplementary angles

$(80° + 35°) + \angle 3 = 180°$

$115° + \angle 3 = 180°$

$\mathbf{\angle 3 = 65°}$

13. $\angle 5$

$\angle 1 = 80°$ Given

$\angle 5 = \angle 1$ Alternate interior angles

$\mathbf{\angle 5 = 80°}$

14. $\angle 6$

$\angle 2 + \angle 5 + \angle 6 = 180°$ Interior angles of a triangle

$35° + 80° + \angle 6 = 180°$

$\angle 6 = 65°$

or

$\angle 6 = \angle 3 = 65°$ Alternate interior angles

$\mathbf{\angle 6 = 65°}$

15. $\angle 7$

$\angle 6 + \angle 7 = 180°$ Supplementary angles

$65° + \angle 7 = 180°$

$\mathbf{\angle 7 = 115°}$

exercise set 1.1

A. exercises

Find the third side in Exercises 1–10, if the given sides are from right triangle ABC.

See Example 1.

1. $a = 15, c = 17$
2. $b = 12, c = 13$
3. $AC = 16, CB = 63$
4. $BC = 9, CA = 40$
5. $a = 11, b = 60$
6. $a = 10, b = 24$
7. $AB = 9, BC = \sqrt{17}$
8. $AC = \sqrt{21}, BA = 11$
9. $a = \sqrt{2}, b = 4\sqrt{2}$
10. $b = 4\sqrt{5}, c = 12$

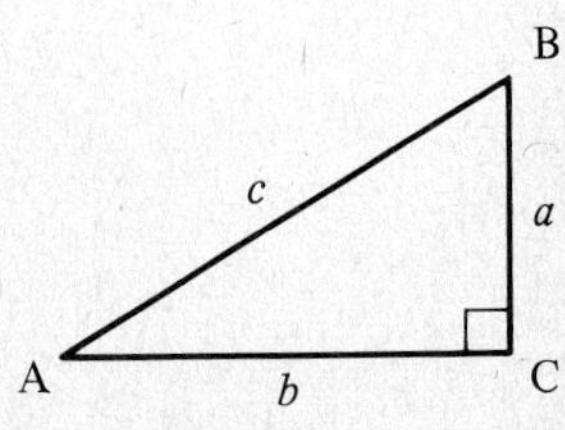

Determine if the numbers in Exercises 11–20 are the sides of a right triangle.

11. 7, 24, 25
12. 12, 35, 37

See Example 2.

13. 10, 15, 18
14. 13, 14, 19
15. 13, 84, 85
16. 20, 99, 101
17. 2, 3, $\sqrt{5}$
18. 3, 4, $\sqrt{7}$
19. $2\sqrt{2}$, $3 + \sqrt{2}$, $1 + 3\sqrt{2}$
20. $\sqrt{6}$, $2\sqrt{3} - 1$, $2 - \sqrt{3}$

In Exercises 21–30, find the distance between the given points. Approximate irrational results to the nearest thousandth.

21. (2, 1), (5, 5)
22. (1, 5), (13, 10)

See Examples 3 and 4.

23. (5, 7), (−3, −8)
24. (3, 9), (−2, −3)
25. (9, 11), (2, −13)
26. (15, 17), (3, −18)
27. $(2\sqrt{2}, 0)$, $(0, \sqrt{17})$
28. $(0, \sqrt{3})$, $(\sqrt{13}, 0)$
29. (2, −3), (−1, 4)
30. (−5, −2), (−1, 2)

Name each of the angles named in Exercises 31–40 in at least one other way.

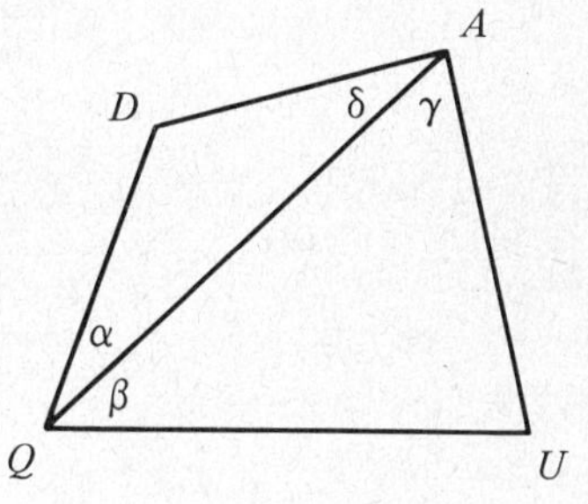

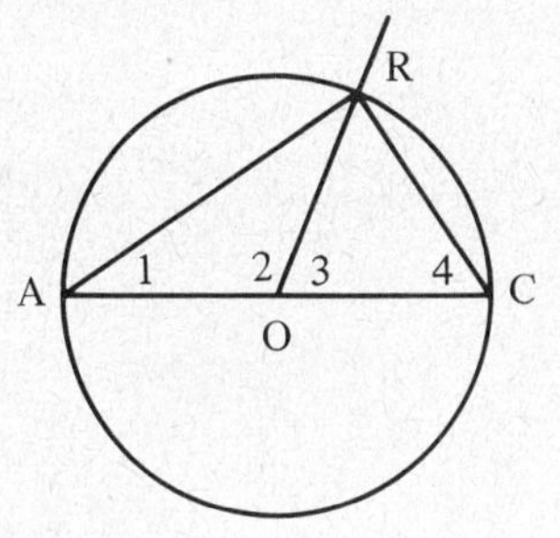

31. α
32. β
33. $\angle QAD$
34. $\angle UAQ$

See Examples 5–8.

35. $\angle U$
36. $\angle D$
37. $\angle AOR$
38. $\angle ROC$
39. $\angle RAC$
40. $\angle OCR$

Determine the degree measure of each of the angles described in Exercises 41–50.

41. $\frac{1}{4}$ of a rotation
42. $\frac{1}{6}$ of a rotation

See Example 9.

43. $\frac{1}{3}$ of a rotation
44. $\frac{1}{2}$ of a rotation
45. $\frac{1}{90}$ of a rotation
46. $\frac{1}{10}$ of a rotation

47. $\frac{3}{4}$ of a rotation

48. $\frac{2}{3}$ of a rotation

49. 1.5 rotations

50. 1.2 rotations

Exercises 51–60 describe as a rotation the angle swept by the minute hand in the time specified. Find the degree measure of each.

See Example 10.

51. 15 min

52. 10 min

53. 20 min

54. 30 min

55. 4 min

56. 6 min

57. 65 min

58. $1\frac{1}{4}$ h

59. 2 h

60. 90 min

Find a positive angle that is coterminal with each of the angles in Exercises 61–70, and less than one rotation.

See Example 11.

61. 500°

62. 450°

63. −800°

64. −750°

65. 1253°

66. −976°

67. −190°35′

68. 421°50′

69. −43°17′20″

70. 391°47′40″

Refer to Figure 1.7 and find the measure of each of the angles in Exercises 71–80. AB ∥ CD and EF ∥ HG.

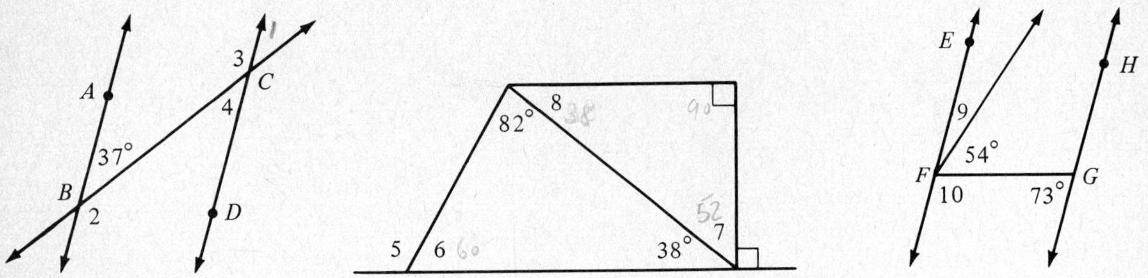

Figure 1.7

See Examples 12–15.

71. ∠1

72. ∠2

73. ∠3

74. ∠4

75. ∠5

76. ∠6

77. ∠7

78. ∠8

79. ∠9

80. ∠10

B. exercises

Given $\triangle ABC$ with $AB \parallel DE$, $\angle 1 = 37°$, and $\angle 2 = 94°$. Find each of the angles specified in Exercises 81–90.

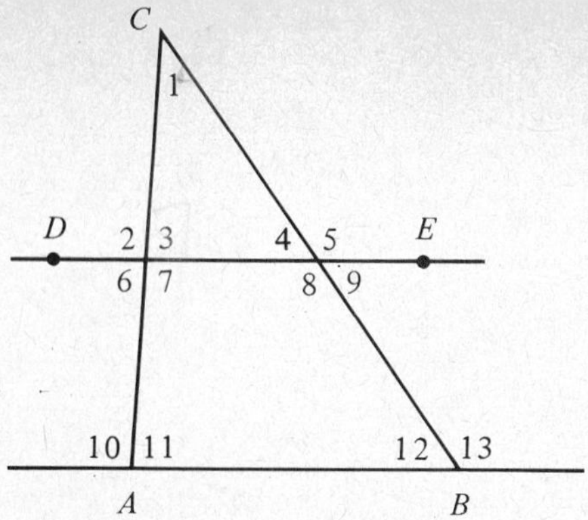

81. $\angle 4$ 82. $\angle 12$ 83. $\angle 3$ 84. $\angle 11$ See Examples 12–15.

85. $\angle 7$ 86. $\angle 9$ 87. $\angle 10$ 88. $\angle 13$

89. $\angle 6$ 90. $\angle 5$

Given $\angle RTW$ with $RT \parallel UV$, $RW \parallel SX$, $\angle W = 57°$, and $\angle UVR = 121°$. Find each of the angles specified in Exercises 91–100.

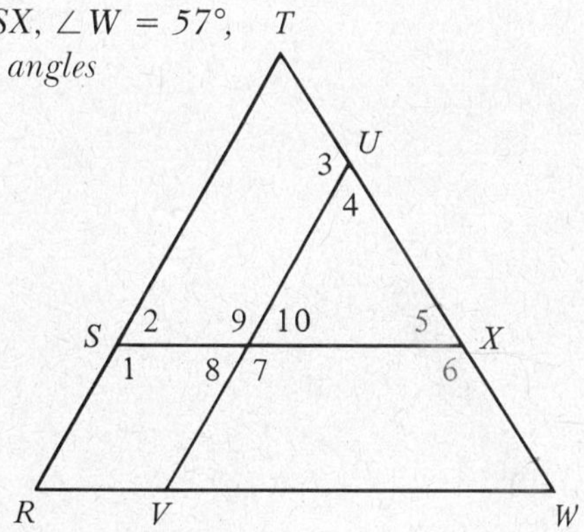

91. $\angle UVW$ 92. $\angle R$ 93. $\angle T$ 94. $\angle VUW$ See Examples 12–15.

95. $\angle 5$ 96. $\angle 8$ 97. $\angle 1$ 98. $\angle 2$

99. $\angle 10$ 100. $\angle 7$

1.2 SIMILARITY AND PROPORTION

The ancient Greeks made some fairly significant calculations with relatively simple concepts. One such concept is that of **similar triangles.** Two triangles are **similar** if their corresponding angles are equal.

SIMILAR TRIANGLES

If two triangles are similar, their corresponding sides are proportional.

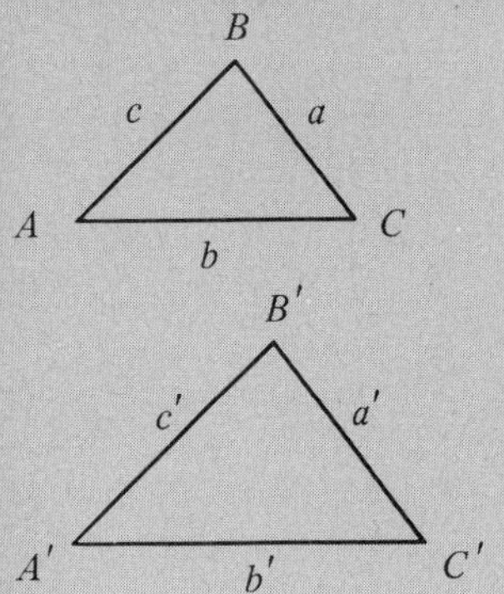

If $\triangle ABC \sim \triangle A'B'C'$, then

$$\frac{a}{a'} = \frac{b}{b'} = \frac{c}{c'}$$

The symbol $\sim$ means "is similar to," and the order of the sides shows their correspondence.

EXAMPLES

1. Given $\triangle NEW \sim \triangle OLD$, $NE = 14$ m, $EW = 49$ m, and $OL = 4$ m. Find LD.

$$\frac{LD}{EW} = \frac{OL}{NE}$$

$$\frac{LD}{49} = \frac{4}{14}$$

$$LD = \frac{4}{14} \cdot 49 = 14$$

Side *LD* is 14 m.

Draw a figure so that you can visualize the correspondence of the sides.

2. A tree casts a 36-ft shadow. At the same time a nearby 4-ft fence post has a 3-ft shadow. Find the height of the tree.

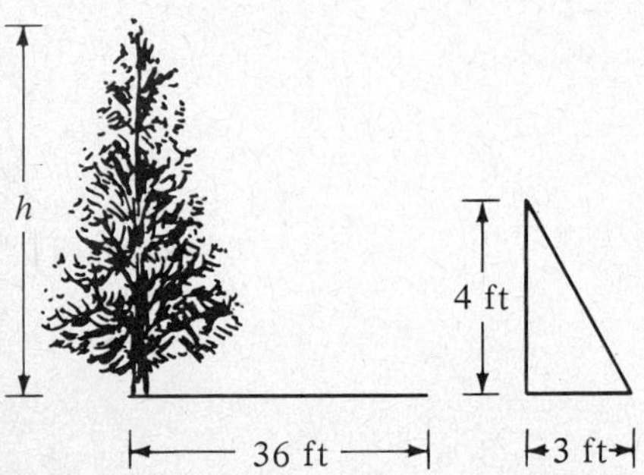

$$\frac{h}{4} = \frac{36}{3}$$

$$h = \frac{4 \cdot 36}{3}$$

$$h = 48$$

The tree is 48 ft tall.

3. $\triangle AGE \sim \triangle EGO$ as shown. Find GE and AE.

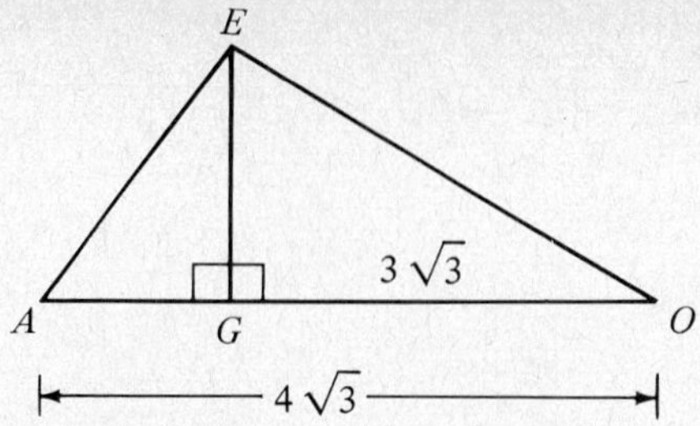

It may help to redraw the triangles as so that the sides correspond.

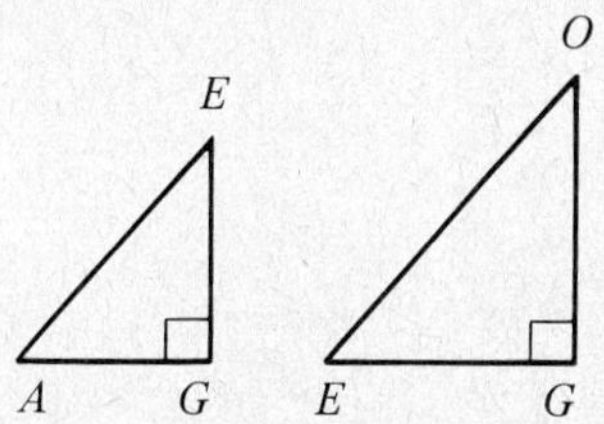

$$AG = AO - GO = 4\sqrt{3} - 3\sqrt{3}$$

$$= \sqrt{3}$$

$$\frac{EG}{OG} = \frac{AG}{EG}$$

$$EG^2 = \text{AG} \cdot \text{OG} = \sqrt{3} \cdot 3\sqrt{3} = 9$$

$$\mathbf{EG = 3}$$

$$EA^2 = AG^2 + EG^2$$

$$= (\sqrt{3})^2 + 3^2 = 3 + 9 = 12$$

$$EA = \sqrt{12}$$

$$\mathbf{EA = 2\sqrt{3}}$$

HISTORICAL ASIDE

When pyramids are mentioned, Egypt is generally assumed. However, the Pyramid of the Sun in the Valley of Teotihuacán, 25 miles northeast of Mexico City, has the same perimeter as the Great Pyramid of Cheops in Egypt. The civilization of early Mexico was brilliant, and their astronomers calculated the solar year at 365.2422 days, only 17 s off.

The Greek Thales (ca. 600 B.C.) is known to have measured the height of the Great Pyramid of Cheops using this concept of similar triangles. He drove a stake at the tip of the shadow of the Great Pyramid and measured the shadow of the pole (see Figure 1.8). He then measured the shadow of the pyramid plus one-half its base and formed the proportion:

$$\frac{H}{h} = \frac{S}{s} \quad \text{or} \quad H = \frac{hS}{s}$$

where h, s, and S are known.

Figure 1.8 THALES MEASUREMENT OF THE GREAT PYRAMID OF CHEOPS.

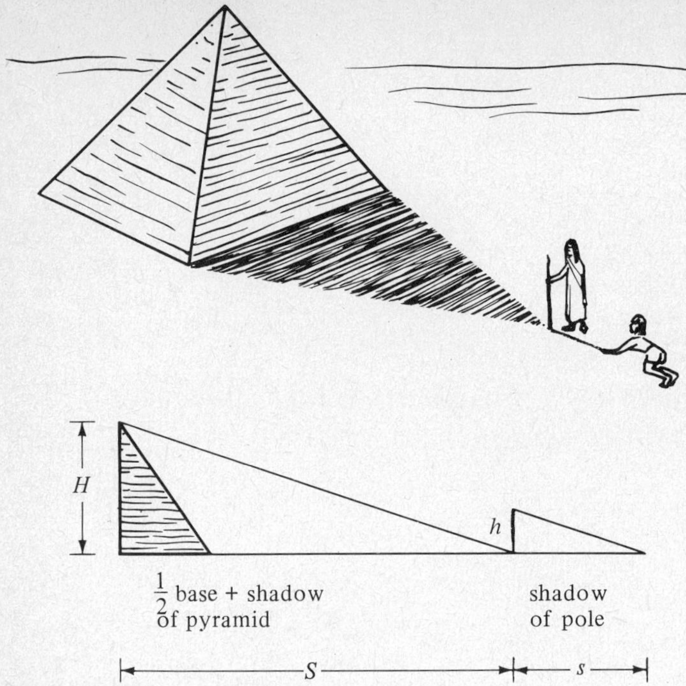

Special Triangles

If you are asked to visualize a draftsman at work, what do you see? Probably someone bent over a large, tilted drawing board, working with a T-square and triangles. These triangles are as familiar an image of the draftsman's work as they are a convenient figure for the mathematician. Both drafting triangles are very special triangles. One is an isosceles right triangle, or 45°-45°-90°. The other is a 30°-60°-90° triangle. The isosceles triangle has two equal sides, so if we give them a value, the Pythagorean theorem can be used to find the third. In Figure 1.9, the equal sides are first set equal to 1 and then given the value x.

$$c^2 = 1^2 + 1^2 \quad \text{or} \quad C^2 = x^2 + x^2$$

$$= 2 \qquad\qquad = 2x^2$$

$$c = \sqrt{2} \qquad\qquad C = x\sqrt{2}$$

The ratio of the sides is then

$$x:x:x\sqrt{2}$$

or

$$\mathbf{1:1:\sqrt{2}}$$

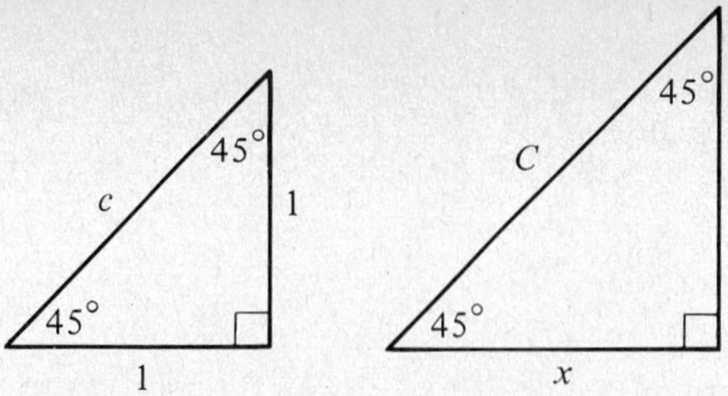

Figure 1.9 ISOSCELES RIGHT TRIANGLE.

The other special triangle can be obtained from an equilateral triangle by erecting an altitude on the base that also bisects the base. The altitude bisects the angle at the vertex, so we have a 30-60-90 triangle. Moreover, the hypotenuse is twice the shorter side. Again the third side can be found by the Pythagorean theorem.

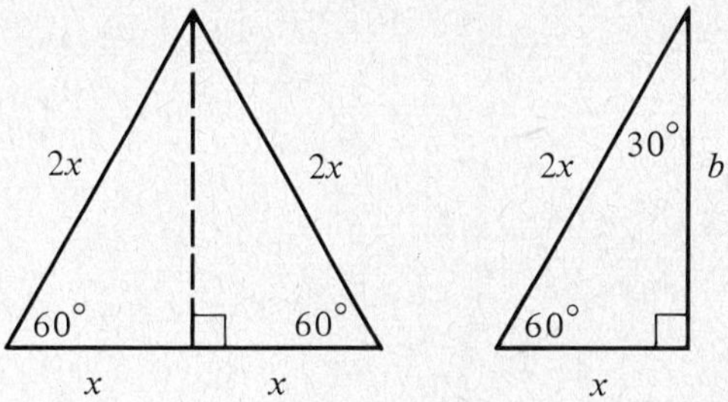

$$x^2 + b^2 = (2x)^2$$

$$= 4x^2$$

$$b^2 = 3x^2$$

$$b = x\sqrt{3}$$

The ratio of the sides is then

$$x{:}2x{:}x\sqrt{3}$$

or

$$\mathbf{1{:}2{:}\sqrt{3}}$$

Knowing the length of any one side of a special triangle, the other sides may be found using ratios peculiar to the figure. Conversely, if a triangle has one of these ratios on its sides, it must have the corresponding special angles.

45-45-90 TRIANGLE, 1:1:$\sqrt{2}$

A right triangle is isosceles if and only if the hypotenuse is $\sqrt{2}$ times one of the equal sides.

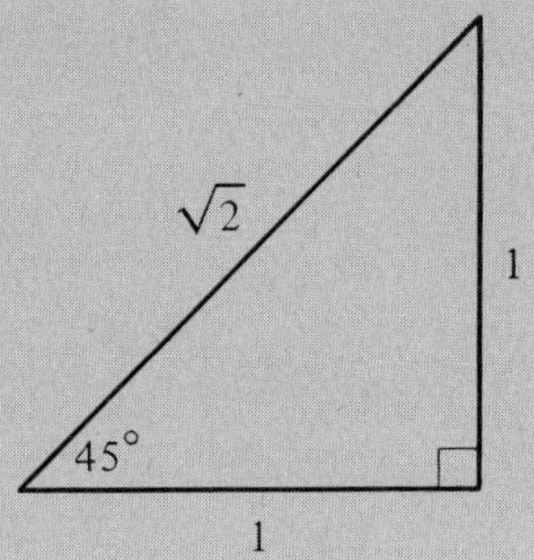

30-60-90 TRIANGLE, 1:2:$\sqrt{3}$

One angle of a right triangle is 30° if and only if the hypotenuse is twice the shorter side.

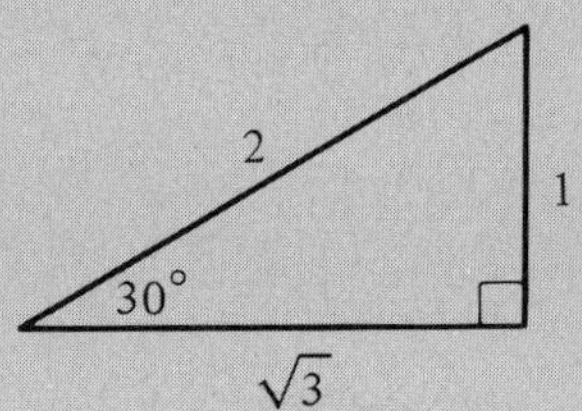

EXAMPLES

4. Find a and b.

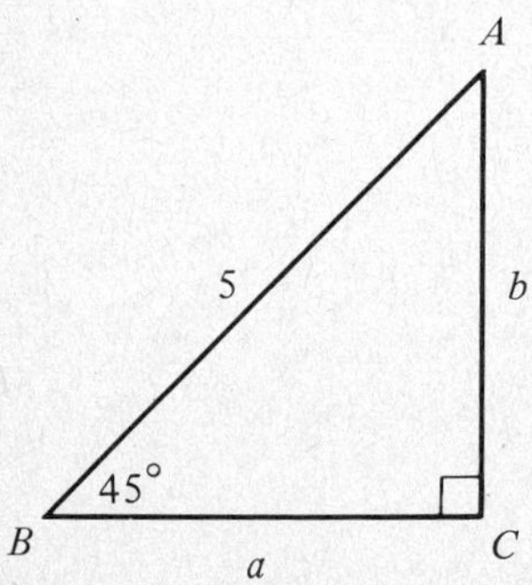

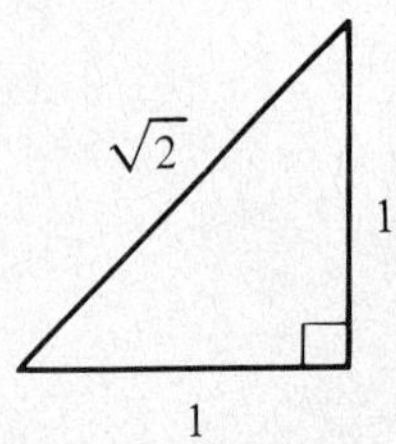

$$\frac{BC}{1} = \frac{AB}{\sqrt{2}}$$

Because $\triangle ABC$ is 45-45-90 sides a and b are equal, and the hypotenuse (5) is $\sqrt{2}$ times side a.

$$5 = (\sqrt{2})(a)$$

$$\frac{5}{\sqrt{2}} = a$$

or

$$\boldsymbol{a = \frac{5}{2}\sqrt{2} = b}$$

5. Find s and t.

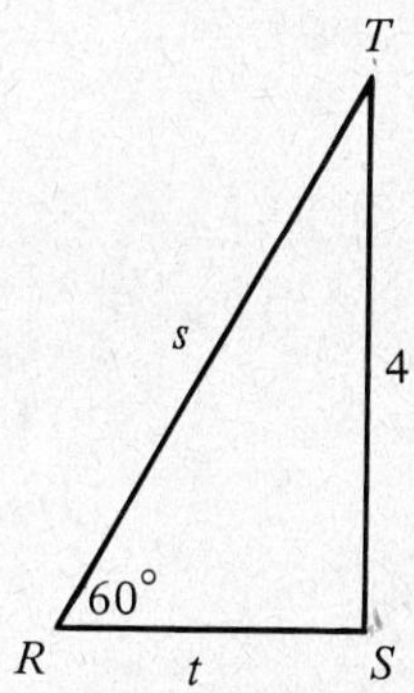

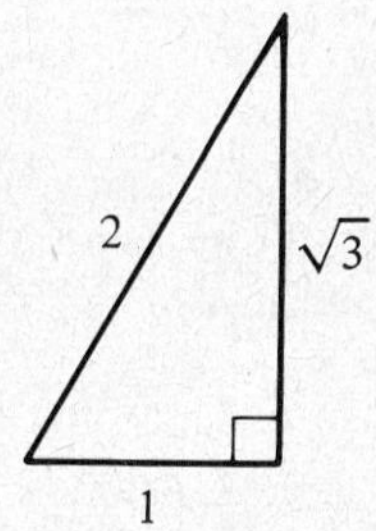

$$\frac{RT}{\sqrt{3}} = \frac{RS}{1}$$

Because $\triangle RST$ is 30-60-90, the hypotenuse s is twice the shorter side t:

$$s = 2t$$

and

$$\sqrt{3}t = 4$$

$$t = \frac{4}{\sqrt{3}} \quad \text{or} \quad \frac{4}{3}\sqrt{3}$$

$$s = 2\left(\frac{4}{3}\sqrt{3}\right)$$

$$= \frac{8}{3}\sqrt{3}$$

$$\boldsymbol{s = \frac{8}{3}\sqrt{3} \quad \textbf{and} \quad t = \frac{4}{3}\sqrt{3}}$$

6. Find w, x, y, and z.
   $\triangle XWZ$ is 30-60-90, so

$$XZ = 2XW$$
$$y = 2\sqrt{2}$$
$$\frac{WZ}{\sqrt{3}} = \frac{XW}{1}$$
$$\frac{w}{\sqrt{3}} = \frac{\sqrt{2}}{1}$$
$$\mathbf{w = \sqrt{6}}$$

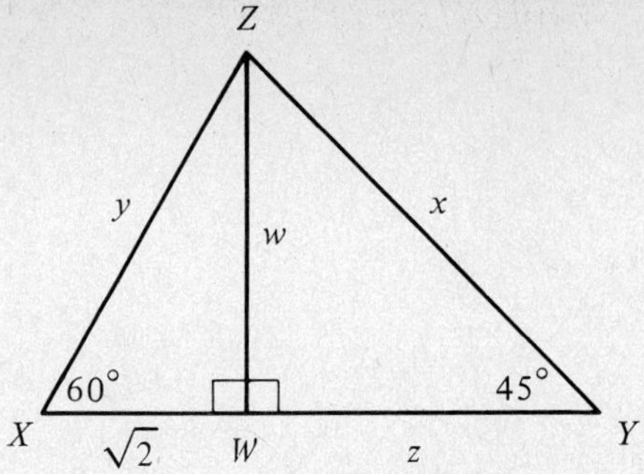

$\triangle WYZ$ is 45-45-90, so

$$WY = WZ$$
$$\mathbf{w = z = 2\sqrt{2}}$$
$$YZ = \sqrt{2}(WZ)$$
$$x = \sqrt{2}(\sqrt{6})$$
$$\mathbf{x = 2\sqrt{3}}$$

7. Identify $\triangle POT$.

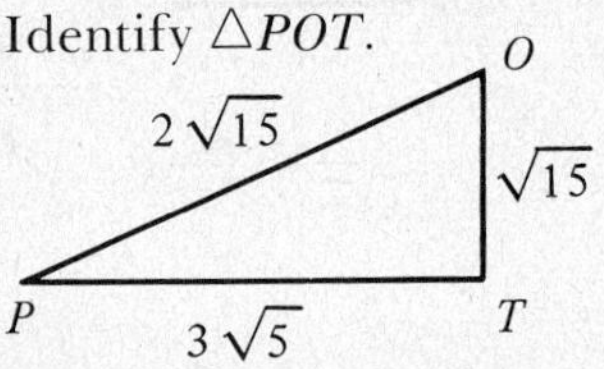

Using the Pythagorean theorem (converse),

$$(\sqrt{15})^2 + (3\sqrt{5})^2 \stackrel{?}{=} (2\sqrt{15})^2$$
$$15 + 45 = 60 \quad \checkmark$$

Thus $\triangle POT$ is a right triangle.

$$PO \stackrel{?}{=} 2(TO) \qquad \text{and} \qquad PT \stackrel{?}{=} \sqrt{3}(TO)$$
$$\sqrt{15} = 2\sqrt{15} \quad \checkmark \qquad\qquad 3\sqrt{5} = \sqrt{3}(\sqrt{15}) \quad \checkmark$$

Thus **$\triangle POT$ is a 30-60-90 triangle.**

exercise set 1.2

A. exercises

For Exercises 1–3, $\triangle PAM \sim \triangle BEV$ and $\triangle PAT \sim \triangle BEN$.

See Examples 1 and 3.

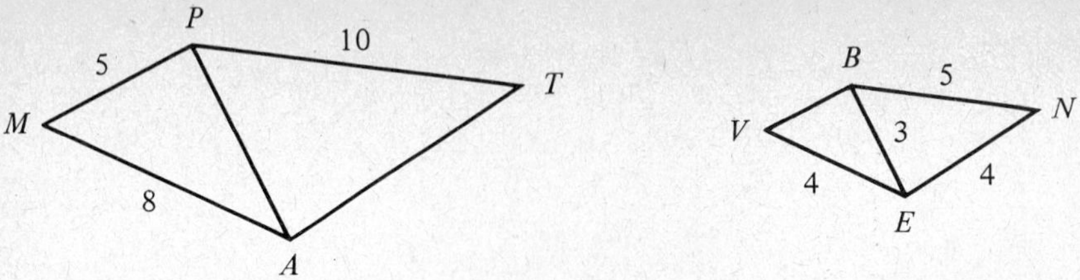

1. $PA = ?$ 2. $BV = ?$ 3. $AT = ?$

In Exercises 4–6, $\triangle IAN \sim \triangle TIM$ and $\triangle TOM \sim \triangle IPN$.

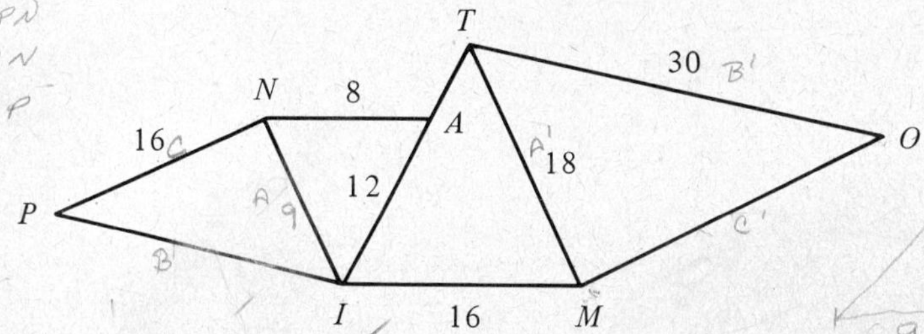

4. $IN = ?$ 5. $MO = ?$ 6. $IT = ?$

In Exercises 7–10, $\triangle HAL \sim \triangle SAM \sim \triangle JAY$.

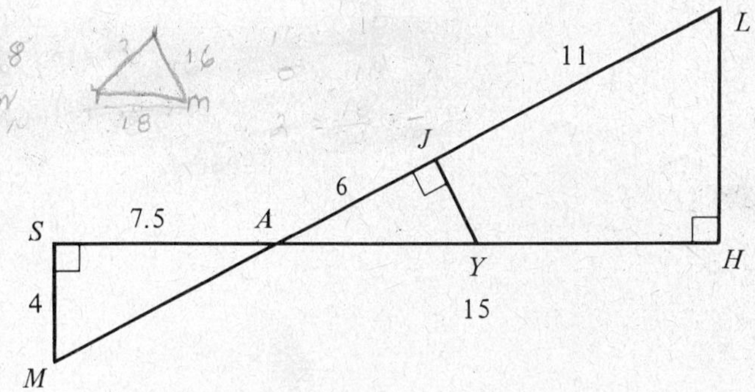

7. $MA = ?$ 8. $JY = ?$ 9. $LH = ?$ 10. $YA = ?$

Find the parts of the special triangles indicated in Exercises 11–20.

See Example 4.

11. $a = ?$

12. $b = ?$

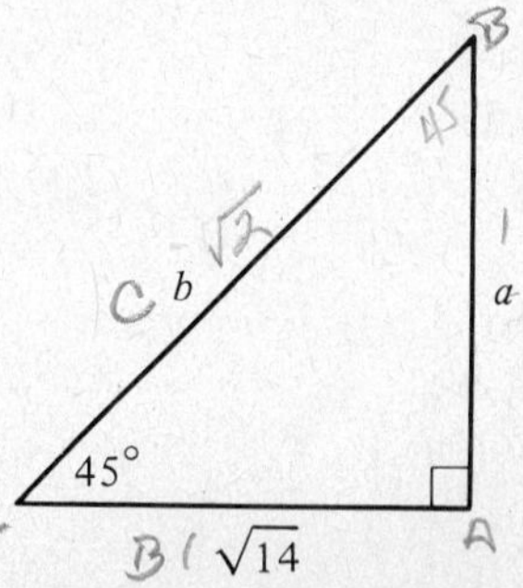

See Examples 4–7.

13. $c = ?$

14. $d = ?$

15. $e = ?$

16. $f = ?$

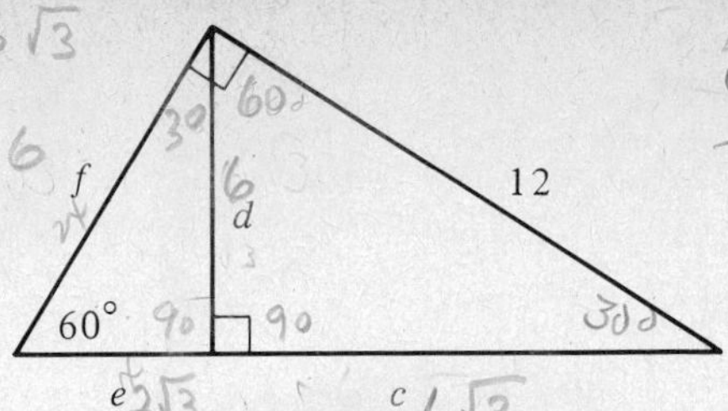

See Examples 4–6.

17. $g = ?$

18. $h = ?$

19. $JH = ?$

20. $GI = ?$

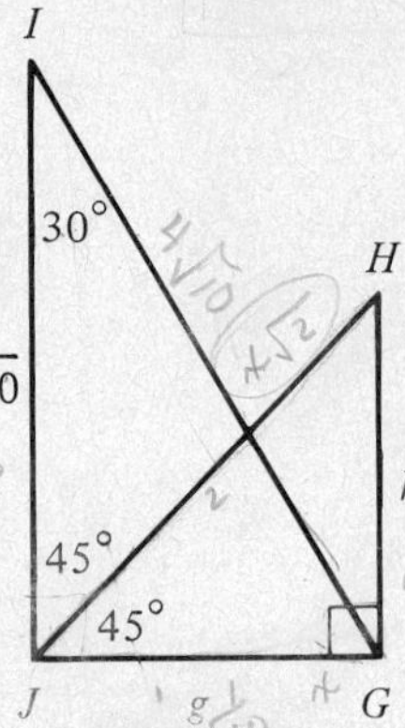

B. exercises

In Exercises 21–24, $\triangle SAM \sim \triangle TAD \sim \triangle TOM$.

See Example 3.

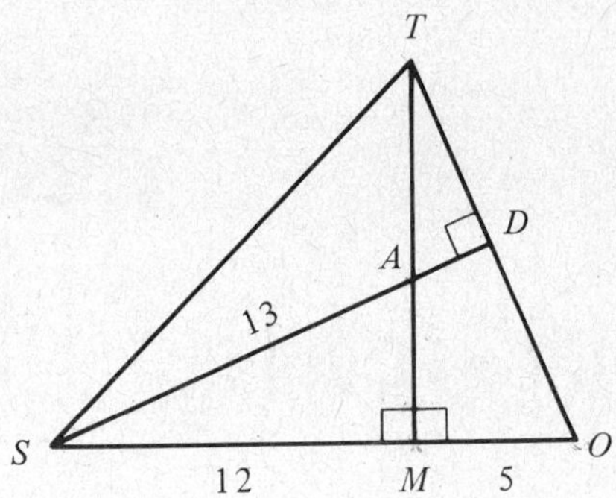

21. $AM = ?$ 22. $AT = ?$ 23. $TO = ?$ 24. $DO = ?$

In Exercises 25–30, $\angle PSI = \angle HIP = 90°$, $\angle PHI = 60°$, $\angle ICH = 45°$, *and* $IS = \sqrt{6}$.

See Example 6.

25. $SH = ?$ 28. $PI = ?$

26. $HI = ?$ 29. $PH = ?$

27. $CI = ?$ 30. $CH = ?$

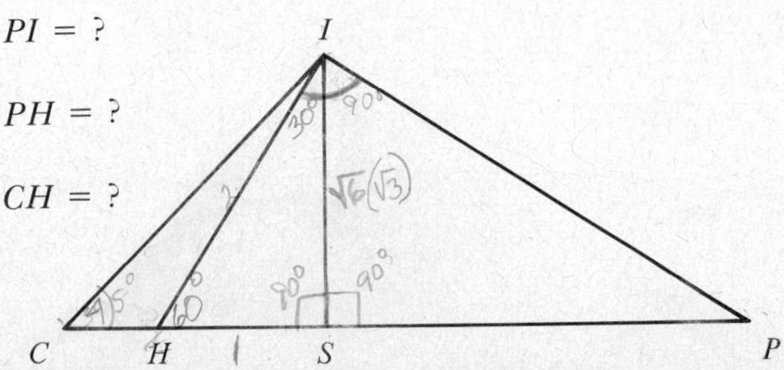

31. A 6-ft man casts a 4-ft shadow at the same time as a nearby building casts a shadow that is 42 ft long. How tall is the building? See Example 2.

32. A tree casts a shadow 18 ft long at the same time as a 4-ft post casts a 16-in. shadow. How tall is the tree? Watch the units!

33. A guy wire 34 ft long is used to secure the top of a tall vertical pole to the ground. Another wire is attached 3 ft above the ground and is 4 ft 3 in. long. How tall is the pole if both wires make the same angle with the ground? Must the ground be horizontal?

34. Standing 72 ft from the base at the back of a building on level ground, one can just see the top of a sign that sits on the flat roof at the front of the building. If the building is known to be 40 ft tall and 27 ft front to back, how tall is the sign on the roof?

35. The diagonal of a rectangle is twice one of its sides. What is its area if the diagonal is 6 yd?

36. It is a 90-ft dash from first to second base on a baseball diamond. How long is the throw from third to first base?

1.3 THE TRIGONOMETRIC RATIOS

The origins of trigonometry are obscure, but it is fairly certain that it arose as a computational tool for the astronomer. Hipparchus (ca. 150 B.C.) was an Alexandrian astronomer who lived a century after Archimedes. He calculated the distance of the moon from the earth. Knowing that distance, many other computations could be made of the distances and sizes of heavenly bodies. As the astronomer on the earth made observations of the moon, sun, planets, and the stars, the only data that could be recorded was the angle of observation. The astronomer needed relationships dependent on that angle in order to make other computations. Similar triangles have proportional sides, and it is equal angles that give similarity.

HISTORICAL ASIDE

Mathematics has long benefited from astronomy—so much so that at one time mathematician meant astronomer. Nicholas Copernicus (1473 – 1543) wrote about trigonometry after the publication of his theory of the universe.

$$\frac{a}{a'} = \frac{b}{b'} = \frac{c}{c'}$$

is equivalent to

$$\frac{a}{c} = \frac{a'}{c'}, \quad \frac{b}{c} = \frac{b'}{c'}, \quad \text{and} \quad \frac{b}{a} = \frac{b'}{a'}$$

Being similar also means that the sides of a given triangle form certain ratios that are constant for all other triangles that are similar (see Figure 1.10). Hipparchus made extensive tables of such ratios for angles at intervals of less than every 1°. Actually, he compiled a table of the ratio of the chord of an arc to the radius or distance (see Figure 1.11). Later

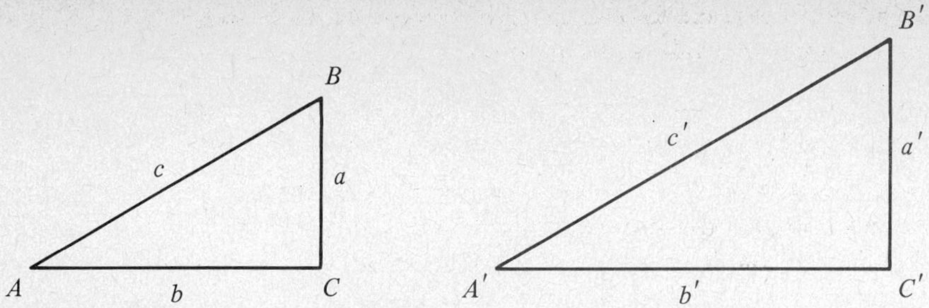

Figure 1.10 SIMILARITY AND RATIOS OF SIDES.

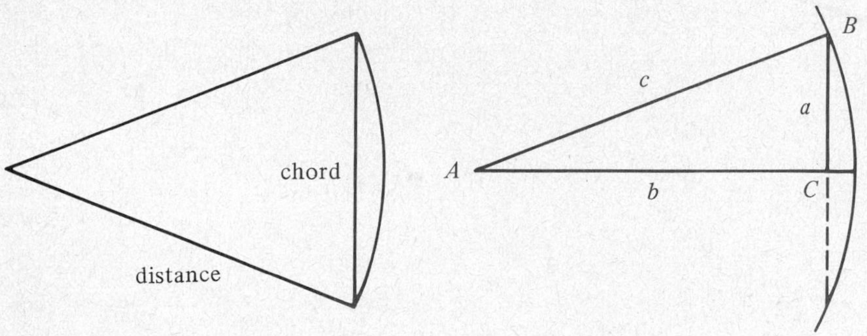

Figure 1.11 HALF-CHORD RATIO a/c.

the ratios were calculated for "half-chords," (see the Historical Aside in this section) or what we now call sines. If the moon's distance is approximately 245,000 mi, Hipparchus' method can be illustrated by Figure 1.12.

$$\text{sine of } \tfrac{1}{4}° = \frac{\text{moon's radius}}{\text{moon's distance}}$$

Figure 1.12 MEASUREMENT OF THE SIZE OF THE MOON.

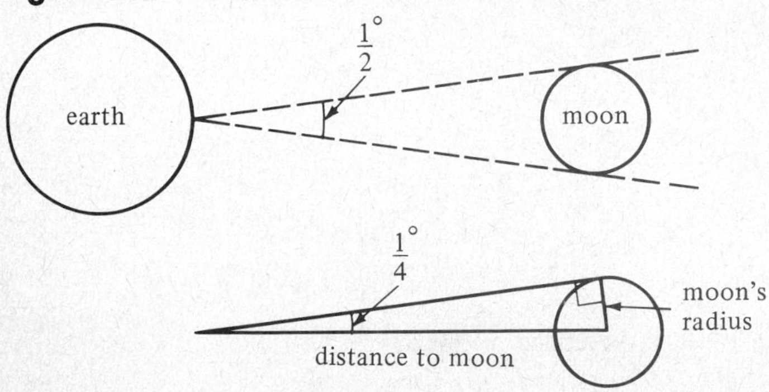

and from his tables, Hipparchus knew that sine of $\frac{1}{4}°$ is approximately 0.0044, so

$$0.0044 \approx \frac{\text{moon's radius}}{245{,}000}$$

$$\text{moon's radius} \approx 0.0044(245{,}000)$$

$$\approx 1100$$

So the radius of the moon was found to be approximately 1100 mi.

There are actually six ratios that can be formed with sides of a right triangle. These six ratios are the **trigonometric ratios,** known historically by the names sine, cosine, tangent, cotangent, secant, and cosecant. These names, however, are abbreviated for utility in the following definition.

TRIGONOMETRIC RATIOS

Given $\triangle ABC$ with a right angle at C (see Figure 1.13),

$$\sin A = \frac{a}{c} \qquad \cot A = \frac{b}{a}$$

$$\cos A = \frac{b}{c} \qquad \sec A = \frac{c}{b}$$

$$\tan A = \frac{a}{b} \qquad \csc A = \frac{c}{a}$$

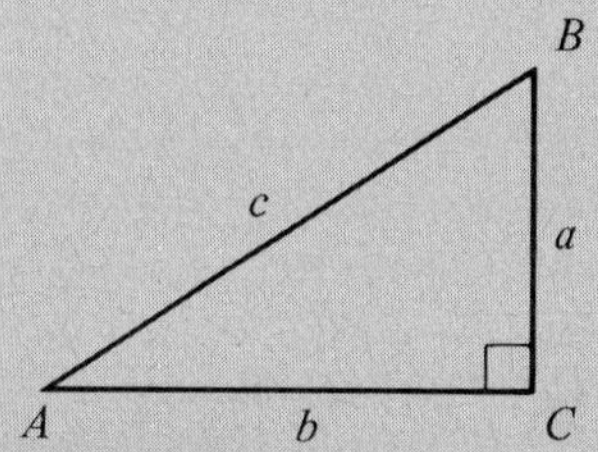

Figure 1.13

Note that in this definition, $\angle A$ must be acute.

EXAMPLES

1. Find all six ratios for angle α in the right triangle shown.

From the triangle, $a = 8$, $b = 15$, and $c = 17$.

$$\sin \alpha = \frac{a}{c} = \frac{8}{17} \qquad \cot \alpha = \frac{b}{a} = \frac{15}{8}$$

$$\cos \alpha = \frac{b}{c} = \frac{15}{17} \qquad \sec \alpha = \frac{c}{b} = \frac{17}{15}$$

$$\tan \alpha = \frac{a}{b} = \frac{8}{15} \qquad \csc \alpha = \frac{c}{a} = \frac{17}{8}$$

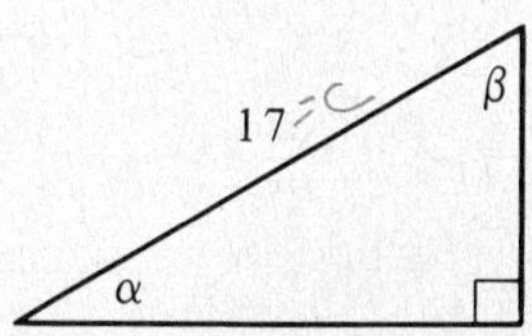

2. Find the sine, cosine, and tangent for angle β in Example 1. It may be helpful if the triangle is reoriented.

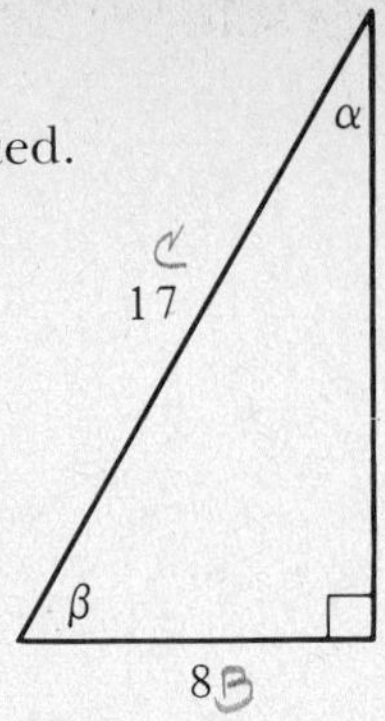

In this orientation, $a = 15$, $b = 8$, and $c = 17$. Thus,

$\sin \beta = \frac{15}{17}$

$\cos \beta = \frac{8}{17}$

$\tan \beta = \frac{15}{8}$

3. If $\sin \theta = \frac{12}{13}$, find the other five ratios.

Because $\sin \theta = a/c$, let $a = 12$ and $c = 13$. The ratios are defined on a right triangle, so $a^2 + b^2 = c^2$, and $12^2 + b^2 = 13^2$ produces $b = 5$.

	$\cot \theta = \frac{5}{12}$
$\cos \theta = \frac{5}{13}$	$\sec \theta = \frac{13}{5}$
$\tan \theta = \frac{12}{5}$	$\csc \theta = \frac{13}{12}$

In Figure 1.13, side c is the *hypotenuse* of the right triangle. Side a is called the side *opposite* $\angle A$ and side b is called the side *adjacent to* $\angle A$. Often the ratios are written with these designations so that they appear as follows:

$$\sin A = \frac{\text{opposite}}{\text{hypotenuse}} \qquad \cot A = \frac{\text{adjacent}}{\text{opposite}}$$

$$\cos A = \frac{\text{adjacent}}{\text{hypotenuse}} \qquad \sec A = \frac{\text{hypotenuse}}{\text{adjacent}}$$

$$\tan A = \frac{\text{opposite}}{\text{adjacent}} \qquad \csc A = \frac{\text{hypotenuse}}{\text{opposite}}$$

The special 30-60-90 and 45-45-90 triangles and the ratios of their sides were reviewed in the last section. With these special triangles, we can now find the trigonometric ratios for the special angles of 30°, 45°, and 60°. If the lengths of the sides are remembered, then sketches like those in Figure 1.14 can be made whenever one of these angles is encountered. With frequent use, many of the trigonometric ratios of these

SPECIAL TRIANGLES

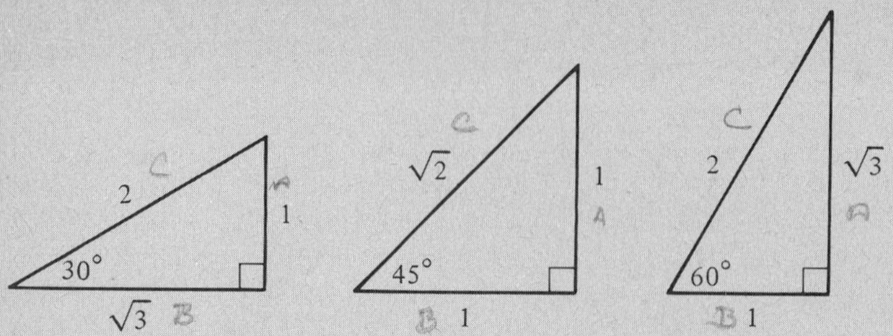

Figure 1.14 SPECIAL TRIANGLES.

angles will become very familiar. The 30-60-90 triangle is placed in two different positions in Figure 1.14. It is the same triangle, but the position emphasizes the angle of interest. Note that the values obtained are exact and not approximate, like those we might obtain on a calculator or in a set of tables.

EXAMPLES

Use Figure 1.14 to obtain each of the following.

4. $\sin 30° = \frac{a}{c} = \frac{1}{2}$

5. $\tan 45° = \frac{a}{b} = \frac{1}{1} = 1$

6. $\sec 60° = \frac{c}{b} = \frac{2}{1} = 2$

7. $\cos 30° = \frac{b}{c} = \frac{\sqrt{3}}{2}$

8. $\cot 60° = \frac{b}{a} = \frac{1}{\sqrt{3}} = \frac{\sqrt{3}}{3}$

All of the ratios for 30°, 45°, and 60° should be found easily by making a small sketch and reading off the sides. In the examples and exercises that follow, you should obtain exact values without the use (or need) of a calculator or tables.

EXAMPLES

Determine if the statement is true or false.

9. $\cos 45° \sin 45° \stackrel{?}{=} \dfrac{1 - \cos^2 45°}{\tan 45°}$. The shorthand notation $\cos^2 \theta$ means $(\cos \theta)^2$.

$$\left(\frac{1}{\sqrt{2}}\right)\left(\frac{1}{\sqrt{2}}\right) \stackrel{?}{=} \frac{1 - (1/\sqrt{2})^2}{1}$$

$$\frac{1}{2} \stackrel{?}{=} 1 - \frac{1}{2}$$

$$\frac{1}{2} \stackrel{?}{=} \frac{1}{2} \qquad \surd \textbf{ True}$$

10. $2 \sec 30° \stackrel{?}{=} \sec 60°$

$$2\left(\frac{2}{\sqrt{3}}\right) \stackrel{?}{=} 2$$

$$\frac{4}{\sqrt{3}} \neq 2 \qquad \textbf{False}$$

11. $\csc 60° \csc 45° \stackrel{?}{=} \dfrac{\tan 30° + \cot 60°}{\cos 45°}$

$$\left(\frac{2}{\sqrt{3}}\right)\left(\frac{\sqrt{2}}{1}\right) \stackrel{?}{=} \frac{(1/\sqrt{3}) + (1/\sqrt{3})}{1/\sqrt{2}}$$

$$\frac{2\sqrt{2}}{\sqrt{3}} \stackrel{?}{=} \left(\frac{2}{\sqrt{3}}\right)\left(\frac{\sqrt{2}}{1}\right)$$

$$\frac{2\sqrt{2}}{\sqrt{3}} = \frac{2\sqrt{2}}{\sqrt{3}} \qquad \surd \textbf{ True}$$

EXAMPLES

12. Find s.

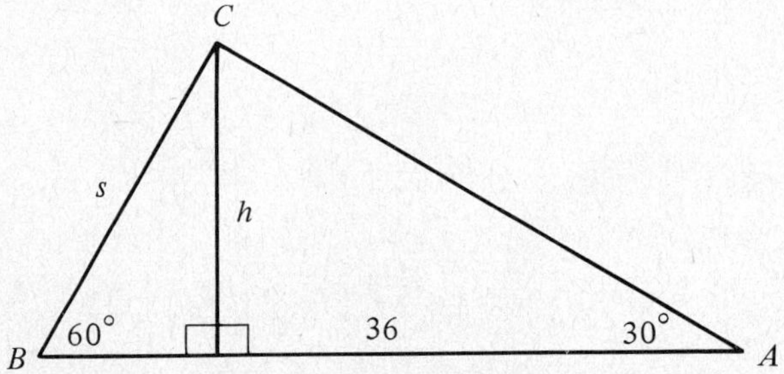

First find h:

$$\frac{h}{36} = \tan A$$

$$h = 36 \tan 30°$$

$$= 36\left(\frac{1}{\sqrt{3}}\right) = \frac{36\sqrt{3}}{3} = 12\sqrt{3}$$

Now find s:

$$\frac{h}{s} = \sin B$$

or

$$\frac{s}{h} = \csc B$$

$$s = h \csc B$$

$$= 12\sqrt{3}\left(\frac{2}{\sqrt{3}}\right)$$

$$\mathbf{s = 24}$$

13. The cross section of a large monument shows that each face is 10 ft and the three faces are at 30°, 45°, and 60° with the horizontal. How high is the structure?

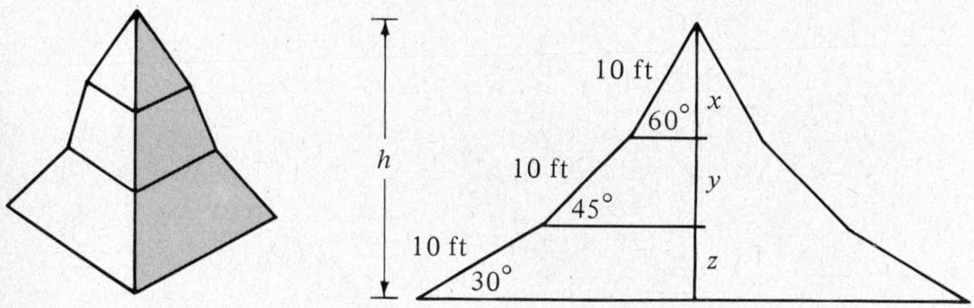

$$\frac{x}{10} = \sin 60° \qquad \frac{y}{10} = \sin 45° \qquad \frac{z}{10} = \sin 30°$$

$$x = 10 \sin 60° \qquad y = 10 \sin 45° \qquad z = 10 \sin 30°$$

and

$$h = x + y + z$$

$$= 10 \sin 60° + 10 \sin 45° + 10 \sin 30°$$

$$= 10\left(\frac{\sqrt{3}}{2} + \frac{1}{\sqrt{2}} + \frac{1}{2}\right) = 10\left(\frac{\sqrt{3}}{2} + \frac{\sqrt{2}}{2} + \frac{1}{2}\right)$$

$$= 5(\sqrt{3} + \sqrt{2} + 1) \approx 20.731322$$

The monument is $5(\sqrt{3} + \sqrt{2} + 1)$ ft tall (or approximately 20.7 ft).

exercise set 1.3

A. exercises

Find all six ratios for each angle in Exercises 1–6.

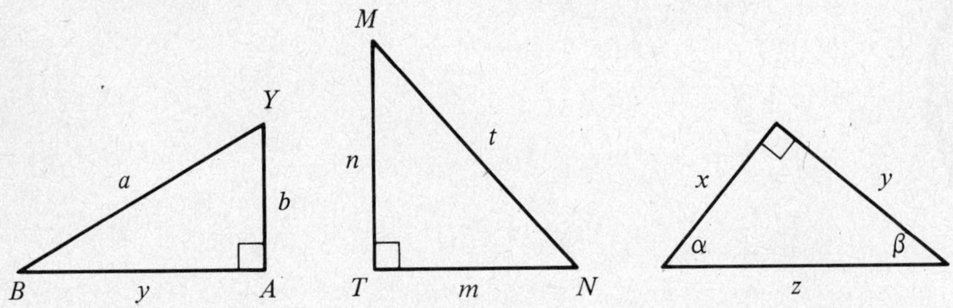

1. B 2. Y 3. N

4. M 5. α 6. β

See the definition. Redraw the triangles if necessary.

In Exercises 7–14, refer to the triangles to find the specified ratio.

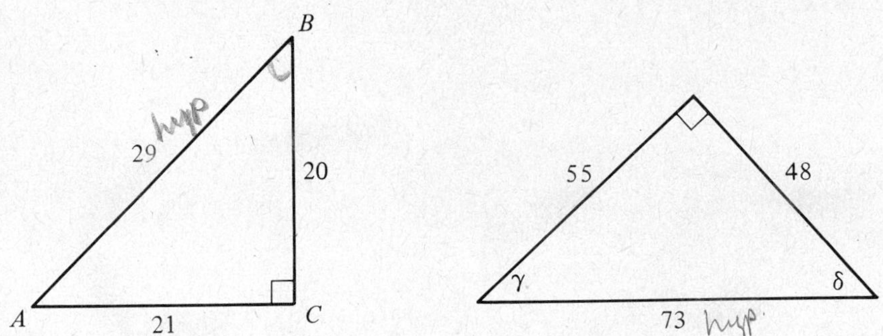

7. $\sin A$ 8. $\cot B$ 9. $\sec \gamma$

10. $\tan \delta$ 11. $\sec B$ 12. $\tan A$

13. $\csc \delta$ 14. $\cos \gamma$

See Examples 1 and 2.

Identify the ratios by name in Exercises 15–20 for the figures on the next page.

15. $\frac{r}{t}$ 16. $\frac{t}{s}$ 17. $\frac{s}{r}$

18. $\frac{z}{x}$ 19. $\frac{y}{z}$ 20. $\frac{x}{y}$

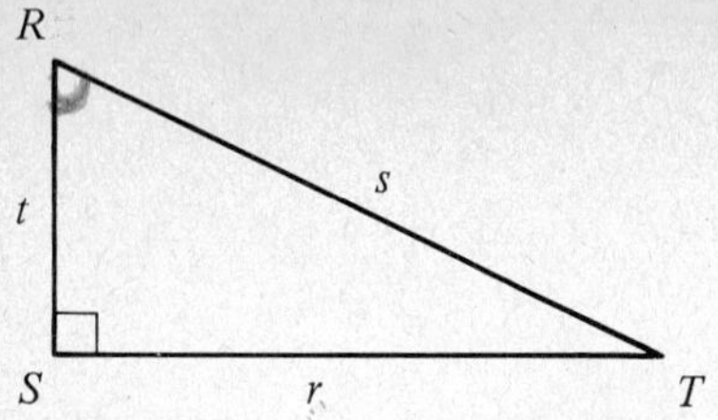

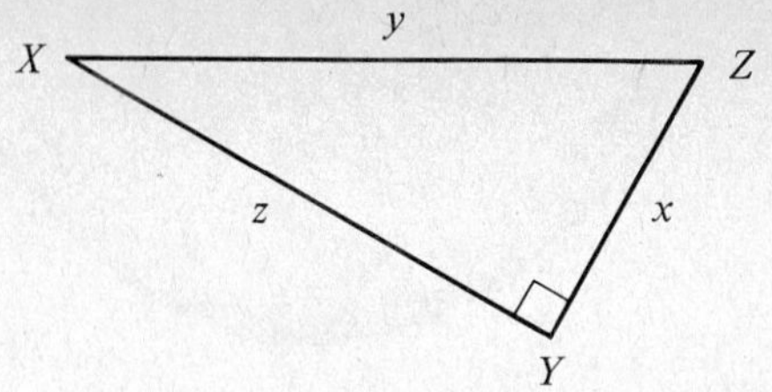

Exercises 21–30 refer to △ABC, where C is the right angle. Find exact results; do not approximate.

See Example 3.

21. Find $\tan A$ if $\sin A = \frac{7}{25}$.

22. Find $\sec A$ if $\tan A = \frac{21}{20}$.

23. If $\csc B = \frac{37}{12}$, find $\cot B$.

24. If $\cot B = \frac{28}{45}$, find $\cos B$.

25. Find $\sin A$ if $\tan B = \sqrt{5}/2$.

Note in Exercises 25–28 that both A and B are involved.

26. Find $\cot A$ if $\cos B = \frac{2}{3}$.

27. Determine $\csc B$ when $\sec A = \sqrt{10}$.

28. Calculate $\tan A$ when $\sin B = \sqrt{2}/10$.

29. If $\cot A = \frac{16}{63}$, find $\csc A$.

30. Find $\sin A$ if $\sec B = \frac{65}{56}$.

Find the exact value of the ratios in Exercises 31–40.

See Examples 4–8.

31. $\cos 45°$
32. $\tan 30°$
33. $\cot 45°$
34. $\sin 60°$
35. $\csc 45°$
36. $\sec 30°$
37. $\tan 60°$
38. $\csc 30°$
39. $\sin 45°$
40. $\cos 30°$

Determine if each of the statements in Exercises 41–50 is true or false.

$\sin^2 \theta = (\sin \theta)^2$
See Examples 9–11.

41. $\sin^2 30° + \cos^2 30° = 1$
42. $\tan^2 60° + 1 = \sec^2 60°$
43. $\tan 30° + \cot 30° = 2 \csc 60°$
44. $\cot 30° - \cot 60° = \csc 60°$
45. $\cos 45°(\sec 45° - \cos 45°) = \sin^2 45°$
46. $\csc 45°(\csc 45° - \sin 45°) = \cot^2 45°$

47. $\dfrac{\cos 30^\circ}{1 - \cos 60^\circ} - \cot 60^\circ = \sec 30^\circ$

48. $\dfrac{\sin^2 60^\circ}{\csc 30^\circ - 1} - \sin 30^\circ = \cos^2 60^\circ$

49. $\dfrac{\sin 30^\circ \cot 60^\circ}{1 - \cos 30^\circ} - \tan 45^\circ = \csc 60^\circ$

50. $\dfrac{\cot 45^\circ}{\sec 30^\circ - \cot 60^\circ} - \tan 30^\circ = \csc 60^\circ$

B. exercises

Find the exact value of each part named in Exercises 51–70 and in the figures shown.

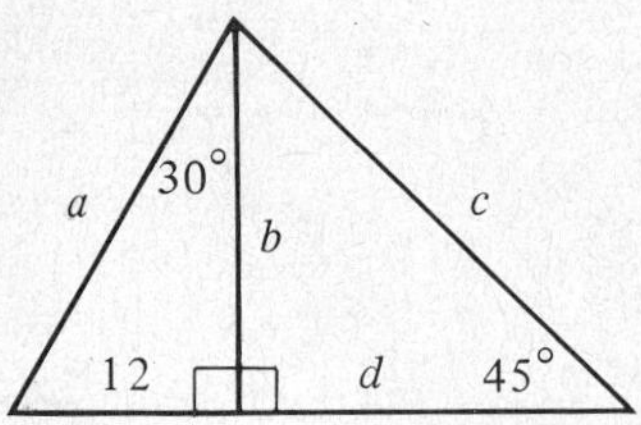

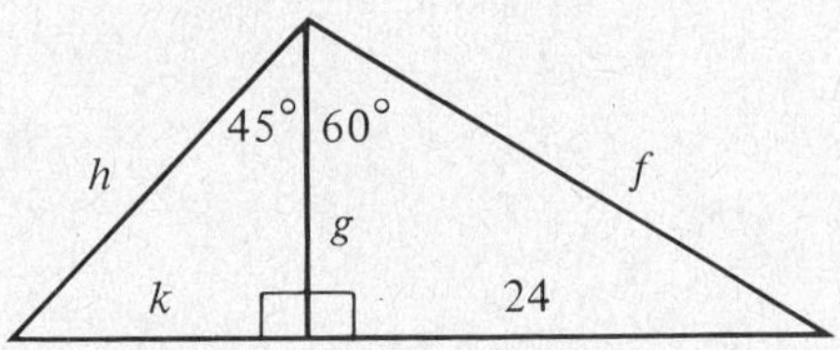

51. a 52. b 53. c 54. d See Example 12.

55. f 56. g 57. h 58. k

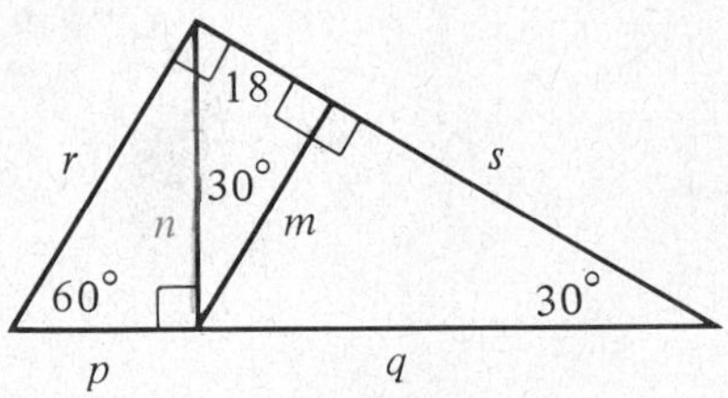

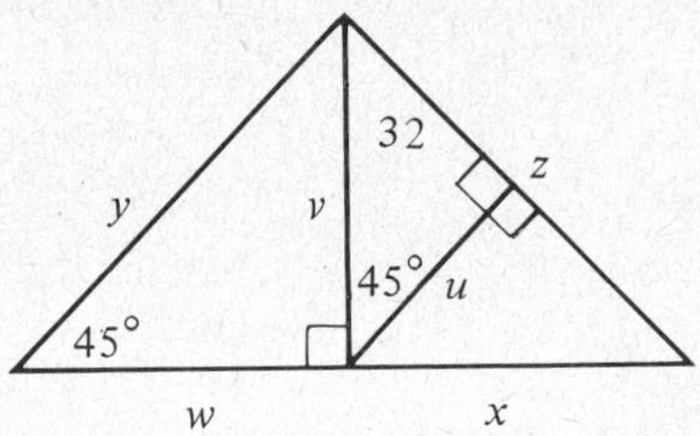

59. m 60. n 61. p 62. q See Example 12.

63. r 64. s 65. u 66. v

67. w 68. x 69. y 70. z

71. The familiar shape of the traditional barn roof is called a *gambrel* roof. If the upper surface (see Figure 1.15) is 20 ft with an elevation of 30° and the lower surface is 12 ft at 60°, what is the total span of the roof? See Example 13.

72. If the gambrel roof in Exercise 71 has an upper surface of 18 ft with an elevation of 30° and the lower surface is 15 ft at 45°, what is the total span of the roof?

73. A camping tent is often shaped like a gambrel roof. If the side is 4 ft and 60° with the ground and the top is 4 ft but at 45°, what is the total height inside the tent?

74. If the side of the tent in Exercise 73 is 8 ft and 45° with the ground and the top is 4 ft at 30°, what is the total height of the tent?

75. The flag of Nepal is red bordered in blue with a white moon above a white sun, as shown in Figure 1.16a. The shape appears to be a 30-60-90 triangle overlapping a 45-45-90 triangle with equal bases. The base of the top triangle is at the midpoint of the height of the banner and the bases of the triangles are 2 m. By how much do the triangles overlap along their vertical sides?

76. The flag of Kuwait has green, white, and red stripes with a black trapezoid at the flagstaff, as shown in Figure 1.16b. The diagonal seam between black and green (or red) is 101 cm on a large state ceremonial model. If the stripes are of equal width and the angle of the diagonal is 45°, how tall is the flag?

77. An equilateral triangle is inscribed in a circle with radius 12 in. What is the perimeter of the triangle?

78. A square is inscribed in a circle with radius 10 cm. What is the perimeter of the square?

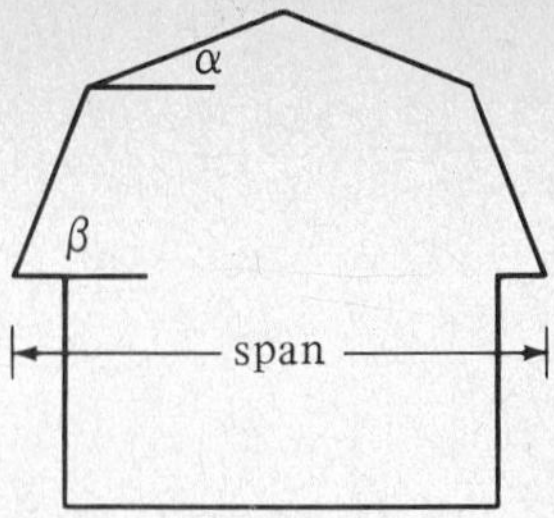

Figure 1.15
GAMBREL ROOF.

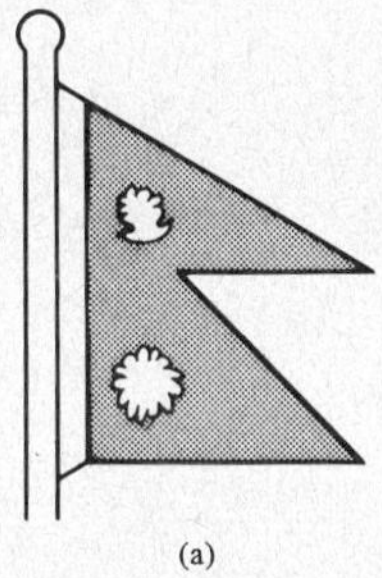

Figure 1.16a
FLAG OF NEPAL.

C. problems

79. **brain teaser** A regular dodecagon can be constructed by the amazing technique illustrated in Figure 1.17. Noticing the 60° and 45° angles (in the detail drawing), compute the exact area of a regular dodecagon inscribed in a circle of radius 2 cm.

A dodecagon is a 12-sided polygon.

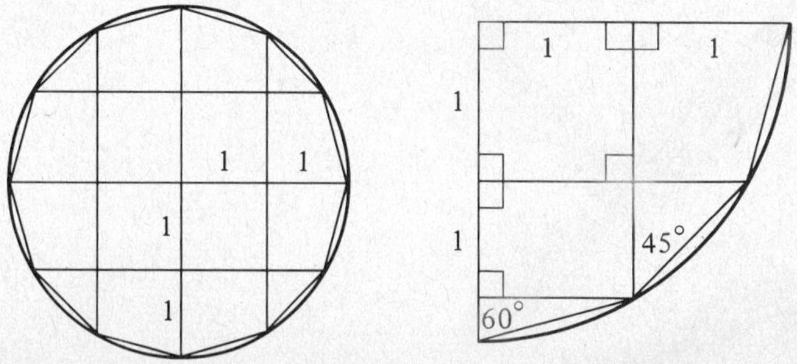

Figure 1.17 REGULAR DODECAGON.

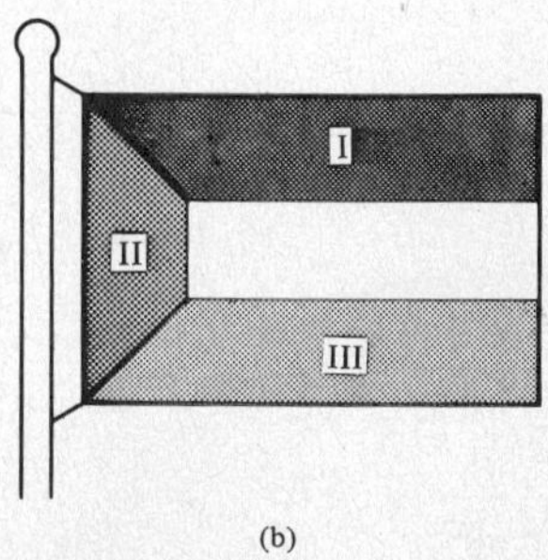

Figure 1.16b
FLAG OF KUWAIT.

80. **brain teaser** Find a, b, and c exactly in Figure 1.18. Use the results to find sin 15°, cos 15°, and tan 15° exactly.

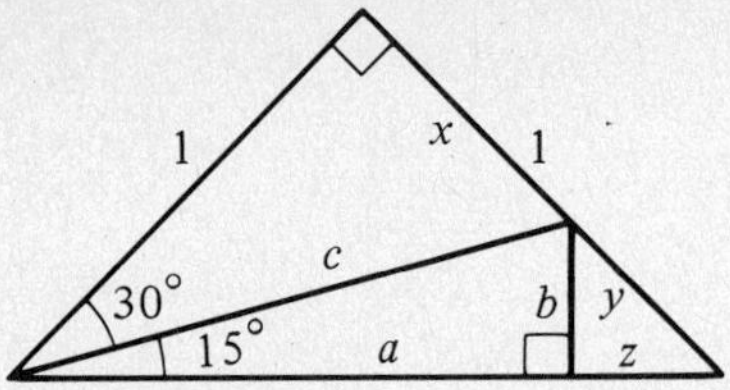

Figure 1.18

1.4 SOLVING RIGHT TRIANGLES I

The exact values of the trigonometric ratios are only practical in very special circumstances. For angles other than 30°, 45°, and 60°, a calculator or tables are much more efficient. Let's first look at a simple form of a table of trigonometric ratios, as shown in Figure 1.19.

The table in Figure 1.19 gives the values of the six trigonometric ratios for angles from 0° to 90° for each degree. The use of the table is best illustrated by example. The table can easily be misread, so notice that 0° to 45° are listed *down* the left-hand column with a legend at the top of the table. However, angles from 45° to 90° are listed *up* the right-hand column and are read from a legend at the bottom of the table. For instance, cos 38° is found by locating 38 in the degree column on the left and moving across that row to the column labeled cosine at the top to find cos 38° ≈ 0.7880. However, tan 76° is determined a little differently, because 76 is in the degree column at the right. Moving across that row to the column labeled tangent at the bottom of the table, we see tan 76° ≈ 4.0108. Double check yourself with the following examples from the table.

EXAMPLES

Use the table in Figure 1.19 to evaluate the following.

1. sin 81°

   sin 81° ≈ 0.9877

2. sec 23°

   sec 23° ≈ 1.0864

3. cot 51°

   cot 51° ≈ 0.8098

   (No, not 1.2349.)

Note the use of the "≈" sign. Results from a table like this or a calculator are approximations.

FIGURE 1.19 TABLE OF TRIGONOMETRIC RATIOS

DEG	SIN	COS	TAN	COT	SEC	CSC	
0	0.0000	1.0000	0.0000	UND	1.0000	UND	90
1	0.0175	0.9998	0.0175	57.2900	1.0002	57.2987	89
2	0.0349	0.9994	0.0349	28.6363	1.0006	28.6537	88
3	0.0523	0.9986	0.0524	19.0812	1.0014	19.1073	87
4	0.0698	0.9976	0.0699	14.3007	1.0024	14.3356	86
5	0.0872	0.9962	0.0875	11.4301	1.0038	11.4737	85
6	0.1045	0.9945	0.1051	9.5144	1.0055	9.5668	84
7	0.1219	0.9925	0.1228	8.1444	1.0075	8.2055	83
8	0.1392	0.9903	0.1405	7.1154	1.0098	7.1853	82
9	0.1564	0.9877	0.1584	6.3138	1.0125	6.3925	81
10	0.1736	0.9848	0.1763	5.6713	1.0154	5.7588	80
11	0.1908	0.9816	0.1944	5.1446	1.0187	5.2408	79
12	0.2079	0.9781	0.2126	4.7046	1.0223	4.8097	78
13	0.2250	0.9744	0.2309	4.3315	1.0263	4.4454	77
14	0.2419	0.9703	0.2493	4.0108	1.0306	4.1336	76
15	0.2588	0.9659	0.2679	3.7321	1.0353	3.8637	75
16	0.2756	0.9613	0.2867	3.4874	1.0403	3.6280	74
17	0.2924	0.9563	0.3057	3.2709	1.0457	3.4203	73
18	0.3090	0.9511	0.3249	3.0777	1.0515	3.2361	72
19	0.3256	0.9455	0.3443	2.9042	1.0576	3.0716	71
20	0.3420	0.9397	0.3640	2.7475	1.0642	2.9238	70
21	0.3584	0.9336	0.3839	2.6051	1.0711	2.7904	69
22	0.3746	0.9272	0.4040	2.4751	1.0785	2.6695	68
23	0.3907	0.9205	0.4245	2.3559	1.0864	2.5593	67
24	0.4067	0.9135	0.4452	2.2460	1.0946	2.4586	66
25	0.4226	0.9063	0.4663	2.1445	1.1034	2.3662	65
26	0.4384	0.8988	0.4877	2.0503	1.1126	2.2812	64
27	0.4540	0.8910	0.5095	1.9626	1.1223	2.2027	63
28	0.4695	0.8829	0.5317	1.8807	1.1326	2.1301	62
29	0.4848	0.8746	0.5543	1.8040	1.1434	2.0627	61
30	0.5000	0.8660	0.5773	1.7321	1.1547	2.0000	60
31	0.5150	0.8572	0.6009	1.6643	1.1666	1.9416	59
32	0.5299	0.8480	0.6249	1.6003	1.1792	1.8871	58
33	0.5446	0.8387	0.6494	1.5399	1.1924	1.8361	57
34	0.5592	0.8290	0.6745	1.4826	1.2062	1.7883	56
35	0.5736	0.8192	0.7002	1.4281	1.2208	1.7434	55
36	0.5878	0.8090	0.7265	1.3764	1.2361	1.7013	54
37	0.6018	0.7986	0.7536	1.3270	1.2521	1.6616	53
38	0.6157	0.7880	0.7813	1.2799	1.2690	1.6243	52
39	0.6293	0.7771	0.8098	1.2349	1.2868	1.5890	51
40	0.6428	0.7660	0.8391	1.1918	1.3054	1.5557	50
41	0.6561	0.7547	0.8693	1.1504	1.3250	1.5243	49
42	0.6691	0.7431	0.9004	1.1106	1.3456	1.4945	48
43	0.6820	0.7314	0.9325	1.0724	1.3673	1.4663	47
44	0.6947	0.7193	0.9657	1.0355	1.3902	1.4396	46
45	0.7071	0.7071	1.0000	1.0000	1.4142	1.4142	45
	COS	SIN	COT	TAN	CSC	SEC	DEG

Now let's put this table to work to solve for the missing parts of a right triangle when certain information is known.

EXAMPLES

4. Find a and b.

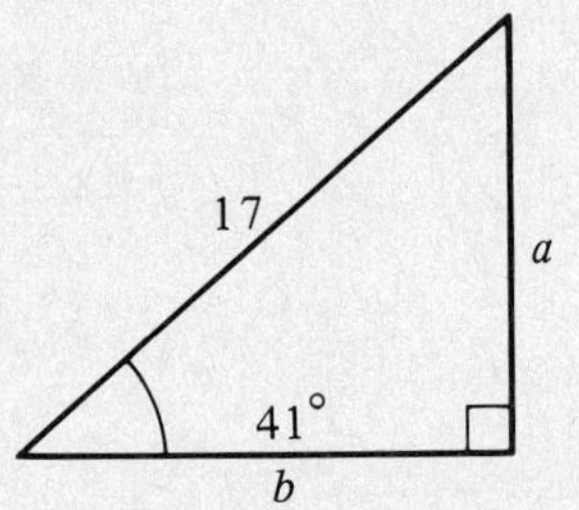

Either sine or cosecant may be used to find a. Sine is more convenient. (Why?) Similarly, cosine is used to find b.

$$\sin 41° = \frac{a}{17} \qquad \cos 41° = \frac{b}{17}$$

$$a = 17 \sin 41° \qquad b = 17 \cos 41°$$

$$\approx 17(0.6561) \qquad \approx 17(0.7547)$$

$$\approx \mathbf{11.2} \qquad \approx \mathbf{12.8}$$

Sin 41° and cos 41° were obtained in the table in Figure 1.19. Did you locate the values?

5. Find b and c.

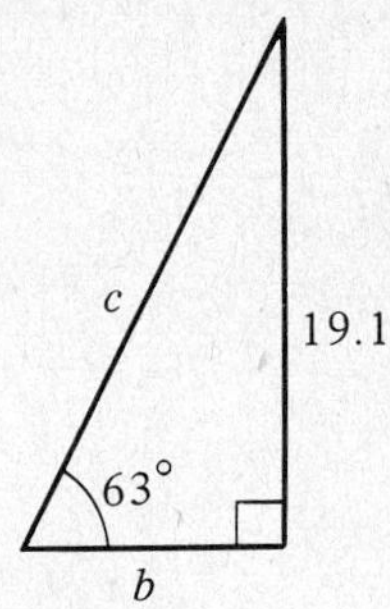

Choose the appropriate ratio and set up a equation to find each.

$$\cot 63° = \frac{b}{19.1} \qquad \csc 63° = \frac{c}{19.1}$$

$$b = 19.1 \cot 63° \qquad c = 19.1 \csc 63°$$

$$\approx 19.1(0.5095) \qquad \approx 19.1(1.122)$$

$$\approx \mathbf{9.7} \qquad \approx \mathbf{21.4}$$

Without the aid of such tables, the ancient astronomers had to resort to scale drawings, which meant much less accuracy and certainly much more time. Today the hand-held calculator offers the same contrast. Although tables are usually accurate to four decimal places, calculators are most commonly accurate to eight, with many having 10- and even 12-place accuracy. This increase in accuracy may be "overkill" for many of our problems, but the time factor enables one to do many more problems than if tables, long multiplication, and logarithms are used. However, just as tables are easily misread, calculators also offer many ways to obtain incorrect results. In fact, before the calculator may be used efficiently and effectively, the operator must have a good understanding of the trigonometric ratios and many of their properties.

The trigonometric ratios are all possible ratios of the three sides of a right triangle. As a result, they are closely related. Notice first that each is the reciprocal of one of the others.

$$\sin A = \frac{1}{\csc A} \qquad \cos A = \frac{1}{\sec A} \qquad \tan A = \frac{1}{\cot A}$$

Further, consider the ratio of sin A to cos A.

$$\frac{\sin A}{\cos A} = \frac{a/c}{b/c} = \frac{a}{c} \cdot \frac{c}{b} = \frac{a}{b} = \tan A$$

These relationships are true for all values of A for which they are defined, so they are called the ratio and reciprocal **identities.**

RECIPROCAL AND RATIO IDENTITIES

$$\sin A = \frac{1}{\csc A} \qquad \cot A = \frac{1}{\tan A} = \frac{\cos A}{\sin A}$$

$$\cos A = \frac{1}{\sec A} \qquad \sec A = \frac{1}{\cos A}$$

$$\tan A = \frac{1}{\cot A} = \frac{\sin A}{\cos A} \qquad \csc A = \frac{1}{\sin A}$$

In smaller tables of trigonometric tables, only sine, cosine, and tangent are given, because the others can be found as reciprocals. This also explains why your calculator has keys for only these three.

To find cot A, for example, you need only find tan A and take the reciprocal by pressing the $1/x$ key. The following examples show how

HISTORICAL ASIDE

Tracing the origin of a word often means a good deal of guesswork, especially when several languages are involved. The word sine is a fanciful example. The Arab writer Āryabhâta (ca. 530 A.D.) used ardha-jyjya (or half-chord) for what we now call sine. The Arabs had an interesting habit of omitting vowels, so jyā was interpreted phonetically as jiba by the Arabs and as jaib by Latin scholars. As a result, jyā or chord became jaib, which means bosom, for which the Latin word is sinus. Thereby hangs the tale. Āryabhata is also known to have given 62,832/20,000 (or 3927/1250) as an approximation for π.

the calculator is used to perform the calculations in Examples 4 and 5 for side b. Blondie admits that she must ". . . find out what it is," before she uses her new calculator. The calculator only "does basic trigonometry" in the hands of a skilled operator, who must set the problem up carefully *before* a calculator can be of any use.

EXAMPLES

6. (from Example 4)

$$b = 17 \cos 41°$$

First be sure that the calculator is set for degrees, if the model has a selection switch for angle measures.

ALGEBRAIC

Key	Display
41	41
cos	0.75470958
×	0.75470958
17	17
=	12.830063

$\mathbf{b = 17 \cos 41° \approx 12.8}$

7. (from Example 5)

$$b = 19.1 \cot 63°$$

ALGEBRAIC

Key	Display
63	63
tan	1.9626105
1/x	0.50952545
×	0.50952545
19.1	19.1
=	9.7319361

Notice how the $1/x$ key is used to compute cotangent.

$$b = 19.1 \cot 63° \approx 9.7$$

Whether you use tables or a calculator, be certain that you carefully set up the problem before attempting to compute an answer. A sketch will be helpful until the ratios are fixed firmly in your mind. Commit them to memory and practice identifying the six by name on an arbitrarily labeled figure. These six ratios form the foundation of the rest of the course, and the most expensive calculator won't remember them for you.

Few calculators allow the direct input of an angle in degrees, minutes, and seconds. The input and display is generally in decimal form. For instance, 34°30′ must be entered as $34\frac{30}{60}° = 34\frac{1}{2}° = 34.5°$. Thus an angle such as 73°49′ must first be converted to decimal form, as shown in the following example.

EXAMPLE

Evaluate the following to the nearest ten-thousandth.

8. cos 73°49′

First, find the fraction of a degree as a decimal. Second, add the whole number of degrees. Finally, find the cosine of the angle in decimal form.

ALGEBRAIC

Key	Display
49	49
÷	49

Key	Display
[60]	60
[+]	0.81666666
[73]	73
[=]	73.81666666
[cos]	0.27871175

cos 73°49′ ≈ 0.2787

9. cot 23°17′

You must find the tangent first, then use the $1/x$ key to get cotangent.

ALGEBRAIC

Key	Display
[17]	17
[÷]	17
[60]	60
[+]	0.28333333
[23]	23
[=]	23.28333333
[tan]	0.43032324
[$1/x$]	2.3238345

cot 23°17′ ≈ 2.3238

In certain applied problems, angles of elevation and depression are relative to the horizontal. The angle of elevation and the angle of depression are actually alternate interior angles and therefore always equal.

EXAMPLE

10. The Eiffel Tower casts a 1275-ft shadow when the elevation of the sun is 37°40′. How tall is this famous Paris landmark?

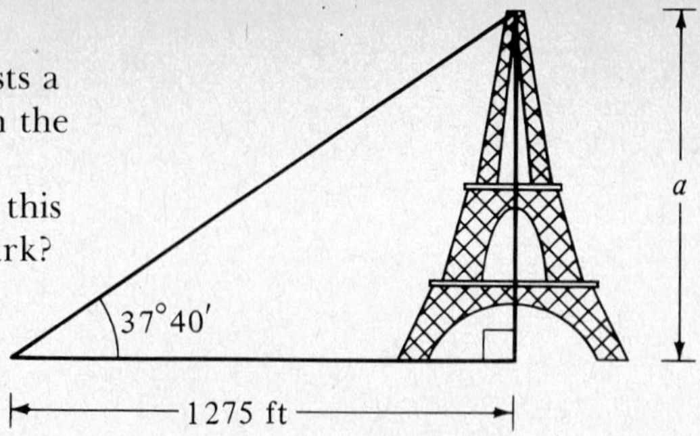

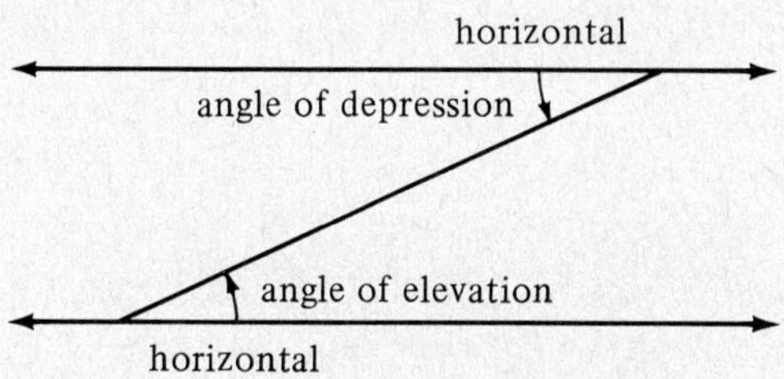

$$\frac{a}{1275} = \tan 37°40'$$

$$a = 1275 \tan 37°40'$$

First enter $\frac{40}{60}$ for degrees. Then add 37° to fraction. Then find the tangent, and multiply by 1275.

ALGEBRAIC

Key	Display
40	40
÷	40
60	60
+	0.66666667
37	37
=	37.66666667
tan	0.77195895
×	0.77195895

1275 1275

= 984.24765

Now, round off the answer.

$$a \approx 984$$

The Eiffel Tower is 984 ft tall.

exercise set 1.4

A. exercises

Find the value of each trigonometric ratio in Exercises 1–20 to the nearest ten-thousandth.

1. $\sin 15°$
2. $\cos 75°$
3. $\cos 69°$ See Examples 1–3.
4. $\sin 21°$
5. $\tan 38°$
6. $\tan 52°$
7. $\sec 43°$
8. $\cot 34°$
9. $\cot 79°$
10. $\csc 54°$
11. $\cos 31°40'$
12. $\tan 6°20'$ See Examples 8 and 9.
13. $\sin 65.5°$
14. $\cos 70.1°$
15. $\cos 5°38'$
16. $\sin 13°54'$
17. $\tan 43.12°$
18. $\cot 27.41°$
19. $\csc 62°48'$
20. $\sec 34°8'$

B. exercises

For Exercises 21–26, choose the appropriate ratio, set up an equation, and find the sides in the triangles shown.

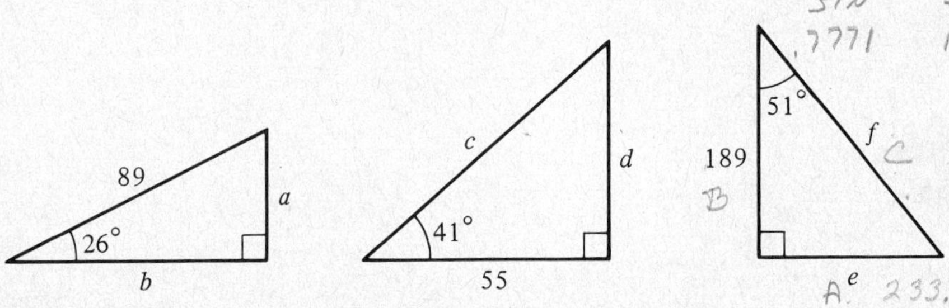

21. a
22. b
23. c See Examples 4 and 5.
24. d
25. e
26. f

Exercises 27–40 refer to △ABC, where C is the right angle.

See Examples 6 and 7.

27. $A = 24°$, $c = 15.7$, $a = ?$
28. $B = 35°$, $c = 23.6$, $b = ?$
29. $B = 61°$, $a = 9.65$, $c = ?$
30. $A = 74°$, $b = 8.76$, $c = ?$
31. $B = 15°$, $a = 234$, $b = ?$
32. $A = 51°$, $a = 197$, $b = ?$
33. $A = 21°30'$, $b = 20.25$, $a = ?$
34. $B = 33°40'$, $b = 41.75$, $a = ?$
35. $B = 39°20'$, $a = 3.875$, $c = ?$
36. $B = 57°50'$, $b = 9.125$, $c = ?$
37. $A = 69.5°$, $a = 0.034$, $c = ?$
38. $A = 75.5°$, $b = 0.738$, $c = ?$
39. $B = 2.45°$, $c = 1247$, $b = ?$
40. $A = 84.18°$, $c = 27.67$, $a = ?$

Read Exercises 41–50 carefully, make a sketch, select the appropriate ratio, and set up an equation before attempting to calculate the result.

See Example 10.

41. A missile is fired at an angle of 62° with the horizontal. If we assume that it travels in a straight line, how high is it when it has gone 557 m?

42. A missile travels 1 km (1000 m) after being launched at an angle of 56° with the horizontal. If it travels in a straight line, what is its height?

43. The shadow of a building is 89 ft when the elevation of the sun is 38°10′. What is the height of the building to the nearest foot?

44. When the elevation of the sun is 23°40′, the shadow of a tower is 64 m. What is the height of the tower to the nearest meter?

45. A kite trails 200 ft of taut line in a stiff breeze. If the string makes an angle of 32.18° with the level ground, what is the altitude of the kite?

46. A tower is secured by guy wires. If one line 105 ft long makes an angle of 62.24° with the horizontal, how high on the tower is the line attached?

47. A metal rod must be tapered at the angle $\theta = 42.25°$ as shown in Figure 1.20. If the tapered portion is 2.54 cm in length, what is the radius of the rod to the nearest hundredth of a centimeter?

Figure 1.20

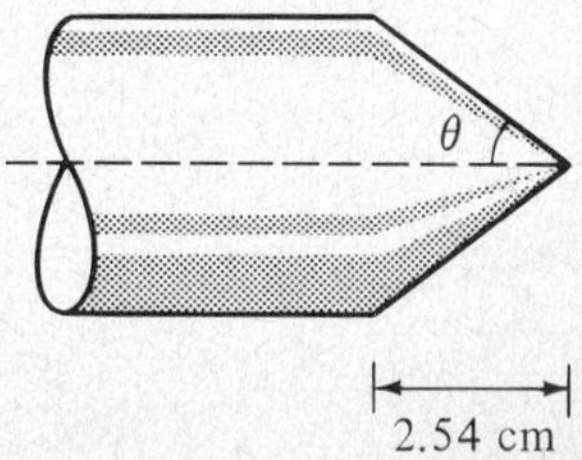

48. A certain pencil sharpener produces a taper with an angle of 7.32° as shown in Figure 1.21. If the diameter of the pencil is 0.65 cm, what is the length along the taper to the nearest hundredth of a centimeter?

49. The world's highest bridge crosses the Colorado River at Royal Gorge, Colorado. If you stand on the riverbank 120 ft from the

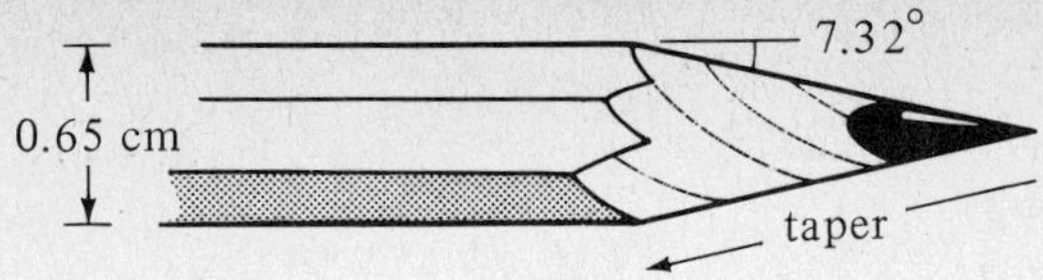

Figure 1.21

base of the bridge, the angle of elevation is 83°30′. How high is this bridge?

50. Russia has the world's tallest dam at Rojunsky. From a point in the river bed 450 m from the base of the structure, the angle of elevation is 35°50′. How tall is the dam?

What is the height in feet? Is the dam as tall as the Royal Gorge Bridge?

C. problems

51. **cagey calculation** You may have noticed that certain numerals in the calculator display look like letters of the alphabet when it is inverted (turned upside down).

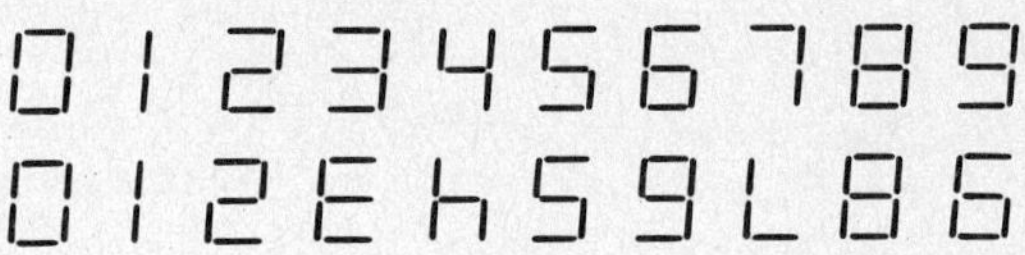

Calculator display

Display inverted

0 I Z E h S g L B G Letters of alphabet

Little Hugh Packer was planning a tenth birthday party for six of his friends, and he kept pestering his mother about taking everyone to a movie as part of the party. She was reluctant to commit herself, so she used her calculator to determine a fair judgment. "Let's see," she said, "the show starts at 1:10 and is out at 4:36. Let's multiply the times and add 3 for the number of features, times seven boys, plus 30 min travel time, and divide by 10 for your tenth birthday." WHAT DID THE CALCULATOR SAY? Just what mothers always say, right?

52. **cagey calculation** Our shrewd and intrepid calculator, David Digit, has a problem. Whenever he computes the following on his trusty HP-41C,

$$\sin 22.222° - 2.4966 \cdot 10^{-6}$$

he contends that his result is illegible. What do you get? Be advised that you must have a 10-digit display for this experiment.

1.5 SOLVING RIGHT TRIANGLES II

As we saw in the previous section, the trigonometric ratios are paired: sine and cosine, tangent and cotangent, secant and cosecent. What is the

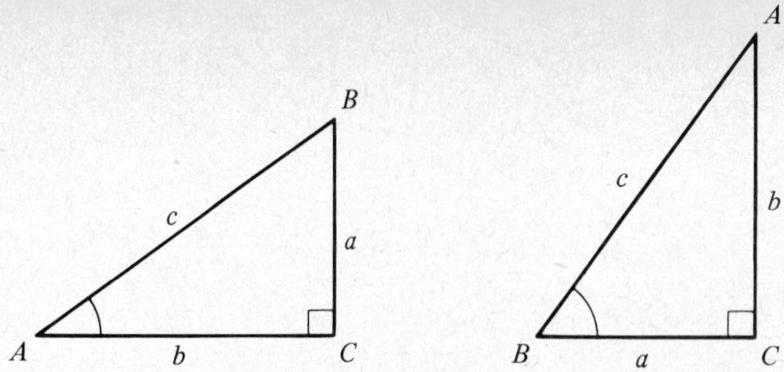

$$\sin A = \frac{a}{c} = \cos B$$
$$\cos A = \frac{b}{c} = \sin B$$
$$\tan A = \frac{a}{b} = \cot B$$
$$\cot A = \frac{b}{a} = \tan B$$
$$\sec A = \frac{c}{b} = \csc B$$
$$\csc A = \frac{c}{a} = \sec B$$

Figure 1.22 RATIOS FOR A AND B.

significance of these *co*ratios? To see the meaning of these names, consider $\triangle ABC$ and the ratios for both angles A and B. To make that easier to see, Figure 1.22 shows the triangle in two positions with all the ratios listed.

One notices immediately the relationship between the coratios of angles A and B. Because A and B are the acute angles in a right triangle, A and B are complementary. The sine of an angle is then equal to the *co*-sine of its *co*mplement. The same is true for tangent and cotangent, and the secant and cosecant. These relationships may be written as follows.

CORATIOS OF COMPLEMENTARY ANGLES

$$\sin A = \cos(90° - A)$$
$$\tan A = \cot(90° - A)$$
$$\sec A = \csc(90° - A)$$

EXAMPLES

1. Find $\cos 63°$ if $\sin 27° = 0.4540$.

$$\cos(90° - 27°) = \sin 27°$$
$$\mathbf{\cos 63° = 0.4540}$$

2. If $\tan\theta = \cot 51°$, then $\theta = ?$

$$\cot 51° = \tan(90° - 51°)$$

$$\tan\theta = \tan 39°$$

Thus

$$\boldsymbol{\theta = 39°}$$

This complementary relationship explains the design of the trigonometric tables. Recall in the tables that 0°–45° runs *down* the left and 45°–90° *up* the right. Notice that angles directly opposite on the left and right are complements and that the legends at the top and bottom simply interchange coratios. This relationship reduces the table to just half the size it might be without it.

In Section 1.4, the angle in each problem was given and one of the sides was computed. Consider now a situation in which the sides are known and the angle must be found. Suppose that your roof has a "pitch" of 5:12. The pitch of a roof is the ratio of its "rise" to its "span" as illustrated in Figure 1.23. What angle does the roof make with the horizontal? This angle is θ and the tangent ratio relates the rise and the span.

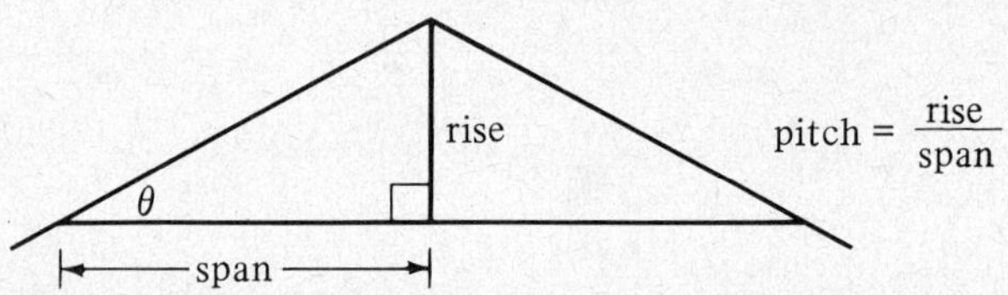

Figure 1.23 THE PITCH OF A ROOF.

$$\tan\theta = \frac{\text{rise}}{\text{span}}$$

$$\tan\theta = \frac{5}{12}$$

$$\tan\theta \approx 0.4167$$

Look in the tangent column of the table (see Figure 1.19) for the value 0.4167. It occurs between 0.4040 and 0.4245, the tangents of 22° and 23°. The closer value is the tangent of 23°, so

$$\tan 23° \approx 0.4167$$

$$\theta \approx 23°$$

This approximation could be obtained more accurately in a better table or using a calculator. Essentially what we found in the tables could be written as follows:

$$\tan \theta = \tfrac{5}{12}$$

$$\theta = \text{angle whose tangent is } \tfrac{5}{12}$$

This is abbreviated as

$$\theta = \arctan \tfrac{5}{12}$$

or

$$\theta = \tan^{-1} \tfrac{5}{12}$$

$$\theta \approx 23°$$

A calculator will have an [ARC], [INV], or [2nd] key to compute this "arctan." It will be a two-key operation: either [ARC] [tan] or [INV] [tan] or [2nd] [$\tan^{-1}$] or a similar combination depending on make and model. Thus, in the examples that follow, some adaptations may have to be made.

EXAMPLES

3. Compute $\theta = \arctan \frac{5}{12}$ from the preceding.

ALGEBRAIC

Key	Display
[5]	5
[÷]	5
[12]	12
[=]	0.41666667
[arc]	0.41666667
[tan]	22.619865

Be certain that the calculator is set on degrees if the model has a selection switch for angle measure.

$$\boldsymbol{\theta \approx 22.619865°}$$

The decimal degree result is usually the preferred form, but it may be converted to degrees-minutes-seconds if necessary. That is, we can subtract the whole number of degrees and convert the decimal part to minutes.

Key	Display
[−]	22.619865
[22]	22

$\boxed{=}$ 0.619865

$\boxed{\times}$ 0.619865

$\boxed{60}$ 60

$\boxed{=}$ 37.191897

$$\boldsymbol{\theta \approx 22°37'}$$

Similarily, we can compute to the nearest second.

$\boxed{-}$ 37.191897

$\boxed{37}$ 37

$\boxed{=}$ 0.191897

$\boxed{\times}$ 0.191897

$\boxed{60}$ 60

$\boxed{=}$ 11.51382

$$\boldsymbol{\theta \approx 22°37'12''}$$

4. A 20-ft ladder reaches 12 ft up the side of a building. What angle will the ladder make with the level ground?
   First, make a sketch!

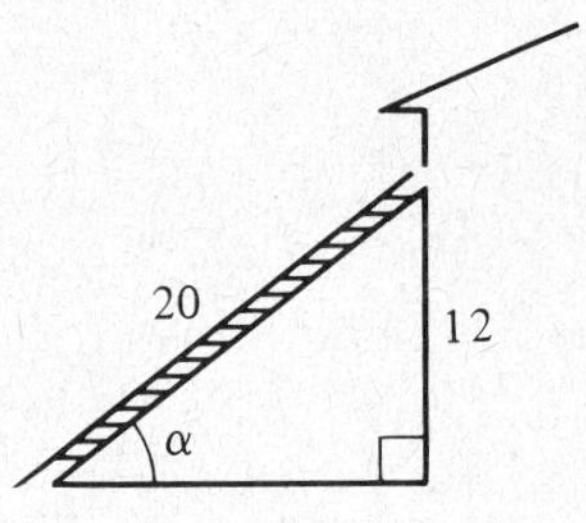

$\sin \alpha = \frac{12}{20}$

$\alpha = \arcsin \frac{12}{20}$

$\approx 36.869898°$

$\approx 36.870°$

Remember, this is read: "α is the angle whose sine is $\frac{12}{20}$."

The ladder makes an angle of approximately 36.870° with the ground.

5. A 30-ft guy wire is to be attached to a TV mast 18 ft from its base. What angle will it make with the horizontal rooftop, and how high up the mast will the wire be attached?

$$\cos\beta = \tfrac{18}{30}$$

$$\beta = \arccos \tfrac{18}{30}$$

$$\approx 53.130102°$$

$$\approx 53.130°$$

$$h^2 + 18^2 = 30^2$$

$$h^2 = 30^2 - 18^2$$

$$= 576$$

$$h = 24$$

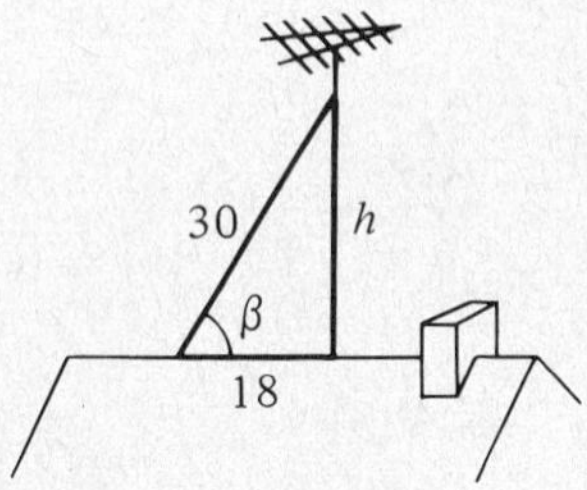

The guy wire is attached 24 ft up the pole and makes an angle of approximately 53.130° with the roof.

6. A tunnel is to be dug under a mountain on a line through A and B. Point C is chosen so that both A and B are visible.

If C is 2.1 mi from A, 1.2 mi from B, and the angle at C is 90°, find the angles that diggers must follow at A and B in order to meet under the mountain and complete the tunnel.

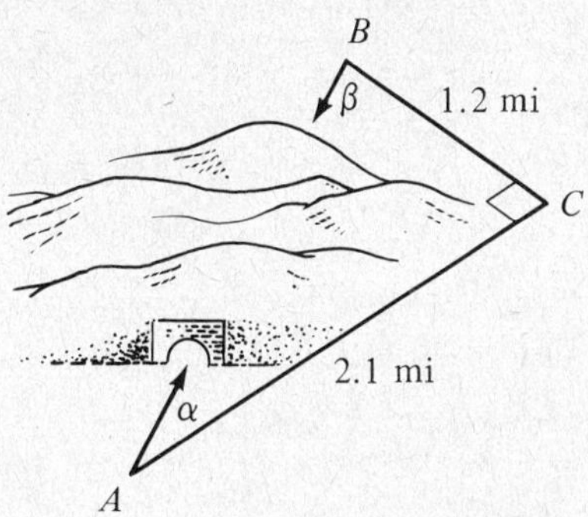

$$\tan\alpha = \frac{1.2}{2.1}$$

$$\alpha = \arctan\frac{1.2}{2.1}$$

$$\approx \arctan 0.57142857$$

[1.2] [÷] [2.1] [=] [ARC] [tan]

$$\alpha \approx 29.74488°$$

$$\beta = 90° - \alpha$$

$$\approx 90° - 29.74488°$$

$$\approx 60.25512°$$

The diggers must stay on 29.745° from A and 60.255° from B.

exercise set 1.5

A. exercises

Express each of the ratios in Exercises 1–10 in terms of the coratio of the ratio given.

1. $\sin 37^\circ$ 2. $\tan 73^\circ$ 3. $\cos 22^\circ$ See Example 1.

4. $\csc 81^\circ$ 5. $\sec 15^\circ$ 6. $\cot 64^\circ$

7. $\cos 54^\circ$ 8. $\cot 45^\circ$ 9. $\sin 56^\circ$

10. $\sec 65^\circ$

Find the angle in Exercises 11–20.

11. $\tan A = \cot 28^\circ$ 12. $\sin B = \cos 78^\circ$ See Example 2.

13. $\csc \alpha = \sec 51^\circ$ 14. $\cot \beta = \tan 41^\circ$

15. $\sec C = \csc 33^\circ$ 16. $\tan D = \cot 63^\circ$

17. $\sin(90^\circ - \alpha) = \cos 17^\circ$ 18. $\sec(90^\circ - \beta) = \csc 87^\circ$

19. $\cos A = \dfrac{1}{\sec 49^\circ}$ 20. $\tan 59^\circ = \dfrac{1}{\cot B}$

Find the value of the ratios in Exercises 21–30, stating the answer to the nearest ten-thousandth.

21. $A = \sec 23.7^\circ$ 22. $B = \sin 19.4^\circ$

23. $C = \cot 62^\circ 53'$ 24. $D = \csc 59.15^\circ$

25. $E = \tan 30^\circ 37'$ 26. $F = \sec 40.95^\circ$ See Exercise 65.

27. $G = \cos 5^\circ 29'$ 28. $H = \csc 84^\circ 31'$

29. $I = \sin 39.8^\circ$ 30. $L = \tan 28.05^\circ$

Find the angle in Exercises 31–40 to the nearest minute.

31. $\sec M = 1.4335$ 32. $\sin N = 0.8342$ See Example 3.

33. $\phi = \arccos 0.2210$ 34. $P = \text{arccot}\ 4.4676$

35. $\csc R = 1.3369$ 36. $\cos S = 0.8736$ See Exercise 65.

37. $T = \arctan 0.8708$ 38. $X = \arcsin 0.9958$

39. $0.4046 = \sin Y$ 40. $1.2624 = \sec Z$

B. exercises

41. The shadow of a vertical post is 31.5 ft long when the angle of elevation of the sun is 18.78°. What is the height of the post? See Problems 4–6.

42. When the angle of elevation of the sun is 29.57°, the shadow of a vertical pole is 21.3 m. What is the height of the pole?

43. The shadow of a 15 ft 8 in. vertical post is 22 ft 3 in. What is the angle of elevation of the sun?

44. A 42 ft 6 in. telephone pole casts a shadow that is 73 ft 9 in. long. What is the angle of elevation of the sun?

45. A vertical post is 9 ft 4 in. tall and the angle of elevation of the sun is 58°7′. What is the length of the shadow of the post?

46. The angle of elevation of the sun is 62.9° and the height of a vertical fence post is 2.3 m. What is the length of the shadow?

47. Standing at the top of a 150-ft cliff, the angle of depression to a ship off shore is 9.07°. How far is the ship from the foot of the cliff?

48. The observation platform of a lighthouse is 135 ft above the water. If the angle of depression of a boat at sea is 3.28°, how far away is the boat?

49. A thousand feet away on level ground, the angle of elevation would be 40°28′ for the Transamerica Pyramid in San Francisco. How tall is this remarkable 48-story building?

50. A thousand feet back from the base of the Sears Tower in Chicago, the angle of elevation is 55°29′. What is the height of this 110-story mammoth?

Exercises 51–60 refer to the figure shown below.

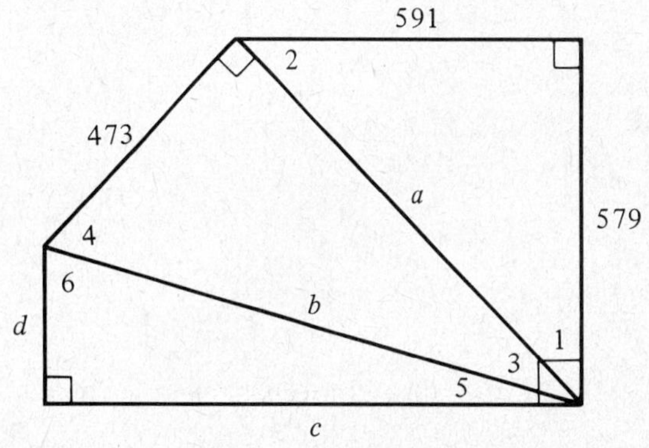

51. $\angle 1$ 52. $\angle 2$ 53. a 54. b

55. $\angle 3$ 56. $\angle 4$ 57. $\angle 5$ 58. $\angle 6$

59. c 60. d

Note that this set is sequential, since a must be found before $\angle 3$, $\angle 4$, or b can be determined.

61. A plane 2100 ft over Staten Island lines up with the Verrazano Narrows bridge to Brooklyn. The angles of depression to the ends of the span are 21°50′ and 12°28′. Find the length of this New York structure to the nearest 10 ft.

62. The Goodyear blimp is hovering over a spot a mile from the Peachtree Plaza in Atlanta. The angles of depression to the base and the top of the Plaza are 13°50′ and 6°14′. How tall is this Georgia complex?

63. A plane is 2 mi from being directly over the MLC Building in Sidney, Australia. The angles of depression to the top and base of the 70-story building are 3°52′ and 8°5′. Find the height of the largest building "down under."

64. An airplane maintains an altitude of 600 m over Lisbon, Portugal. As it approaches the Salazar Bridge, the angles of depression to the ends of the bridge are 16°36′ and 30°58′. Find the length of this span over the Tagus River.

C. exercises

65. **trig talk** Use Exercises 21–40 to fill the blanks with the capital letter that corresponds to the value below the blank. When all the blanks are filled, the quotation from Plato may be read.

“ ___ ___ ___   ___ ___   ___ ___ ___

0.5328 0.5918 41°3′   56°32′ 77°14′   77°14′ 56°32′ 0.5918

___ ___ ___ ___ ___ ___ ___ ___   ___ ___

0.6401 0.9954 56°32′ 77°14′ 48°25′ 1.0921 56°32′ 41°3′   77°14′ 1.3240

___ ___ ___ ___ ___ ___ ___ ___

0.9954 0.5918 77°14′ 45°46′ 0.5918 41°3′ 48°25′ 23°52′

___ ___ ___ ___ ___   ___ ___

0.5918 56°32′ 41°3′ 0.5918 48°25′   45°46′ 23°52′

___ ___ ___ ___.”

1.1648 77°14′ 77°14′ 48°25′   (Plato)

66. **brain teaser** If $a^2 + b^2 = c^2$, then $\angle ABC$ is a right triangle.

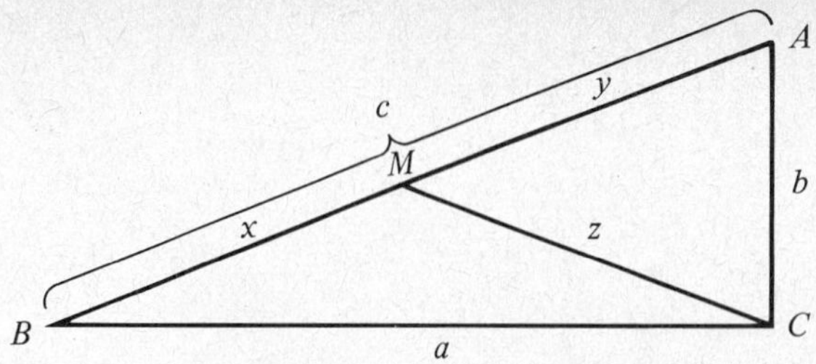

But instead, suppose that $x = y = z$; then what do you know about $\triangle ABC$? Why?

67. **brain teaser** If s is the length of a side of a regular polygon of n sides inscribed in a circle of radius r, show that

$$s = 2r \sin \left(\frac{180}{n}\right)^\circ$$

1.6 CHAPTER REVIEW

Fill in the symbol, word, or words that make the statements in Exercises 1–8 complete and correct.

1. (1.1) Angles with the same initial and terminal sides are called ________________________ angles.
2. (1.1) Two angles whose sum is 90° are called ________________________ angles.
3. (1.2) If two triangles are similar, then their sides are ________________________.
4. (1.2) The ratio of the sides of a 45-45-90 triangle is ________: ________:________.
5. (1.2) The hypotenuse of a 30-60-90 triangle is ________ times the shortest side.
6. (1.3) In $\triangle RAP$ with right angle at A, ________________________ $= p/a$.
7. (1.4) The angle of elevation or depression is always measured relative to the ________________________.

8. (1.5) The cosecant of an angle is equal to the ________________

________________ of its complement.

9. (1.1) Given $\triangle AXE$ with $DZ \parallel XE$, $\angle 1 = 29°$, and $\angle 2 = 106°$.

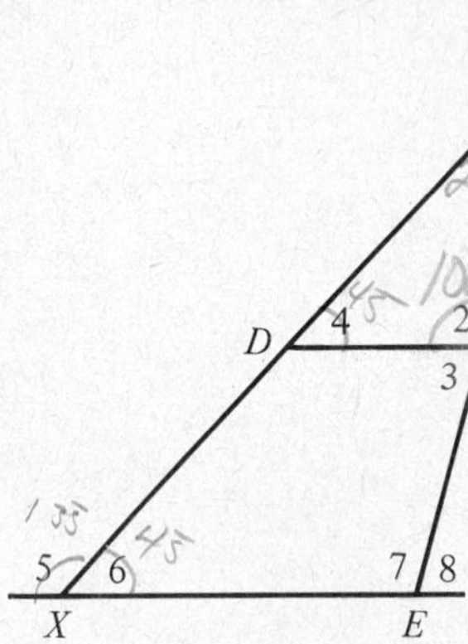

Find each of the following.

(a) $\angle 3$
(b) $\angle 4$
(c) $\angle 5$
(d) $\angle 6$
(e) $\angle 7$

10. (1.2) Given $\triangle TRY \sim \triangle ART$, find each of the following.

(a) RA
(b) RT
(c) AT
(d) TY

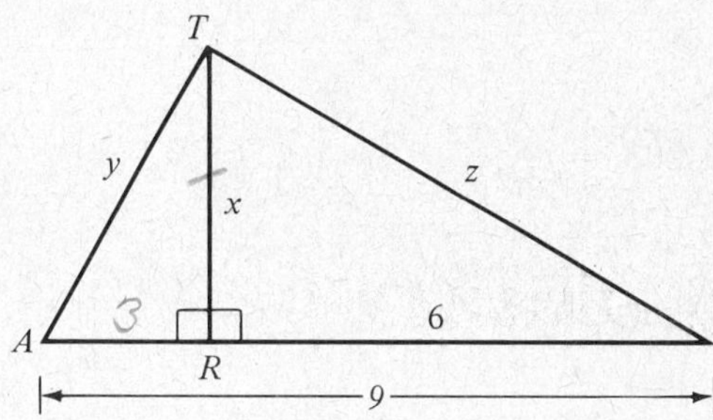

11. (1.3) Find the ratios for the triangle in Figure 1.24.

(a) $\sin \alpha$
(b) $\cos \alpha$
(c) $\tan \alpha$
(d) $\cos \beta$
(e) $\sec \beta$

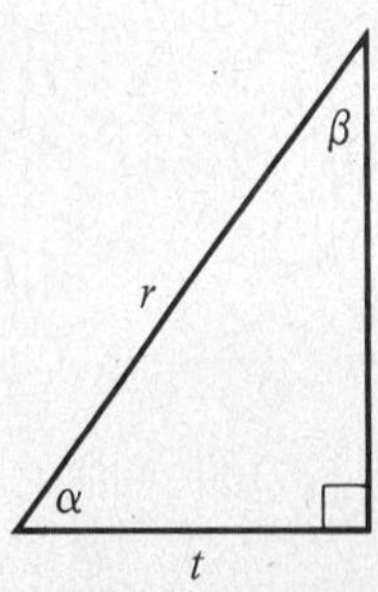

Figure 1.24

12. (1.3) Name the ratios for the triangle in Figure 1.25.

(a) $\frac{p}{y}$

(b) $\frac{y}{h}$

(c) $\frac{h}{p}$

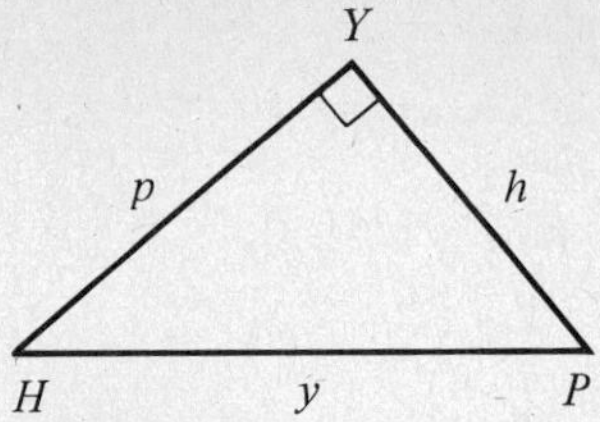

Figure 1.25

Find the exact value in Exercises 13 and 14. Do not approximate.

13. (1.3) Find

(a) $\tan \alpha$ if $\cos \alpha = \frac{9}{41}$
(b) $\sec \beta$ if $\cot \beta = \frac{12}{35}$
(c) $\sin \gamma$ if $\cos \gamma = 3/\sqrt{13}$

14. (1.3, 1.4) Find each part named in Figure 1.26.

(a) α
(b) b
(c) c
(d) d
(e) e

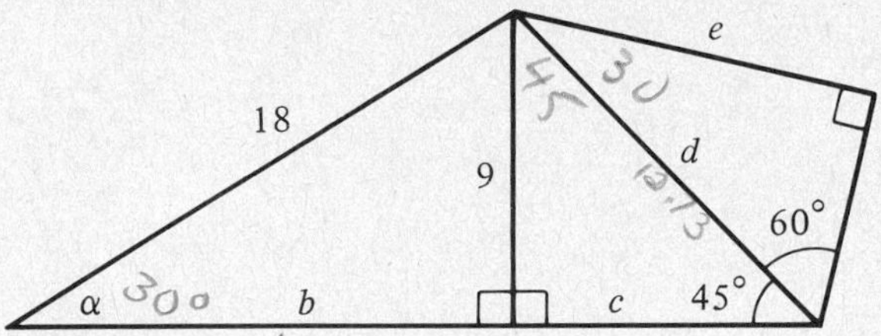

Figure 1.26

In Exercises 15–17, state the answer to the nearest ten-thousandth or to the nearest minute.

15. (1.3, 1.4, 1.5) Find the value of each unknown.

(a) $\sin A = \csc 49.74°$
(b) $1/\cos B = \sec 31°28'$
(c) $\sec 32.14° = C$
(d) $\delta = \text{arccsc } 2.07795$
(e) $\tan \text{E} = 0.6305$

16. (1.4) C is a right angle in $\triangle ABC$.

(a) $A = 23°$, $c = 14.8$, $b = ?$
(b) $B = 58.5°$, $a = 7.56$, $c = ?$
(c) $A = 3°51'$, $c = 2485$, $a = ?$

17. (1,4) A is a right angle in $\triangle NAP$.

(a) $n = 14.8$, $a = 29.6$, $N = ?$
(b) $n = 2344$, $a = 3556$, $P = ?$
(c) $n = 79.82$, $p = 113.56$, $N = ?$

18. (1.4) The Diamond Head Tower in Honolulu is 350 ft tall. At a certain distance the angle of elevation is 49.4°. If you walk 150 ft

directly toward the tower, the angle at that point is 66.8°. How far back were you originally?

19. (1.5) The Ala Moana Hotel in Honolulu is 390 ft tall. If you stand back 150 ft, what is the angle of elevation to the top of the building?

20. (1.5) A ramp is constructed inclined at 30° with the horizontal. If it is 22 ft long, what is the vertical rise along its length?

the trigonometric functions

two

2.1 THE TRIGONOMETRIC FUNCTIONS

Imagine the hands of a large clock. Watch the minute hand as it moves. As the tip of the hand traces its circular path, the *height* of the tip of the hand changes. As a matter of fact, the tip of the hand is continually raised and lowered all day long. Its height depends on the time or the position of the hand, which can be measured by an angle with the horizontal. It is this notion that we wish to formalize and study.

The trigonometric ratios of the preceding chapter were defined for *acute* angles only. The hand of the clock may make *any* angle with the horizontal. In order to extend our definitions, it is necessary to define carefully the position of the angle in the coordinate plane.

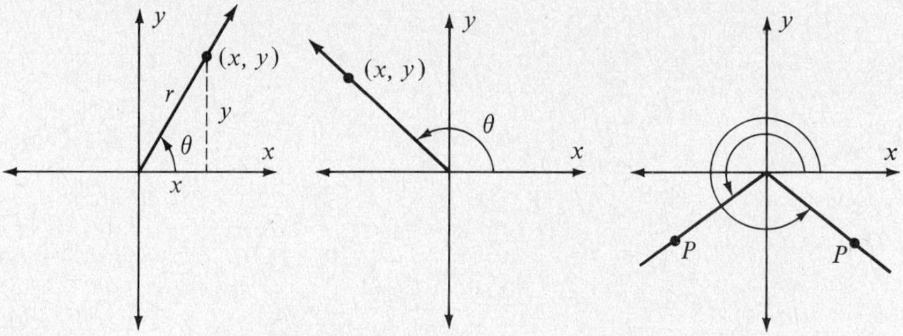

Figure 2.1 Angle in standard position.

An angle in **standard position** has its vertex at the origin and its initial side along the positive x axis. See Figure 2.1. Let θ be in standard position and let P be any point on the terminal side of θ. If P is r units from the origin and has coordinates (x, y), then the distance $r = \sqrt{x^2 + y^2}$ is always nonnegative. The trigonometric functions may now be defined for any angle θ.

TRIGONOMETRIC FUNCTIONS OF ANY ANGLE

$$\sin\theta = \frac{y}{r} \qquad \cot\theta = \frac{x}{y}$$

$$\cos\theta = \frac{x}{r} \qquad \sec\theta = \frac{r}{x}$$

$$\tan\theta = \frac{y}{x} \qquad \csc\theta = \frac{r}{y}$$

where θ is an angle in standard position and (x, y) is the point on the terminal side r units from the origin.

It is important to note that this definition includes as a special case the trigonometric ratios for a right triangle whenever θ is restricted to the first quadrant. That is, if θ is acute, the trigonometric ratios of a right triangle are a special case of the definition.

EXAMPLES

Find sine, cosine, and tangent for the angle specified in standard position.

1. (7, 12) lies on the terminal side of θ.

$$x = 7 \qquad y = 12 \qquad r = \sqrt{x^2 + y^2}$$
$$= \sqrt{49 + 144}$$
$$= 13$$

$$\sin\theta = \frac{y}{r} = \mathbf{\frac{12}{13}}$$

$$\cos\theta = \frac{x}{r} = \mathbf{\frac{7}{13}}$$

$$\tan\theta = \frac{y}{x} = \mathbf{\frac{12}{7}}$$

2. The terminal side of α contains $(-15, 8)$.

$$x = -15 \qquad y = 8 \qquad r = \sqrt{225 + 64}$$
$$= \sqrt{289}$$
$$= 17$$

$$\sin\alpha = \frac{y}{r} = \mathbf{\frac{8}{17}}$$

$$\cos\alpha = \frac{x}{r} = \frac{-15}{17} = \mathbf{-\frac{15}{17}}$$

$$\tan\alpha = \frac{y}{x} = \frac{8}{-15} = \mathbf{-\frac{8}{15}}$$

As we see in Example 2, the values of the functions may now be negative. That is, r is always positive, but the signs of x and y change as the terminal side of the angle moves through the quadrants. The signs of the functions depend on the quadrant of the angle. For instance, sine (and cosecant) will be positive in QI and QII and negative in QIII and QIV, because y is positive above the x axis and negative below. Table 2.1 summarizes the results for all functions.

TABLE 2.1 SIGNS OF THE TRIGONOMETRIC FUNCTIONS

(x_0, y_0)	QI $(+, +)$	QII $(-, +)$	QIII $(-, -)$	QIV $(+, -)$
$\sin\theta$, $\csc\theta$	+	+	−	−
$\cos\theta$, $\sec\theta$	+	−	−	+
$\tan\theta$, $\cot\theta$	+	−	+	−

HISTORICAL ASIDE

The words cosine and cotangent are credited to Edmund Gunter (1581–1626). In 1620 he published a book of seven-place tables for sine and tangent. He was, however, originally a theological student. His classmates recalled that his sermons were just terrible.

EXAMPLES

Name the quadrant containing the terminal side of the angle specified.

3. $\cos\alpha > 0$ and $\tan\alpha < 0$

$\cos\alpha > 0$ for α in QI and QIV.

$\tan\alpha < 0$ for α in QII and QIV.

The conditions are met simultaneously only in QIV; thus

α is in QIV.

4. $\sin\beta < 0$ and $\sec\beta < 0$

$\sin\beta < 0$ for β in QIII and QIV.

$\sec\beta < 0$ for β in QII and QIII.

β is in QIII.

Find $\sin\beta$, $\cos\beta$, and $\tan\beta$ if:

5. $\sec\beta = \frac{13}{5}$ and β is in QIV

$$\sec\beta = \frac{13}{5} = \frac{r}{x}$$

so let $x = 5$ and $r = 13$. Then

$$x^2 + y^2 = r^2$$

$$25 + y^2 = 169$$

$$y^2 = 144$$

$$y = -12$$

since β is in QIV.

$$\sin \beta = \frac{y}{r} = \frac{-12}{13} = -\mathbf{\frac{12}{13}}$$

$$\cos \beta = \frac{x}{r} = \mathbf{\frac{5}{13}}$$

$$\tan \beta = \frac{y}{x} = \frac{-12}{5} = -\mathbf{\frac{12}{5}}$$

This discussion of angles in one of the four quadrants has excluded angles whose terminal side lies on a coordinate axis. Such angles are called *quadrantal* angles, for instance, 0°, 90°, 180°, and so on. One of the coordinates of a point on an axis is zero. Because division by zero is undefined, certain functions of each quadrantal angle will be undefined, for which we use the symbol ∞.

EXAMPLE

6. Find the values of the trigonometric functions of 270°.

First, $(0, -1)$ is a point on the terminal side of 270°, so $x = 0$, $y = -1$, and $r = 1$.

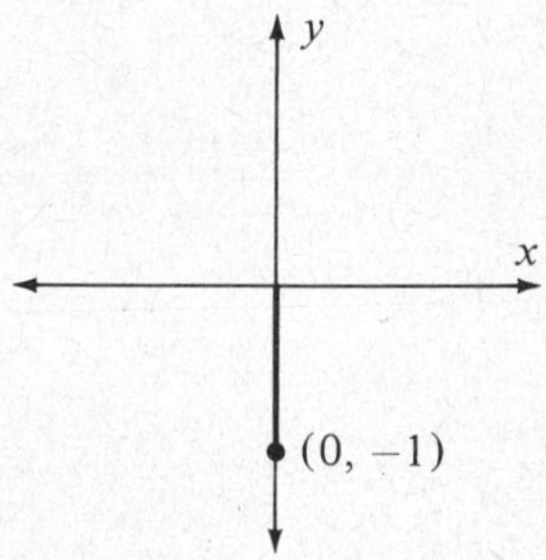

$$\sin 270° = \frac{-1}{1} = -1 \qquad \cot 270° = \frac{0}{-1} = 0$$

$$\cos 270° = \frac{0}{1} = 0 \qquad \sec 270° = \frac{1}{0} = \infty$$

$$\tan 270° = \frac{-1}{0} = \infty \qquad \csc 270° = \frac{1}{-1} = -1$$

Similarly, the other quadrantal angles can be found. The results are summarized in Table 2.2.

TABLE 2.2 QUADRANTAL ANGLES

θ	SIN θ	COS θ	TAN θ	COT θ	SEC θ	CSC θ
0°	0	1	0	∞	1	∞
90°	1	0	∞	0	∞	1
180°	0	−1	0	∞	−1	∞
270°	−1	0	∞	0	∞	−1

exercise set 2.1

A. exercises

Find all six functions in Exercises 1–18. The given point is on the terminal side of an angle in standard position.

See Examples 1 and 2.

1. (5, 12)
2. (15, 8)
3. (−3, 4)
4. (4, −3)
5. $(-\sqrt{5}, -2)$
6. $(1, -\sqrt{3})$
7. $(-\sqrt{6}, \sqrt{3})$
8. $(\sqrt{2}, -\sqrt{7})$
9. (−12, −35)
10. (63, 16)
11. $\left(\frac{\sqrt{3}}{2}, -\frac{1}{2}\right)$
12. $\left(-\frac{\sqrt{2}}{2}, -\frac{\sqrt{2}}{2}\right)$
13. $\left(\frac{1}{\sqrt{5}}, -\frac{2}{\sqrt{5}}\right)$
14. $\left(-\frac{3}{\sqrt{10}}, \frac{1}{\sqrt{10}}\right)$
15. $\left(-\frac{40}{41}, -\frac{9}{41}\right)$
16. $\left(-\frac{1}{2}, 0\right)$
17. (0, 1)
18. $(\sqrt{3}, 0)$

See Example 6.

19. The point (p, q) is on the terminal side of γ. If γ is in standard position, find all six trigonometric functions of γ.
20. State the trigonometric functions of δ if δ is in standard position with point (h, k) on its terminal side.

Name the quadrant containing the terminal side of the angle in Exercises 21–30.

See Examples 3 and 4.

21. $\sin \alpha > 0, \cos \alpha > 0$
22. $\sin \alpha > 0, \sec \alpha < 0$
23. $\csc \beta < 0, \sec \beta < 0$
24. $\cos \beta > 0, \csc \beta > 0$
25. $\cos \gamma > 0, \sin \gamma < 0$
26. $\tan \gamma < 0, \cos \gamma > 0$
27. $\sin \delta < 0, \cot \delta > 0$
28. $\csc \delta < 0, \cot \delta < 0$
29. $\tan \lambda > 0, \sin \lambda < 0$
30. $\cos \lambda < 0, \tan \lambda > 0$

B. exercises

Find sine, cosine, and tangent for the angles in standard position that satisfy the conditions given in Exercises 31–50.

See Example 5 and Table 2.1 with discussion.

31. $\sin \theta = \frac{12}{13}$, θ in QII
32. $\cos \theta = -\frac{9}{41}$, θ in QIII
33. $\cot \theta = \frac{15}{8}$, θ in QIII
34. $\tan \theta = -\frac{24}{7}$, θ in QII
35. $\sec \theta = \frac{5}{3}$, θ in QIV
36. $\csc \theta = \frac{5}{4}$, θ in QI

37. $\cos\alpha = -\frac{11}{61}$, $\tan\alpha > 0$

38. $\sin\alpha = \frac{12}{37}$, $\sec\alpha < 0$

39. $\tan\beta = \frac{84}{13}$, $\csc\beta < 0$

40. $\cot\beta = -\frac{16}{63}$, $\sin\beta < 0$

41. $\csc\gamma = \frac{113}{15}$, $\cot\gamma > 0$

42. $\sec\gamma = \frac{101}{99}$, $\csc\gamma > 0$

43. $\sin\delta = 1$

44. $\cos\delta = -1$

45. $\tan\theta = -1$, θ in QII

46. $\cot\theta = 1$, θ in QI

47. $\csc\alpha = -1$

48. $\sec\alpha = 1$

49. $\cot\beta = 0$, $180° < \beta < 360°$

50. $\tan\beta = 0$, $90° < \beta < 270°$

C. exercises

51. **brain teaser** Imagine a semicircular disk turning on an axis in front of a screen. A strong light source projects the shadow of the disk on to the screen. As the disk rotates, the height h of the shadow changes. If the height is the vertical distance from a horizontal axis of rotation, the shadow will have a negative height when it is below its axis. Figure 2.2 shows a simple technique that can be used to plot points of the function $f(\theta) = h$, where θ is the angle of rotation from the horizontal. Use Figure 2.2 to plot sufficiently many points of $f(\theta) = h$ to sketch its graph over the interval 0° to 360°.

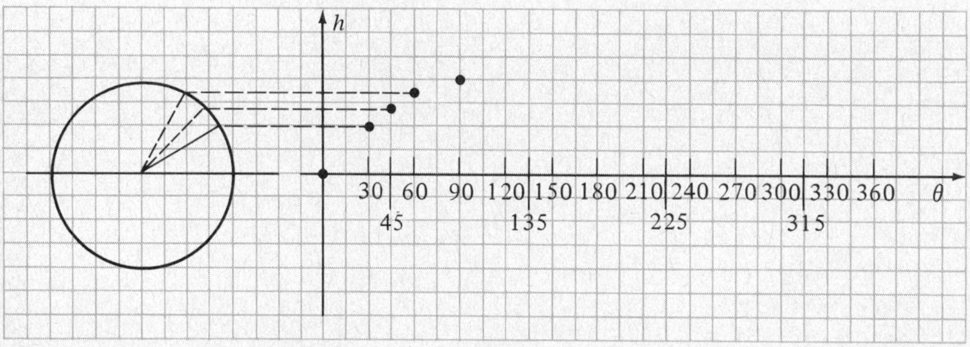

Figure 2.2 Graphing the height of a disk.

2.2 ANGLE MEASURE

We take our units of measure much for granted. They are often very arbitrary and sometimes only accidental. Why do we use an angle measure that divides a rotation into 360 units? There are many supposed explanations, but one seems most reasonable. The Babylonians used an especially large unit of distance, actually several of our miles. It was natural that this huge unit be subdivided into more manageable pieces. The distance was divided into 30 subunits. It was so great a distance that the time needed to travel it was also used as a unit of time. A complete day or one rotation of the earth turned out to be 12 of these time units in duration; hence $12 \cdot 30$ or 360 units in a complete rotation.

Degree measure is not necessarily the best way to measure angles. For one thing, separate units of measure must be used for angular and linear measure. When you set up a coordinate system, the axes are marked off in "units." The "units" are rarely specified, but the angles of the previous work were each designated as in degrees. If the same unit used for linear measure could be adapted to measure angles, our analysis would be independent of the actual unit used. This is, in fact, what is done in the calculus, scientific work, and advanced mathematics, using a system called **radian measure.**

HISTORICAL ASIDE

King Edward II of England "standardized" the inch in 1324 as the length of "three barley corns, round and dry, and laid lengthwise." The rod was likewise specified in 1514 as the sum of the lengths of the left feet of the first 16 men out of church on a certain feast. An acre was originally the area plowed by a team of oxen in a day's work.

BY PERMISSION OF JOHN HART AND FIELD ENTERPRISES, INC.

By permission of John Hart and Field Enterprises, Inc.

A protractor is the usual instrument used to measure angles. It is simply an arc from a circle that has been divided into 360 equal units called degrees. Protractors come in many sizes, from small plastic models to the large wooden chalkboard versions you've seen in classrooms. If a standard-size protractor were used, then the circumference could be computed and marked off in linear units that depend only on the radius chosen. Radian measure uses a circle with radius 1, the unit circle. The circumference is 2π, so 2π represents one rotation (360°). One-half of a rotation (180°) is π, and similarly any angle can be represented in this way. For a given angle, the arc subtended is proportional to the circumference. The measure of any angle should be proportional to the measure of one rotation.

$$\frac{\text{Radian Measure of angle } \theta}{2\pi} = \frac{\text{Degree Measure of angle } \theta}{360}$$

$$\frac{R}{2\pi} = \frac{D}{360}$$

or

$$\frac{R}{\pi} = \frac{D}{180}$$

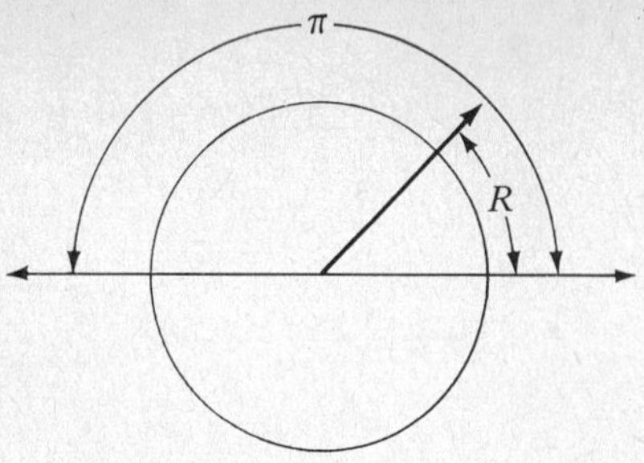

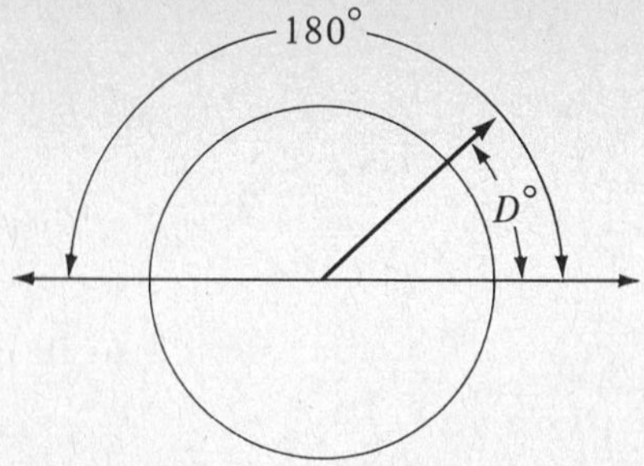

Any real number is the radian measure of an angle. Frequently, **radians** is used to designate the radian measure of an angle. For instance, 180° is referred to as π radians, and $\pi/2$ radians is 90°. If no unit of measure is specified, it is assumed to be radian measure.

Conversion of Angle Measures

Using the proportion above, degrees-to-radian measure and radian measure-to-degrees conversions are simple matters, as illustrated by the following examples.

EXAMPLES

1. Convert 45° to radian measure.

$$\frac{R}{\pi} = \frac{D}{180}$$

$$= \frac{45}{180}$$

$$R = \frac{45\pi}{180} = \frac{\pi}{4}$$

45° is equivalent to $\pi/4$.

2. Convert $2\pi/5$ to degree measure.

$$\frac{R}{\pi} = \frac{D}{180}$$

$$\frac{2\pi/5}{\pi} = \frac{D}{180}$$

$$D = \frac{2}{5} \cdot 180 = 72$$

$2\pi/5$ is equivalent to 72°.

3. Convert 130° to radian measure.

$$\frac{R}{\pi} = \frac{130}{180}$$

$$R = \frac{13\pi}{18}$$

130° is equivalent to $13\pi/18$.

4. Convert $11\pi/9$ to degree measure.

$$\frac{11\pi/9}{\pi} = \frac{D}{180}$$

$$D = \frac{11 \cdot 180}{9}$$

$$= 220$$

$11\pi/9$ is equivalent to 220°.

5. Convert $\frac{5}{9}$ to degree measure.

$$\frac{R}{\pi} = \frac{D}{180}$$

$$\frac{\frac{5}{9}}{\pi} = \frac{D}{180}$$

$$D = \frac{\frac{5}{9} \cdot 80}{\pi} = \frac{100}{\pi}$$

Notice that any real number can be used; radian measure is not necessarily a multiple of π.

$\frac{5}{9}$ is equivalent to $(100/\pi)$° or approximately 31.83°.

Special Angles

A number of special angles are used often enough that their radian measure should be familiar. See Figure 2.3. Complete Table 2.3 and observe the examples that follow using their radian measure.

Coterminal Angles

Recall that an angle may designate more than a complete rotation or may rotate in either a clockwise or counterclockwise direction. Given an angle in standard position, many other angles may be coterminal with it. Each additional rotation of 360° or 2π produces another coterminal angle. For example, consider $7\pi/4$ or 315°.

Recall that coterminal angles have the same initial and terminal sides.

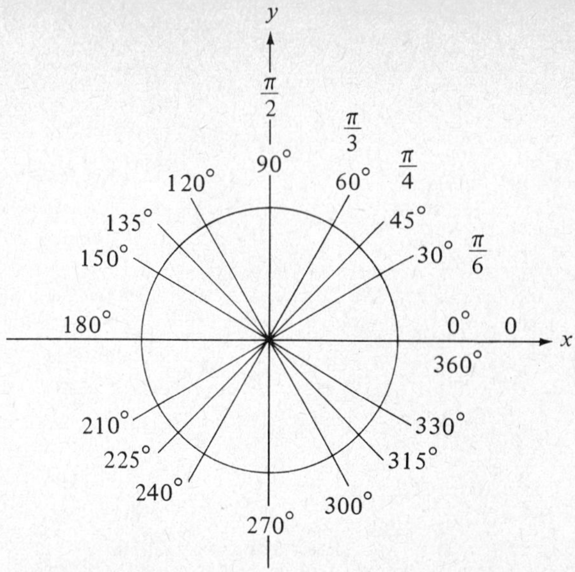

Figure 2.3 Special angles.

$$\frac{7\pi}{4} + 2\pi = \frac{15\pi}{4} \qquad 315° + 360° = 675°$$

$$\frac{7\pi}{4} - 2\pi = -\frac{\pi}{4} \qquad 315° - 360° = -45°$$

$$\frac{7\pi}{4} + 4\pi = \frac{23\pi}{4} \qquad 315° + 720° = 1035°$$

4π or 720° represents two complete rotations.

$$\frac{7\pi}{4} - 4\pi = -\frac{9\pi}{4} \qquad 315° - 720° = -405°$$

$\frac{15\pi}{4}, -\frac{\pi}{4}, \frac{23\pi}{4}, -\frac{9\pi}{4}, \ldots, 675°, -45°, 1035°, -405°, \ldots$ are all coterminal with $7\pi/4$ or 315°. In general, it may be stated that:

COTERMINAL ANGLES

θ is coterminal with $(\theta + 360k)°$ for any integer k.	θ is coterminal with $(\theta + 2\pi k)$ for any integer k.

If two angles are coterminal, then they share the terminal side and all of its points. A point (x, y) on the terminal side of θ is also on the ter-

TABLE 2.3 SPECIAL ANGLES

DEGREES	RADIANS	DEGREES	RADIANS	DEGREES	RADIANS
$0°$	0	$135°$		$270°$	
$30°$	$\frac{\pi}{6}$	$150°$		$300°$	
$45°$	$\frac{\pi}{4}$	$180°$	π	$315°$	
$60°$	$\frac{\pi}{3}$	$210°$		$330°$	
$90°$	$\frac{\pi}{2}$	$225°$		$360°$	2π
$120°$		$240°$			

minal side of $(\theta + 2\pi k)$ or $(\theta + 360k)°$. Thus the trigonometric functions of two coterminal angles are equal. For instance, $\sin \theta = \sin(\theta + 2\pi)$.

EXAMPLES

6. $\sin 765° = ?$

$\sin 765° = \sin(765° - 360°) = \sin 405°$

$\sin 405° = \sin(405° - 360°) = \sin 45°$

$\sin 765° = \sin 45°$

$\mathbf{\sin 765° = \frac{1}{\sqrt{2}}}$ or $\frac{\sqrt{2}}{2}$

7. $\sec 5\pi = ?$

$\sec 5\pi = \sec(5\pi - 2\pi) = \sec 3\pi$

$= \sec(3\pi - 2\pi)$

$= \sec \pi$ or $\sec 180°$

$\mathbf{\sec 5\pi = -1}$

8. $\cos 5100° = ?$

$$\begin{aligned}\cos 5100° &= \cos(5100° - 14 \cdot 360°)\\ &= \cos(5100° - 5040°)\\ &= \cos 60° \quad \text{or} \quad \cos \frac{\pi}{3}\end{aligned}$$

$$\mathbf{\cos 5100° = \frac{1}{2}}$$

Reference Angles

The table of values of the trigonometric functions lists only angles in the interval 0 to $\pi/2$ or 0° to 90°. This certainly reduces the size of the tables, but how is it possible to list only functions on such a small interval? Consider an angle θ_{II}, in standard position with terminal side in the second quadrant. As shown in Figure 2.4, the terminal side of θ_{II} makes an acute angle with the negative x axis. We call θ_R the **reference angle** for θ_{II}. If θ_R is placed in standard position, $(-x, y)$ is on the terminal side if (x, y) is on the terminal side of θ_{II}. Because $|-x| = |x|$ and $y = y$, the functions of θ_R and θ_{II} will have the same absolute values. That is, the functions of an angle and its reference angle will differ only in sign.

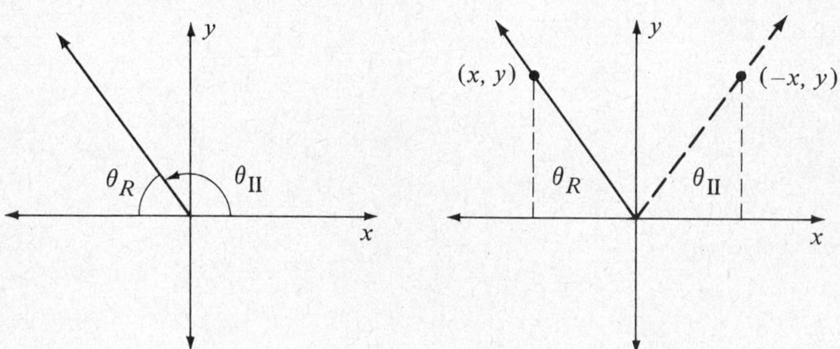

$$\theta_R = 180° - \theta_{\mathrm{II}}$$
$$\theta_R = \pi - \theta_{\mathrm{II}}$$

Figure 2.4 Second-quadrant reference angle.

EXAMPLE

9. Express cos 110° in terms of its reference angle.

110° is in QII, so $\theta_R = 180° - 110° = 70°$, and cosine is negative in QII. Thus.

$$\mathbf{\cos 110° = -\cos 70°}$$

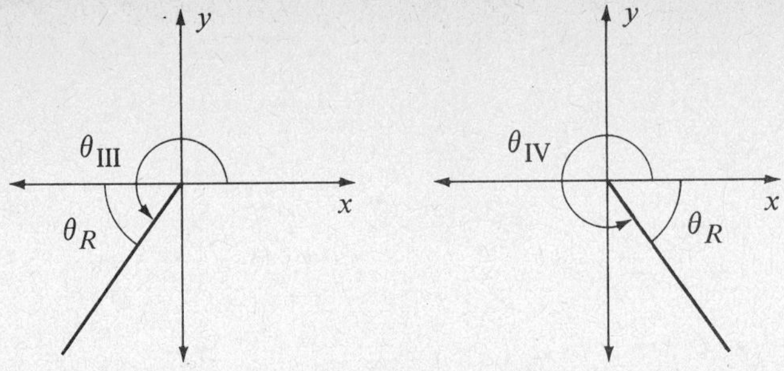

$\theta_R = \theta_{III} - 180°$ $\qquad$ $\theta_R = 360° - \theta_{IV}$

$\theta_R = \theta_{III} - \pi$ $\qquad$ $\theta_R = 2\pi - \theta_{IV}$

Figure 2.5 Reference angles in QIII and QIV.

The angles θ_{III} and θ_{IV} with terminal sides in the third and fourth quadrants, respectively, have reference angles as shown in Figure 2.5.

REFERENCE ANGLES

Given that θ is in standard position and $0 \leq \theta < 2\pi$, the reference angle of θ is the positive acute angle that the terminal side of θ makes with the x axis.

θ IN QII	θ IN QIII	θ IN QIV
$\theta_R = 180° - \theta_{II}$	$\theta_R = \theta_{III} - 180°$	$\theta_R = 360° - \theta_{IV}$
$\theta_R = \pi - \theta_{II}$	$\theta_R = \theta_{III} - \pi$	$\theta_R = 2\pi - \theta_{IV}$

So not only can we reduce θ to a coterminal angle of less than one rotation, now we can examine a reference angle that is acute and write a function in terms of this reference angle. The function of any angle can be expressed in terms of that function for an acute angle.

EXAMPLES

Express each function in terms of that function for an acute angle.

10. $\csc 7\pi/4 = ?$

$7\pi/4$ is in QIV, so

$$\theta_R = 2\pi - \frac{7\pi}{4} = \frac{\pi}{4}$$

and cosecant is negative in QIV. Thus,

$$\mathbf{csc\,\frac{7\pi}{4} = -csc\,\frac{\pi}{4}}$$

11. sin 150° = ?
150° is in QII, so

$$\theta_R = 180° - 150° = 30°$$

and sine is positive in QII. Thus,

$$\mathbf{sin\ 150° = sin\ 30°}$$

12. tan 970° = ?
First reduce 970° to a positive coterminal angle between 0° and 360°.

$$\tan 970° = \tan 610° = \tan 250°$$

Then 250° is in QIII, so

$$\theta_R = 250° - 180° = 70°$$

and tangent is positive in QIII. Thus,

$$\mathbf{tan\ 970° = tan\ 70°}$$

13. sin 9 = ?
$9 > 2\pi$, so first reduce to a positive coterminal angle between 0 and 2π.

$$\sin 9 = \sin(9 - 2\pi)$$

$9 - 2\pi$ is in QII, so

$$\theta_R = \pi - (9 - 2\pi) = 3\pi - 9$$

and because sine is positive in QII,

$$\mathbf{sin\ 9 = sin(3\pi - 9)}$$

Once a function is rewritten in terms of an acute angle, the value of the function may be found. Exact values can be found for special angles, as in the following examples.

EXAMPLES

Find the exact value of the trigonometric functions given.

14. tan 300° = ?

$$\tan 300° = -\tan(360° - 300°)$$

$$= -\tan 60^\circ \qquad \text{300° is in QIV and tangent is negative in QIV.}$$

$$= -(\sqrt{3})$$

$$\mathbf{\tan 300^\circ = -\sqrt{3}}$$

15. $\cos\left(-\frac{8\pi}{3}\right) = ?$

$$\cos\left(-\frac{8\pi}{3}\right) = \cos\left(-\frac{8\pi}{3} + 2\pi\right) = \cos\left(-\frac{2\pi}{3}\right)$$

$$= \cos\left(-\frac{2\pi}{3} + 2\pi\right) = \cos\frac{4\pi}{3}$$

$$= -\cos\left(\frac{4\pi}{3} - \pi\right)$$

$$= \cos\frac{\pi}{3}$$

$$= -\frac{1}{2}$$

$$\mathbf{\cos\left(-\frac{8\pi}{3}\right) = -\frac{1}{2}}$$

Throughout this discussion we have ignored the use of the calculator. The calculator will, of course, give tan 300° or $\cos(-8\pi/3)$ directly without recourse to reference angles. However, the calculator is unable to simplify an expression such as $\cos(x + 3\pi/2)$ unless the value of x is known. The principles of this section are an important part of our ability to work with trigonometric statements, and we continue to develop this facet in this chapter.

exercise set 2.2

A. exercises

In Exercises 1–20, convert to radian measure. See Examples 1 and 3.

1. 30°	2. 45°	3. −60°	4. 90°
5. −25°	6. 40°	7. −70°	8. 85°
9. 110°	10. −140°	11. 200°	12. 205°
13. 280°	14. −305°	15. 720°	16. 900°
17. 1000°	18. −2000°	19. π°	20. $(1/\pi)$°

In Exercises 21–40, convert to degrees.

21. $\frac{\pi}{4}$ 22. $\frac{\pi}{6}$ 23. $-\frac{2\pi}{3}$ 24. $\frac{3\pi}{4}$

See Examples 2 and 4.

25. $\frac{5\pi}{6}$ 26. $-\frac{7\pi}{6}$ 27. $-\frac{5\pi}{4}$ 28. $\frac{11\pi}{6}$

29. $\frac{\pi}{5}$ 30. $\frac{\pi}{12}$ 31. 3π 32. -4π

33. $\frac{9\pi}{2}$ 34. $\frac{7\pi}{3}$ 35. $\frac{2}{3}$ 36. $\frac{3}{4}$

37. 3 38. 4 39. π^2 40. $\frac{1}{\pi}$

Watch out! Exercises 35–40 are not integral multiples of π. See Example 5.

Find a positive coterminal angle of less than one rotation for each angle in Exercises 41–50.

41. $400°$ 42. $440°$ 43. $-55°$ 44. $-110°$

See Examples 6–8.

45. $-\frac{13\pi}{3}$ 46. $\frac{12\pi}{5}$ 47. $1940°$ 48. $-\frac{23\pi}{2}$

49. -23π 50. $3250°$

Find the reference angle for the angles in Exercises 51–60.

51. $95°$ 52. $115°$ 53. $\frac{3\pi}{4}$ 54. $\frac{5\pi}{6}$

See Examples 9–13.

55. $229°$ 56. $283°$ 57. $301°$ 58. $342°$

59. $\frac{5\pi}{3}$ 60. $\frac{11\pi}{6}$

B. exercises

Find the exact values of the trigonometric functions in Exercises 61–80.

61. $\sin 30°$ 62. $\cos 45°$ 63. $\tan \frac{\pi}{3}$

See Examples 9–15.

64. $\cot \frac{\pi}{2}$ 65. $\sec 120°$ 66. $\csc 135°$

67. $\sin \frac{5\pi}{6}$ 68. $\cos \pi$ 69. $\tan 210°$

70. $\cot 240°$ 71. $\sec \frac{3\pi}{2}$ 72. $\csc 2\pi$

73. $\sin 315°$ 74. $\cos 330°$ 75. $\tan 2\pi$

76. $\cot 450°$ 77. $\sec \frac{8\pi}{3}$ 78. $\csc 900°$

79. $\sin\left(-\frac{\pi}{6}\right)$ 80. $\cos\left(-\frac{\pi}{4}\right)$

Express the functions for the angles given in Exercises 81–100 in terms of the reference angle.

See Examples 9–13.

81. $\sin 122°$ 82. $\cos 137°$ 83. $\tan 204°$

84. $\cot 253°$ 85. $\sec \frac{4\pi}{7}$ 86. $\csc \frac{9\pi}{8}$

87. $\sin \frac{13\pi}{7}$ 88. $\cos \frac{10\pi}{11}$ 89. $\tan 302°$

90. $\cot 356°$ 91. $\sec 491°$ 92. $\csc 657°$

93. $\sin(-29°)$ 94. $\cos(-73°)$ 95. $\tan(-701°)$

96. $\cot 1046°$ 97. $\sec 10.5\pi$ 98. $\csc \pi^2$

99. $\sin 2$ 100. $\cos 3$

C. exercises

The military uses an angle measure that was adopted because it would lend itself to easy, rapid, accurate computation in measurement at great distances, such as artillery fire. One **mil** *was intended to be that which would subtend an arc of one linear unit at a distance of 1000 units. That would mean $2\pi \cdot 1000$ such units, but 6283.18 . . . was not convenient, easy, or rapid. So 6400 was selected so that*

$$1 \text{ mil} = \frac{1}{6400} \text{ of a rotation}$$

Thus, 6400 mil = 2π, 3200 mil = π, and 1600 mil is $\pi/2$, a right angle, or 90°. Use this background in Exercises 101–110.

HISTORICAL ASIDE

The Swiss army artillery introduced the mil in 1864. The French adopted it in 1879 and the U.S. army in 1900. The army's use of the unit is in directing artillery fire, determining range, and making corrections to fire. Forward observers often use binoculars with built-in scales for measuring in mils between objects in their field of vision.

101. 800 mil = ? degrees 102. $\frac{\pi}{4}$ = ? mil

103. $30° = ?$ mil

104. 160 mil $= ?$

105. 500 mil $= ?$

106. $400° = ?$ mil

107. $\frac{3\pi}{4} = ?$ mil

108. $? = 2400$ mil

109. $5\pi = ?$ mil

110. 480 mil $= ?$ degrees

Most scientific calculators allow the user to choose degree or radian measure, and a few have a deg-rad-grad switch. The third choice, unused in this country, is a system proposed in France and adopted to some extent in Europe. The right angle is divided into 100 equal parts, each called a ***grad.*** *The grad is subdivided into 100 parts each called a minute, and the minute is divided into 100 parts each called a second.*

$$1 \text{ grad} = \frac{1}{100} \text{ of a right angle}$$

$$= \frac{1}{400} \text{ of a rotation}$$

Thus 100 grads, written 100^g, *is equal to* $90°$, *and* $30^g73'19''$ *is*

$$\frac{30}{100} + \frac{73}{100^2} + \frac{19}{100^3}$$

or 0.307319 right angles. Use this information in Exercises 111–120.

111. $50^g = ?$ degrees

112. $?$ grad $= 135°$

113. $?$ grad $= \frac{\pi}{2}$

114. $250^g = ?$

115. $630° = ?$ grad

116. $?$ grad $= 5\pi$

117. $81^g37'50'' = ?$ degrees

118. $81°37'50'' = ?$ grad

119. $?$ mil $= 75^g$

120. $?$ grad $= 400$ mil

2.3 FUNDAMENTAL IDENTITIES

There are many fundamental relationships among the trigonometric functions. All are derived directly from the definitions. They are useful in making certain computations, but primarily they provide a needed flexibility in working with complicated trigonometric expressions. Such expressions may often be simplified considerably by employing these fundamentals. This process is a major concern in later work, notably the calculus.

Reciprocal Identities

The very definition of the functions provides three pairs of reciprocals. That is,

$$\sin \theta = \frac{y}{r} \qquad \frac{r}{y} = \csc \theta$$

$$\cos \theta = \frac{x}{r} \qquad \frac{r}{x} = \sec \theta$$

$$\tan \theta = \frac{y}{x} \qquad \frac{x}{y} = \cot \theta$$

Thus, if sin 53.14° is known, csc 53.14° is simply the reciprocal of that value. Many tables have only sine, cosine, and tangent entries, just as the slide rule had only S, C, and T scales. Today's scientific calculator has only sin, cos, and tan keys. The other functions are obtained by applying these reciprocal identities.

RECIPROCAL IDENTITIES

$$\sin \theta = \frac{1}{\csc \theta} \qquad \csc \theta = \frac{1}{\sin \theta}$$

$$\cos \theta = \frac{1}{\sec \theta} \qquad \sec \theta = \frac{1}{\cos \theta}$$

$$\tan \theta = \frac{1}{\cot \theta} \qquad \cot \theta = \frac{1}{\tan \theta}$$

We use the word "identity" here to mean that the statement is true for all values of the variables for which it is defined. In general, an equation is simply a statement of equality. An equation may be true always, sometimes, or never, as the scientist in the cartoon remarks. An equation that is always true is called an **identity,** one that is never true is called a **contradiction.** The equations you solved in algebra are called **conditional equations** and are true for some value or values of the variable and false for others. Note that numerical equations are always identities such as $1 + 1 = 2$ or $\sin \pi = 0$, or contradictions such as $\sin 0 = \pi$.

Courtesy of Sidney Harris, from What's So Funny About Science? (William Kaufman, Inc., Los Altos, CA, 1978).

EXAMPLES

1. Verify that cos 60° sec 45° = sin 135°.

$$(\cos 60°)(\sec 45°) = (\sin 135°)$$

$$\left(\frac{1}{2}\right)(\sqrt{2}) = \left(\frac{1}{\sqrt{2}}\right)$$

$$\frac{\sqrt{2}}{2} = \frac{1}{\sqrt{2}}\frac{\sqrt{2}}{\sqrt{2}}$$

$$\mathbf{\frac{\sqrt{2}}{2} = \frac{\sqrt{2}}{2}} \quad \checkmark$$

2. Prove that the statement is an identity.

$$\cos \alpha \sec \alpha = \tan \alpha \cot \alpha$$

In this example we change *both sides* to show that each is equal to a third expression. That is, if $a = b$ and $b = c$, then $a = c$—the transitive property.

$$\cos\alpha\left(\frac{1}{\cos\alpha}\right) = \tan\alpha\left(\frac{1}{\tan\alpha}\right)$$

$$\frac{\cos\alpha}{\cos\alpha} = \frac{\tan\alpha}{\tan\alpha}$$

$$\mathbf{1 = 1} \quad \checkmark$$

Ratio Identities

Also related directly to the definitions are the so-called ratio identities.

$$\frac{\sin\theta}{\cos\theta} = \frac{y/r}{x/r} = \frac{y}{x} = \tan\theta$$

and similarily for cot θ.

RATIO IDENTITIES

$$\tan\theta = \frac{\sin\theta}{\cos\theta} \qquad \cot\theta = \frac{\cos\theta}{\sin\theta}$$

EXAMPLES

3. Verify that sin 60° csc 30° = tan 60°.

(sin 60°)(csc 30°) = (tan 60°)

$$\left(\frac{\sqrt{3}}{2}\right)(2) = (\sqrt{3})$$

$$\boldsymbol{\sqrt{3} = \sqrt{3}} \quad \checkmark$$

4. Use the identities to simplify the expression cos x csc x.

$$\cos x \csc x = \cos x \frac{1}{\sin x}$$

$$= \frac{\cos x}{\sin x}$$

$$\mathbf{\cos x \csc x = \cot x}$$

5. Prove that the statement is an identity.

$$\tan\beta \sec\beta = \frac{\sin\beta}{\cos^2\beta}$$

Unlike Example 2, here we make substitutions on *only one side*, showing that it is in fact identical to the other side. This is the classical form of proving identities.

$$\frac{\sin\beta}{\cos\beta}\frac{1}{\cos\beta} =$$

$$\frac{\sin\beta}{\cos^2\beta} = \frac{\sin\beta}{\cos^2\beta} \quad \checkmark$$

Pythagorean Identities

Recall that in defining the functions in terms of x, y, and r, $x^2 + y^2 = r^2$. Beginning with this, observe what happens when we divide by r^2.

$$x^2 + y^2 = r^2$$

$$\frac{x^2}{r^2} + \frac{y^2}{r^2} = \frac{r^2}{r^2}$$

$$\left(\frac{x}{r}\right)^2 + \left(\frac{y}{r}\right)^2 = 1$$

$$\cos^2\theta + \sin^2\theta = 1$$

Similarly, dividing by x^2 and y^2, respectively, we obtain

$$1 + \tan^2\theta = \sec^2\theta$$

and

$$\cot^2\theta + 1 = \csc^2\theta$$

which we call the square or **Pythagorean identities.**

PYTHAGOREAN IDENTITIES

$$\cos^2\theta + \sin^2\theta = 1$$

$$1 + \tan^2\theta = \sec^2\theta$$

$$\cot^2\theta + 1 = \csc^2\theta$$

These three groups of fundamental identities provide ample means to simplify many complicated expressions and develop other interesting relationships. For instance, now all six functions may be expressed strictly in terms of any one of the functions, as shown for sine in the following example.

HISTORICAL ASIDE

The greatest mathematician of the nineteenth century was Karl Friedrich Gauss (1777–1855). In spite of his preeminence he worried about such basic concerns as the clarity of notation. He remarked that "$\sin^2 \phi$ is odious to me" because he was concerned that it would be confused with $\sin \phi^2$, $(\sin \phi)^2$, and $\sin(\sin \phi)$. [We use $\sin^2 \phi$ to mean $(\sin \phi)^2$.]

EXAMPLES

6. Express each of the other functions in terms of $\sin \phi$.

$$\cos^2 \phi + \sin^2 \phi = 1$$

$$\cos^2 \phi = 1 - \sin^2 \phi$$

$$\mathbf{\cos \phi = \pm\sqrt{1 - \sin^2 \phi}}$$

The sign of the radical depends on the quadrant of the angle.

$$\tan \phi = \frac{\sin \phi}{\cos \phi}$$

$$\mathbf{\tan \phi} = \frac{\sin \phi}{\pm\sqrt{1 - \sin^2 \phi}} \qquad \mathbf{\cot \phi} = \frac{\pm\sqrt{1 - \sin^2 \phi}}{\sin \phi}$$

$$\mathbf{\sec \phi} = \frac{1}{\cos \phi} = \frac{1}{\pm\sqrt{1 - \sin^2 \phi}}$$

$$\mathbf{\csc \phi} = \frac{1}{\sin \phi}$$

7. Use the fundamental identities to simplify $\cos x + \tan x \sin x$.

$$\begin{aligned}\cos x + \tan x \sin x &= \cos x + \frac{\sin x}{\cos x} \sin x \\ &= \cos x + \frac{\sin^2 x}{\cos x} \\ &= \frac{\cos^2 x}{\cos x} + \frac{\sin^2 x}{\cos x} \\ &= \frac{\cos^2 x + \sin^2 x}{\cos x} \\ &= \frac{1}{\cos x}\end{aligned}$$

$$\mathbf{\cos x + \tan x \sin x = \sec x}$$

8. Prove:

$$\cos^2 \frac{13\pi}{5} + \sin^2 \frac{3\pi}{5} + \sin \frac{7\pi}{5} \csc \frac{2\pi}{5} = 0$$

First express each function in terms of its reference angle. Notice that $3\pi/5$ is in QII, $7\pi/5$ is in QII, and $13\pi/5$ is in QIV. All have the same reference angle.

$$\left(-\cos\frac{2\pi}{5}\right)^2 + \left(\sin\frac{2\pi}{5}\right)^2 + \left(-\sin\frac{2\pi}{5}\right)\left(\csc\frac{2\pi}{5}\right)$$

Then simplify using the reciprocal and Pythagorean identities.

$$\left(\cos^2\frac{2\pi}{5} + \sin^2\frac{2\pi}{5}\right) + \left(-\sin\frac{2\pi}{5}\,\frac{1}{\sin 2\pi/5}\right) =$$

$$(1) + (-1) =$$

$$\mathbf{0 = 0} \quad \checkmark$$

9. Prove the identity:

$$\cot\theta(\cot\theta + \tan\theta) = \csc^2\theta$$

$$\frac{1}{\tan\theta}\left(\frac{1}{\tan\theta} + \tan\theta\right) =$$

Make use of the reciprocal identity $\cot\theta = \dfrac{1}{\tan\theta}$.

$$\frac{1}{\tan^2\theta} + 1 =$$

$$\frac{1 + \tan^2\theta}{\tan^2\theta} =$$

$$\frac{\sec^2\theta}{\tan^2\theta} =$$

$$\frac{1/(\cos^2\theta)}{(\sin^2\theta)/(\cos^2\theta)} =$$

Then express in terms of sine and cosine.

$$\frac{1}{\sin^2\theta} =$$

$$\mathbf{\csc^2\theta = \csc^2\theta} \quad \checkmark$$

10. Prove Exercise 9 in a different way.

$$\cot\theta(\cot\theta + \tan\theta) = \csc^2\theta$$

Try all sines and cosines to begin. Add the fractions in parentheses.

$$\frac{\cos\theta}{\sin\theta}\left(\frac{\cos\theta}{\sin\theta} + \frac{\sin\theta}{\cos\theta}\right) =$$

$$\frac{\cos\theta}{\sin\theta}\left(\frac{\cos^2\theta + \sin^2\theta}{\sin\theta\cos\theta}\right) =$$

$$\frac{1}{\sin\theta}\left(\frac{1}{\sin\theta}\right) =$$

$$\frac{1}{\sin^2\theta} =$$

$$\mathbf{\csc^2\theta = \csc^2\theta}$$

11. Prove Exercise 9 differently again.

$$\cot\theta(\cot\theta + \tan\theta) = \csc^2\theta$$

$$\cot\theta\left(\cot\theta + \frac{1}{\cot\theta}\right) =$$

$$\cot\theta\left(\frac{\cot^2\theta + 1}{\cot\theta}\right) =$$

$$\cot^2\theta + 1 =$$

$$\boldsymbol{\csc^2\theta = \csc^2\theta} \quad \checkmark$$

As illustrated in Examples 7–9, there are many ways of approaching the proof of an identity. Some are just smoother than others. By attempting such proofs in this section you will come to know the identities well and prepare for later work in Chapter 5 when a broad array of relationships are developed.

exercise set 2.3

A. exercises

In Exercises 1–10, determine if the statement is true or false, an identity or a contradiction.

See Examples 1, 3, and 8.

1. $\sin 45° \cot 45° = \cos 45°$
2. $\tan 30° \cos 30° = \sin 30°$
3. $2 \sin 30° = \sin 2(30°)$
4. $\tan 2(45°) = 2 \tan 45°$
5. $\sin^2\left(\frac{3\pi}{4}\right) + \cos^2\left(\frac{3\pi}{4}\right) = 1$
6. $\tan^2\left(\frac{\pi}{6}\right) + \cot^2\left(\frac{\pi}{6}\right) = 1$
7. $\sec\frac{5\pi}{6} = -\sqrt{\tan^2\left(\frac{5\pi}{6}\right) + 1}$
8. $\cot\left(\frac{2\pi}{3}\right) = \sqrt{\csc^2\left(\frac{2\pi}{3}\right) - 1}$
9. $\cos 120° \tan 60° = \sin 240° \cot 180°$
10. $\cos^2\left(\frac{\pi}{4}\right) + \sin^2\left(\frac{3\pi}{4}\right) = \tan^2\left(\frac{5\pi}{4}\right)$

Use the fundamental identities to simplify the expressions in Exercises 11–20 to a single function if possible.

See Examples 4 and 7.

11. $\cos\alpha(\tan\alpha + \cot\alpha)$
12. $\sec\alpha - \sin\alpha\tan\alpha$
13. $(\sin^2\beta + \cos^2\beta)^2$
14. $(\sec^2\beta - \tan^2\beta)^3$
15. $\tan\gamma(\sin\gamma + \cot\gamma\cos\gamma)$
16. $(\tan\gamma + \cot\gamma)\cos\gamma$

17. $\csc \delta - \cot \delta \cos \delta$

18. $\sin \delta + \cot \delta \cos \delta$

19. $(\cos \lambda + \sin \lambda)^2 - (\sin \lambda - \cos \lambda)^2$

20. $\sin^2 \lambda + \sin \lambda \csc \lambda + \cos^2 \lambda + \cos \lambda \sec \lambda$

Express each of the other functions in terms of the function given in Examples 20–24.

See Example 6.

21. $\cos \alpha$

22. $\tan \beta$

23. $\cot \gamma$

24. $\csc \delta$

B. exercises

In Exercises 25–30, make the indicated substitution and simplify the expression. Assume that all angles are acute.

See Examples 4 and 7.

25. $y = \sqrt{4 - x^2}$; let $x = 2 \sin \alpha$.

26. $y = \sqrt{25 - x^2}$; let $x = 5 \cos \beta$.

27. $y = 49 + 25x^2$; let $x = \frac{7}{5} \tan \gamma$.

28. $y = 16 + 36x^2$; let $x = \frac{2}{3} \cot \delta$.

29. $y = \dfrac{\sqrt{x^2 - 9}}{x}$; let $x = 3 \sec \lambda$.

30. $y = \dfrac{\sqrt{4 + x^2}}{x}$; let $x = 2 \tan \theta$.

Prove the identities stated in Exercises 31–50.

See Examples 2, 5, and 9–11.

31. $\tan \alpha = \sin \alpha \sec \alpha$

32. $\cos \alpha \csc \alpha = \cot \alpha$

33. $\cos \beta \tan \beta \csc \beta = 1$

34. $\sin \beta \cot \beta \sec \beta = 1$

35. $\tan \gamma(\tan \gamma + \cot \gamma) = \sec^2 \gamma$

36. $\csc^2 \gamma(1 - \sin^2 \gamma) = \cot^2 \gamma$

37. $\tan \delta + \cot \delta = \sec \delta \csc \delta$

38. $\sec \delta - \cos \delta = \tan \delta \sin \delta$

39. $\csc \lambda - \sin \lambda = \cot \lambda \cos \lambda$

40. $\csc \lambda - \cot \lambda \sec \lambda = 0$

41. $\dfrac{\tan x}{\sin^2 x + \cos^2 x} = \dfrac{1}{\cot x}$

42. $\dfrac{\sec x}{1 + \tan^2 x} = \dfrac{1}{\sec x}$

43. $(\sin y + \cos y)^2 = 1 + 2 \sin y \cos y$

44. $(1 - \tan y)^2 = \sec^2 y - 2 \tan y$

45. $\sec^2 z + \csc^2 z = \sec^2 z \csc^2 z$

46. $\tan^2 z - \sin^2 z = \tan^2 z \sin^2 z$

47. $\dfrac{\cos^2 \theta - \sin^2 \theta}{\sin \theta \cos \theta} = \cot \theta - \tan \theta$

48. $\dfrac{\cot^2 \theta - \cos^2 \theta}{\cos^2 \theta} = \cot^2 \theta$

49. $\dfrac{1}{1 - \sin \sigma} + \dfrac{1}{1 + \sin \sigma} = 2 \sec^2 \sigma$

50. $\dfrac{1}{\csc \sigma - \cot \sigma} - \dfrac{1}{\csc \sigma + \cot \sigma} = 2 \cot \sigma$

C. exercises

51. **brain teaser** Where is the fallacy in the following argument?

$$\begin{aligned}
\cos^2 \lambda + \sin^2 \lambda &= 1 \\
\cos^2 \lambda &= 1 - \sin^2 \lambda \\
\cos \lambda &= (1 - \sin^2 \lambda)^{1/2} \\
1 + \cos \lambda &= 1 + (1 - \sin^2 \lambda)^{1/2} \\
(1 + \cos \lambda)^2 &= [1 + (1 - \sin^2 \lambda)^{1/2}]^2
\end{aligned}$$

Let $\lambda = \pi$.

$$\begin{aligned}
(1 + \cos \pi)^2 &= [1 + (1 - \sin^2 \pi)^{1/2}]^2 \\
(1 - 1)^2 &= [1 + (1 - 0^2)^{1/2}]^2 \\
(0)^2 &= (1 + 1)^2 \\
\mathbf{0} &= \mathbf{4} \qquad \text{Hmmmmmmm?}
\end{aligned}$$

52. **brain teaser** Find one value of θ as a counterexample to show that

$$8 \sin^4 \theta - 4(\sqrt{3} + \sqrt{2} + 1) \sin^3 \lambda + 2(\sqrt{6} + \sqrt{3} + \sqrt{2}) \sin^2 \theta - \sqrt{6} \sin \theta = 0$$

is *not* an identity.

53. Prove the logarithmic identity:

$$\log(\cos \sigma) + \log(\tan \sigma) + \log(\csc \sigma) = 0$$

Not all students will be familiar with the topics or notation of Exercises 53 and 54.

54. Prove the determinant identity:

$$\begin{vmatrix} \cos\sigma & \sin\sigma \\ -\sin\sigma & \cos\sigma \end{vmatrix} = \begin{vmatrix} \sec\sigma & \tan\sigma \\ \tan\sigma & \sec\sigma \end{vmatrix}$$

2.4 APPLICATIONS AND CALCULATION

The trigonometric functions have wide applications, especially in physical problems involving circular motion. There are extensive tables available to aid in the calculations needed in such work, but the hand-held calculator makes many more calculations practical. The calculator does not, however, set up these calculations; you do!

"Math is easy with an electronic calculator. What does a 7 look like?"

You may be satisfied if it is shown that trigonometry can, in principle, be applied in a particular instance. Once a problem is set up, the answer may require lengthy calculation. The uses illustrated in this section are chosen to indicate the scope of application *and* the extent of the

2. If θ_a is 34° for Kodak Optical Glass EK-448, then θ_m is 17.3°. Find the index of refraction for this material.

$$D = \frac{\sin 34°}{\sin 17.3°}$$

ALGEBRAIC

Key	Display
[34]	34
[sin]	0.55919290
[÷]	0.55919290
[17.3]	17.3
[sin]	0.29737487
[=]	1.8804309

D = 1.8804 for EK-448 to the nearest ten-thousandth.

Diamond has a very high index of refraction, 2.4195. A high index means a greater tendency to bend light. Does this contribute to the "brilliance" of a diamond? How? Suppose that the angle of entry is 21°. What is the angle in the medium?

$$D = \frac{\sin \theta_a}{\sin \theta_m}$$

$$2.4195 = \frac{\sin 21°}{\sin \theta_m}$$

$$\sin \theta_m = \frac{\sin 21°}{2.4195}$$

$$\boldsymbol{\theta_m = \text{angle whose sine is } \frac{\sin 21°}{2.4195}}$$

abbreviated as

$$\boldsymbol{\theta_m = \arcsin\left(\frac{\sin 21°}{2.4195}\right)}$$

computation necessary. Each of these aspects should be given the same attention and effort.

Index of Refraction

Light entering a medium denser than air from air bends toward a perpendicular to the surface of that medium. A measure of this tendency is called the **index of refraction** D and is defined by

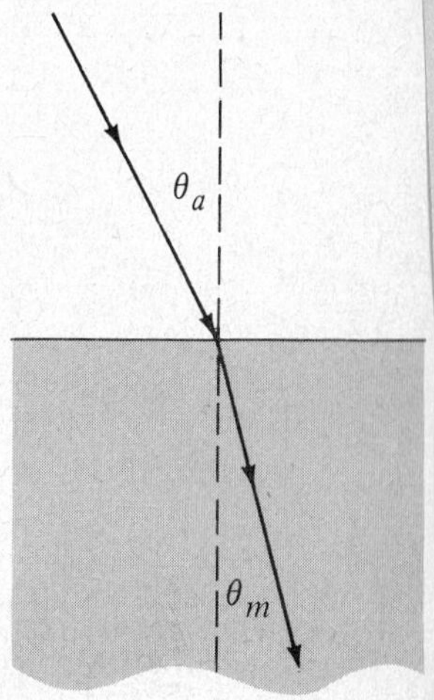

$$\boldsymbol{D = \frac{\sin \theta_a}{\sin \theta_m}}$$

where θ_a and θ_m are the angles with perpendicular to the surface in air and in the medium, respectively. If the angles θ_a and θ_m are known for a particular medium, the index can be found as in the following examples.

EXAMPLES

1. If light encounters ordinary crystalline quartz at 63° to the perpendicular and is refracted to 35° in the quartz, find its index of refraction to the nearest ten-thousandth.

$$D = \frac{\sin \theta_a}{\sin \theta_m}$$

$$= \frac{\sin 63°}{\sin 35°}$$

ALGEBRAIC

Key	Display
[63]	63
[sin]	0.89100652
[÷]	0.89100652
[35]	35
[sin]	0.57357643
[=]	1.5534224

$D \approx 1.5534$

The index of refraction for ordinary crystalline quartz is approximately 1.5534.

or

$$\sin^{-1}\left(\frac{\sin 21°}{2.4195}\right)$$

which is calculated in Example 3.

EXAMPLE

3. Find θ_m to the nearest minute.

$$\theta_m = \arcsin\left(\frac{\sin 21°}{2.4195}\right)$$

ALGEBRAIC

Key	Display
[21]	21
[sin]	0.35836795
[÷]	0.35836795
[2.4195]	2.4195
[=]	0.14811653
[arc]	0.14811653
[sin]	8.5177925

The diamond refracts the 21° light to 8.518°.

Volume

Many common implements are circular: wheels, gears, pipes, tubes, dials, and gauges. Consider a large drum that is a right circular cylinder. Such drums are sometimes cut lengthwise to make a water trough for animals. What is the volume of such a container, called a **segment** of a cylinder?

First let's look at a diagram and label all of the important dimensions. When we sketch the figure, as in Figure 2.6, we see that there are two cases (if we don't simply bisect the cylinder). Either way we have a figure called a prism, whose volume is the product of the area of the base and the height. We need an expression for the area of the segment of the circle that is the base of the prism.

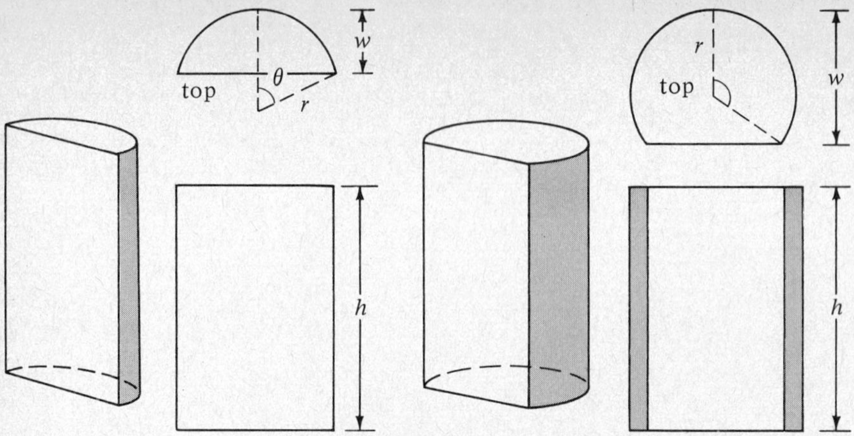

Figure 2.6 Segment of a right circular cylinder.

Observe the segment for θ acute as shown in Figure 2.7a. Notice that the segment (I) is actually a sector (II) of a circle less a triangle (III). That is, the area of the segment is the area of a sector minus the area of a triangle.

$$\begin{pmatrix}\text{area of}\\ \text{segment}\end{pmatrix} = \begin{pmatrix}\text{area of}\\ \text{sector}\end{pmatrix} - \begin{pmatrix}\text{area of}\\ \text{triangle}\end{pmatrix}$$

The volume of the segment of the cylinder is the height of the cylinder times the area of the base, which is the segment of the circle.

$$\text{volume} = \begin{pmatrix}\text{height of}\\ \text{cylinder}\end{pmatrix}\left[\begin{pmatrix}\text{area of}\\ \text{sector}\end{pmatrix} - \begin{pmatrix}\text{area of}\\ \text{triangle}\end{pmatrix}\right]$$

Figure 2.7 Base of cylinder.

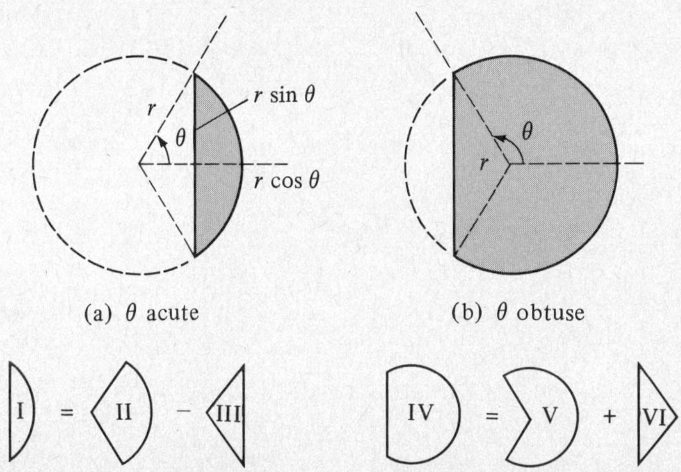

The area of a sector of a circle is to the area of a circle as the angle of the sector is to 2π. That is,

$$\frac{\begin{pmatrix}\text{area of}\\ \text{sector}\end{pmatrix}}{\begin{pmatrix}\text{area of}\\ \text{circle}\end{pmatrix}} = \frac{2\theta}{2\pi}$$

$$\begin{pmatrix}\text{area of}\\ \text{sector}\end{pmatrix} = \frac{\theta}{\pi}\begin{pmatrix}\text{area of}\\ \text{circle}\end{pmatrix}$$

$$\text{volume} = \begin{pmatrix}\text{height of}\\ \text{cyclinder}\end{pmatrix}\left[\frac{\theta}{\pi}\begin{pmatrix}\text{area of}\\ \text{circle}\end{pmatrix} - \frac{1}{2}\begin{pmatrix}\text{base of}\\ \text{triangle}\end{pmatrix}\begin{pmatrix}\text{height of}\\ \text{triangle}\end{pmatrix}\right]$$

$$V = (h)\left[\frac{\theta}{\pi}(\pi r^2) - \frac{1}{2}(2r\cos\theta)(r\sin\theta)\right]$$

$$= h[\theta r^2 - r^2\cos\theta\sin\theta]$$

$$\boldsymbol{V = hr^2(\theta - \sin\theta\cos\theta)}$$

This is satisfactory for an acute θ. What if θ is obtuse? First, $\cos\theta$ is negative for obtuse θ and that would seem to make the base of the triangle and hence the area of the triangle negative. Actually, notice in Figure 2.7b that if θ is obtuse, the area of the segment (IV) is obtained by adding the area of the sector (V) and the area of the triangle (VI) rather than adding as when θ is acute. The change of sign of the cosine then has the effect of adding the area of the triangle for an obtuse θ. The formula gives the volume of the segment for any value of θ such that $0 \leq \theta \leq \pi$. Can you explain what happens when θ is $\pi/2$?

EXAMPLE

4. Find the volume of the segment of an 18-in.-radius cylinder, 3 ft high, if θ is 75°.

First, the radius is 1.5 ft and 75° has radian measure $5\pi/12$. Second, substitute into the formula.

$$V = hr^2(\theta - \sin\theta\cos\theta)$$

$$= 3(1.5)^2\left(\frac{5\pi}{12} - \sin\frac{5\pi}{12}\cos\frac{5\pi}{12}\right)$$

Third, calculate this quantity.

ALGEBRAIC

Key	Display
[5]	5
[×]	5

Key	Display	Comment
π	3.1415926	Compute $5\pi/12$ and store, because θ is used several times.
$\div$	1.5707963	
12	12	
=	1.3089969	
STO	1.3089969	
sin	0.96592582	Check the degree-radian switch!
$\times$	0.96592582	
RCL	1.3089969	
cos	0.25881904	
+/−	−0.25881904	Change signs and add.
+	−0.25000000	
RCL	1.3089969	
=	1.0589969	
$\times$	1.0589969	
1.5	1.5	
x^2	2.25	
$\times$	2.3827431	
3	3	
=	7.1482293	

$$V \approx 7.1482293$$

Finally, state an answer to the original question.

The volume is approximately 7.15 ft^3.

The angle θ might not be one of the given dimensions. While making a water trough from a 50-gallon drum, you would be more likely to measure the width w of the segment as shown in Figure 2.4. However, θ may be found from this width, as we see below.

$$w + r\cos\theta = r$$

$$w = r - r\cos\theta$$

$$\mathbf{w = r(1 - \cos\theta)}$$

$$\frac{w}{r} = 1 - \cos\theta$$

$$\cos\theta = 1 - \frac{w}{r}$$

$$\boldsymbol{\theta = \cos^{-1}\left(1 - \frac{w}{r}\right)}$$

An example should clarify the use of these new results.

EXAMPLE

5. A 4-m log is 45 cm in diameter. If a 15-cm slice is removed lengthwise, what is the volume of the remaining segment?

First convert all measurements to meters.

$$h = 4 \qquad r = 0.225 \qquad w = 0.3$$

$$\theta = \cos^{-1}\left(1 - \frac{w}{r}\right) = \cos^{-1}\left(1 - \frac{0.3}{0.225}\right)$$

$$\approx 1.9106332$$

Be certain to have the degree-radian switch on radian. Store θ and recall as necessary. Do not round off and reenter. Round off only at the final step.

Now that θ is known, the volume may be found.

$$V = hr^2(\theta - \sin\theta\cos\theta)$$

$$= 4(0.225)^2(\theta - \sin\theta\cos\theta)$$

$$\approx 0.45054284$$

The volume is approximately 0.45 m^3.

6. Suppose that the log in Example 5 is white pine, which weighs approximately 1100 lb/m^3. What is the weight of the log?

$$\begin{pmatrix}\text{total}\\\text{weight}\end{pmatrix} = (\text{volume})\begin{pmatrix}\text{weight per}\\\text{unit volume}\end{pmatrix}$$

$$w \approx (0.45054284)(1100)$$

$$\approx 495.59712$$

The log weighs approximately 496 lb.

Piston Application

Many physical problems can be analyzed with trigonometry, especially those that deal with circular motion. Consider the typical piston, connecting rod, and crankshaft in an automobile engine, steam engine, or electric pump. The piston moves back and forth or up and down along a straight path. This linear motion in an engine produces the circular motion of the crankshaft. The rod connects the piston and shaft at pivot points P and A, as shown in Figure 2.8. As the shaft turns in an electric pump, the rod pushes and pulls the piston in its cylinder in a line. In an engine, however, it is the piston that turns the shaft.

A mechanical diagram is shown in a coordinate system in Figure 2.8. Note that if $\theta = 0$, the piston and connecting rod would be fully extended such that

$$r + a + c = OB$$

In this "top-of-the-stroke" position, point P is colinear with O, A, and B. But the crankshaft is turning in a counterclockwise direction, so as θ

Figure 2.8 Piston, connecting rod, and crankshaft.

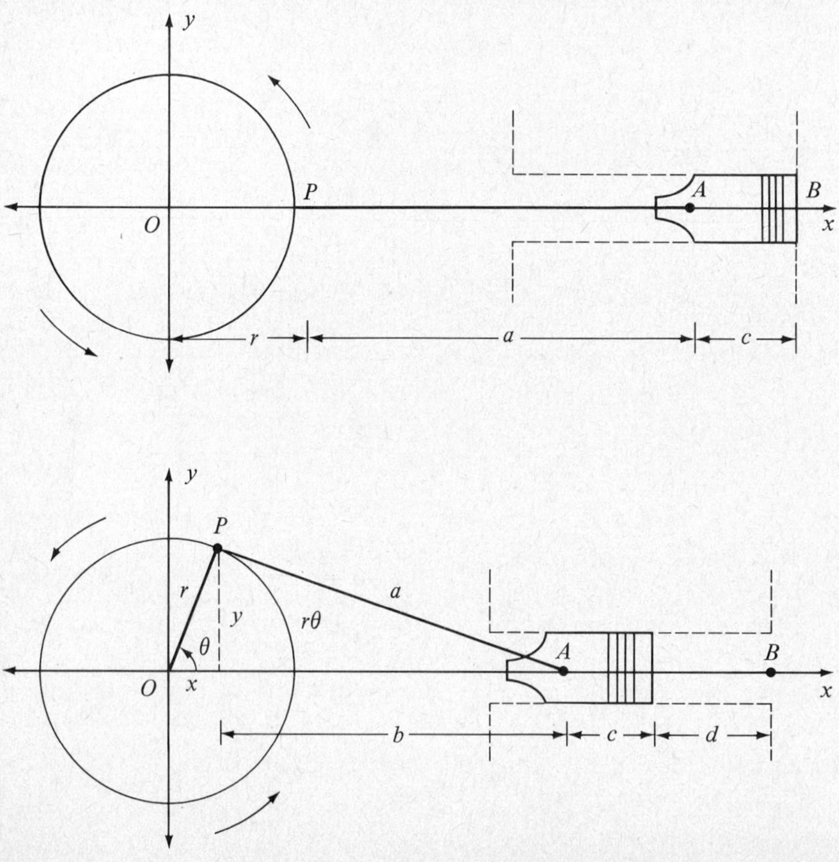

changes, the head of the piston moves away from B a distance d such that

$$OA + c + d = OB$$

$$\begin{aligned} d &= OB - (OA + c) \\ &= (r + a + c) - (x + b + c) \\ &= r + a - x - b \end{aligned}$$

Then notice that $b^2 + y^2 = a^2$ and $b = \sqrt{a^2 - y^2}$, so

$$d = r + a - x - \sqrt{a^2 - y^2}$$

Finally, $x = r \cos \theta$ and $y = r \sin \theta$, so that

$$\mathbf{d = r + a - r \cos \theta - \sqrt{a^2 - r^2 \sin^2 \theta}}$$

The distance depends on θ alone, because a and r are constants. d is then a function of θ, but suppose the crankshaft is turning at 120 r/min and we wish d as a function of time t.

Knowing the speed or velocity v, the arc length $r\theta$ traveled can be expressed in terms of the time t.

$$\begin{pmatrix}\text{arc}\\ \text{length}\end{pmatrix} = \begin{pmatrix}\text{length of}\\ \text{revolution}\end{pmatrix}\begin{pmatrix}\text{number of}\\ \text{revolutions}\end{pmatrix}$$

$$\downarrow \qquad\qquad \downarrow \qquad\qquad \downarrow \searrow$$

$$\begin{pmatrix}\text{arc}\\ \text{length}\end{pmatrix} = \begin{pmatrix}\text{length of}\\ \text{revolution}\end{pmatrix}\begin{pmatrix}\text{revolutions}\\ \text{per minute}\end{pmatrix}\begin{pmatrix}\text{time in}\\ \text{minutes}\end{pmatrix}$$

$$\downarrow \qquad\qquad \downarrow \qquad\qquad \downarrow \qquad\qquad \downarrow$$

$$(r\theta) = (2\pi r) \quad (v) \quad (t)$$

$$\begin{aligned} r\theta &= 2\pi r v t \\ \theta &= 2\pi v t \end{aligned}$$

and in this instance, $v = 120$, so

$$\begin{aligned} \theta &= 2\pi(120)t \\ &= 240\pi t \end{aligned}$$

Now substitute for θ in the expression for d.

$$\mathbf{d = r + a - r \cos 240\pi t - \sqrt{a^2 - r^2 \sin^2 240\,\pi t}}$$

Finally, distance is expressed as a function of time.

Evaluation of these expressions for values of θ and t is complicated. The calculator certainly gives computing power, but it won't do it by itself. Some ingenuity will be necessary at times to compute an involved expression. In the following example, the stages of calculation are featured, and the key scheme may require some adaptation for your calculator.

EXAMPLES

7. Find the position of a piston with a 10-in. connecting rod (a) and 3-in. radius crankshaft (r), at 60° of rotation (θ).

$$\begin{aligned} d &= r + a - r\cos\theta - \sqrt{a^2 - r^2\sin^2\theta} \\ &= 3 + 10 - 3\cos 60^\circ - \sqrt{100 - 9\sin^2 60^\circ} \\ &= 13 - 3\cos 60^\circ - \sqrt{100 - 9(\sin 60^\circ)^2} \end{aligned}$$

ALGEBRAIC	Flow Chart of Stages of Calculation
[60] [sin] [x^2]	$A = \sin^2 60^\circ$
[×] [9] [−] [100] [=] [+/−] [√]	$B = \sqrt{100 - 9A}$
[+]	
[(] [60] [cos] [×] [3] [)]	$C = 3\cos 60^\circ$
[=] [+/−] [+] [13] [=]	$D = 13 - C - B$

$d \approx 1.8433960$
$d \approx 1.84$

The piston is approximately 1.84 in. from the top of its stroke.

8. Find the position of the piston in Example 7 after 0.01 s.

$$\begin{aligned} d &= r + a - r\cos 240\pi t - \sqrt{a^2 - r^2\sin^2 240\pi t} \\ &= 3 + 10 - 3\cos 240\pi(0.01) - \sqrt{100 - 9(\sin 240\pi(0.01))^2} \\ &= 13 - 3\cos 2.4\pi - \sqrt{100 - 9\sin^2 2.4\pi} \end{aligned}$$

(Calculate as in Example 7 except in radian mode.)

≈ 2.4886168

After 0.01 s, the piston is approximately 2.49 in. from the top of the stroke.

The exercises that follow will give you an opportunity to test your calculating power. However, don't confuse this power with the true power of trigonometry. The analysis of the motion of the piston would be *impossible* without trigonometry, but the calculation is only *tedious* without a calculator (and sometimes with one).

exercise set 2.4

A. exercises

Find the value or angle specified in Exercises 1–20 to the nearest ten-thousandth or the nearest minute. All angles are positive and less than 270° or $3\pi/2$.

1. $\sin 171° = A$
2. $\cos 97° = B$
3. $\tan 146° = C$
4. $\cot 102° = D$
5. $\arcsin(-0.3936) = E$
6. $\arccos(-0.4598) = F$
7. $\tan^{-1}(-1.4080) = G$
8. $\cot^{-1}(-0.7102) = J$
9. $\sec 212.25° = I$
10. $\csc 193.7° = H$
11. $\sin(-68°14') = K$
12. $\cos(-21°56') = L$
13. $\operatorname{arccsc} 1.7537 = M$
14. $\operatorname{arcsec} 1.7357 = N$
15. $\cot 324.4° = \phi$
16. $\csc 283.85° = P$
17. $\csc^{-1}(-2.5129) = Q$
18. $\cot^{-1}(3.2561) = R$
19. $\operatorname{arcsec} 1.4142 = S$
20. $\operatorname{arccsc}(-1.7321) = T$

B. exercises

Calculate the quantity in Exercises 21–30.

21. $U = \dfrac{35.2 \sin 21°}{\sin 64°}$
22. $V = \dfrac{\sin 52°}{127.5 \sin 19°}$
23. $W = (1.5821)^2 - 16 \sec 79°$
24. $X = 25 \cot^2 44° - (2.6157)^2$
25. $\alpha = \arcsin\left(\dfrac{0.0147}{\cos 23°}\right)$
26. $\beta = \arctan\left(\dfrac{\sin 29°}{1.4023}\right)$
27. $Y = 0.1014 - \sqrt{1 - \sin^2 79°}$
28. $Z = \sqrt{\cos^2 15° + \dfrac{1}{0.01267}}$

29. $\gamma = \text{arccot}(1.847 - \cos^2 49°)$ 30. $\delta = \text{arcsec}\left(\dfrac{4.4155 + \sin 26°}{\tan 74°}\right)$

Exercises 31–40 deal with the index of refraction D. Find the missing quantities to the nearest ten-thousandth or to the nearest minute.

	MEDIUM	D	θ_a	θ_m
31.	Ice		20°	15.133°
32.	Topaz		17°	10.333°
33.	Garnet		13°30′	7°44′
34.	Acetone		10.417°	7.567°
35.	Benzene	1.5011	22°25′	
36.	Chloroform	1.4467	20°35′	
37.	Nicotine	1.5280		19.55°
38.	Styrene	1.5469		17.367°
39.	Diamond	2.4195	15°43′	
40.	Diamond	2.4195		13°13′

See Examples 1–3.

Exercises 41–48 refer to the volume of a segment of a cylinder, where

$$V = hr^2(\theta - \sin\theta\cos\theta)$$

41. A plastic cylindrical container is 10 cm across and 20 cm high. It is cut lengthwise such that θ is 110°. What is the volume of the resulting section? See Example 4.

42. A cylindrical tank must sit against a wall, so one side is constructed flat. The tank is 8 ft in radius and 10 ft high. What is the volume if θ is 130°?

43. A 45-ft-long tunnel is planned under a new roadway. It is to be made from a 6-ft-radius culvert. If the height inside the tunnel is 9 ft, find the number of cubic yards of fill dirt saved by the tunnel. See Example 5.

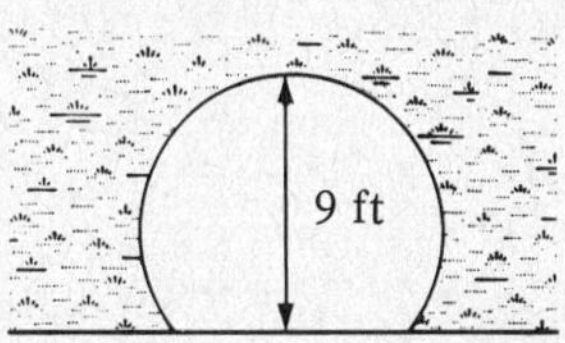

44. A 150-ft-long culvert carries water under a freeway. There is nor-

mally 2 ft of water in the 3-ft-radius pipe. What volume of water is normally in the pipe?

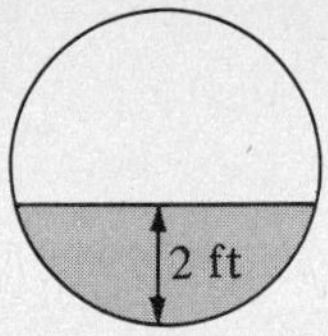

45. A 50-gallon drum is cut lengthwise to create an opening. The drum is 37 in. high, 20 in. in diameter, and petroleum weighs approximately 0.0289 lb/in^3. What is the weight of the oil if the segment is filled to a depth of 12 in. so that it can be used as an oil bath for machine parts?

See Example 6.

lb/in^3 means pounds per cubic inch.

46. A marble column has a flat surface where it abuts a building facade. The column has a 1-ft radius, is 14 in. front to back, and is 12 ft tall. How much does the column weigh if marble weighs approximately 168.55 lb/ft^3?

47. A circular hump across a road is often used to control speeding. If such a speed bump is 4 in. high and has a radius of 16 in., what volume of asphalt is needed to place one across a 30-ft roadway? What is the weight of the asphalt used if it weighs approximately 90 lb/ft^3?

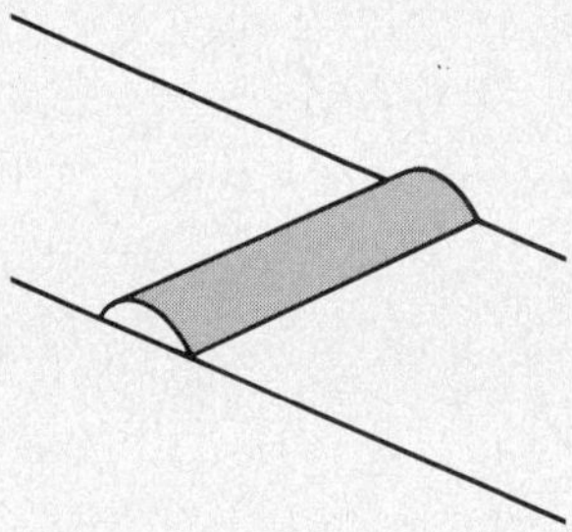

48. A circular depression on a roof has filled with water and frozen. The owner is concerned about the weight over a prolonged period. The depression is 3 in. deep and has an 8-in. radius. What is the volume of the trapped ice? Find the weight in a 20-ft run if ice weighs approximately 0.033 lb/in.^{3}.

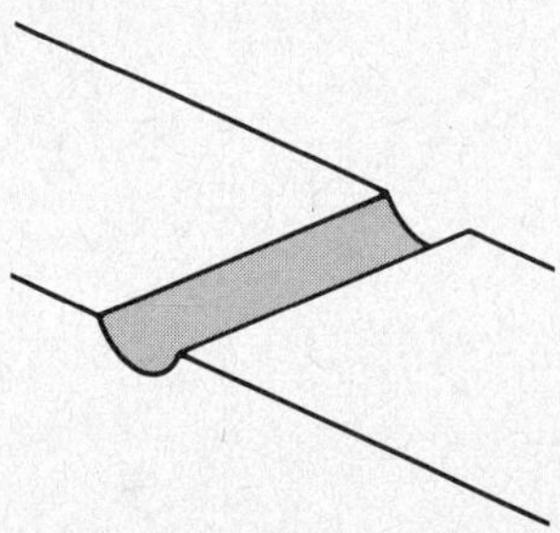

49. If a cylinder is cut lengthwise through two radii, the resulting figure is called a **sector** of a cylinder. Write a formula for a sector of a cylinder in terms of r, h, and θ as shown in Figure 2.9.

50. A hollow sector of a cylinder is shown in Figure 2.10. Write a formula for its volume in terms of r, R, h, and θ.

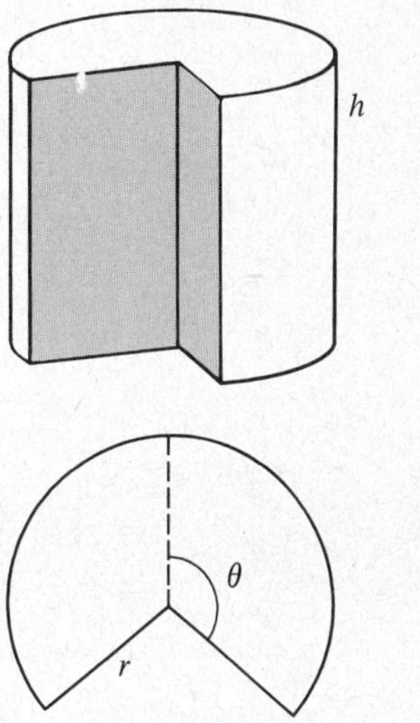

Figure 2.9

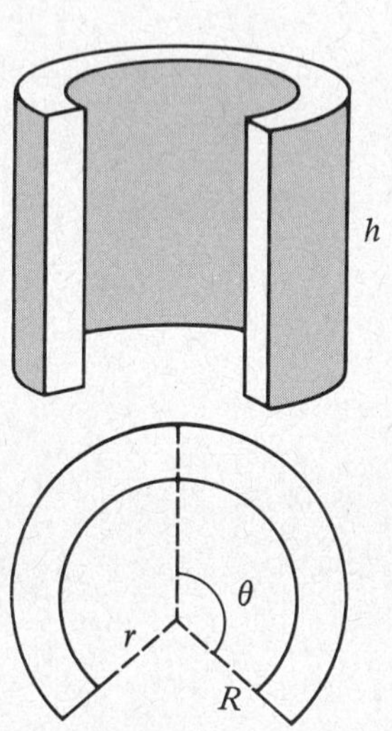

Figure 2.10

Exercises 51–60 deal with the piston mechanical system where

$$d = r + a - r\cos\theta - \sqrt{a^2 - r^2\sin^2\theta}$$

51. $r = 2$, $a = 5$, $\theta = 85°$, $d = ?$ See Example 7.

52. $r = 3$, $a = 8$, $\theta = 125°$, $d = ?$

53. $r = 10.4$, $a = 27.6$, $\theta = 164°25'$, $d = ?$

54. $r = 12.3$, $a = 25.9$, $\theta = 206.2°$, $d = ?$

55. Suppose that the shaft is turning at 30 rpm. Rewrite the equation in terms of time t in minutes, instead of angle θ. See Example 8.

56. Rewrite the equation in terms of time t in seconds instead of angle θ if the shaft turns at 2 r/s (revolutions per second).

57. In Exercise 55, $r = 4$, $a = 9$, $t = 1.15$, $d = ?$

58. In Exercise 56, $r = 21$, $a = 52$, $t = 0.09$, $d = ?$

59. Suppose that the length of the connecting rod is four times the radius of the crankshaft. Rewrite the equation for d and simplify, if possible.

60. Rewrite the equation in Exercise 59 so that the ratio d/r is a function of time if the shaft turns at v r/min.

C. exercises

calculator problems *The constant g, the acceleration due to gravity, often appears in physical formulas. In 1930 the International Union of Geodesy and Geophysics adopted the following expression for g.*

$$\mathbf{g = 978.049(1 + 0.0052884 \sin^2 \phi - 0.0000059 \sin^2 2\phi) - 0.0003086A}$$

where ϕ is the latitude and A is the altitude in meters. 981.274 cm/s² is used in scientific work, whereas the more familiar value is 32 ft/s²(32.194). In Exercises 61–70, use the preceding formula to determine g at the following locations.

	CITY	LATITUDE	ALTITUDE
61.	Honolulu, HI	21°18′N	6.4 m
62.	New Orleans, LA	29°56′N	0 m
63.	Phoenix, AZ	33°27′N	332 m
64.	Dallas, TX	33°47′N	133 m
65.	Denver, CO	39°44′N	1609 m
66.	Pittsburgh, PA	40°26′N	227 m
67.	Chicago, IL	41°52′N	181 m
68.	Boise, ID	43°37′N	824 m
69.	Spokane, WA	47°39′N	576 m
70.	Fairbanks, AL	64°48′N	137 m

71. **trig talk** Use Exercises 1–30 to fill the blanks with the capital letter that corresponds to the value below the blank. When all the blanks are filled, the Felix Klein quotation may be read.

Notice you are given the letter "oh" below.

"

___	___	___	___	___	___	___	___	___	___	___
34°46′	0.1564	215°16′	−4.2223	203°11′	34°46′	0.1564	215°16′	−1.1824	−0.6745	45°0′

___	___		___	___	___
−1.1824	45°0′	. . .	215°16′	−4.2223	203°11′

___	___	___	___	___	O	___
45°0′	215°16′	14.0350	−0.2126	−0.0894		117°22′

______	______	______	______	______	______	______	______	______	______	______
45°0′	203°11′	0.9276	117°22′	203°11′	0.0190	−1.1824	−0.2126	203°11′	54°49′	215°16′

______	______	______	______	______	______."	
215°16′	−4.2223	−1.1824	54°49′	125°23′	45°0′	(FELIX KLEIN)

2.5 CHAPTER REVIEW

Fill in the symbol, word, or words that make the statements in Exercises 1–10 complete and correct.

1. (2.1) If θ is in standard position with (h, k) on the terminal side d units from the origin, $\sin \theta =$ ________.
2. (2.1) Both ________ and ________ are positive in QI and QIII.
3. (2.1) If θ is in standard position with (x, y) on the terminal side r units from the origin, ________ $= x/r$.
4. (2.2) Any real number is the measure of an angle in ________ measure.
5. (2.2) θ is coterminal with ________ for any integer k and θ in radian measure.
6. (2.2) If θ is in QII, then its reference angle θ_R is ________.
7. (2.3) An equation that is true for all values of the variables is called an ________.
8. (2.3) $1 + \tan^2 \theta = \sec^2 \theta$ is one of the so-called ________ identities.
9. (2.3) $\sin \theta$, $\cos \theta$, and $\tan \theta$ are related by one of the so-called ________ identities.
10. (2.4) The ________ identities are used to find cotangent, secant, and cosecant on a calculator.
11. (2.1) Find sine, cosine, and tangent for the angle θ in standard position with the given point on the terminal side.

    (a) $\left(\frac{15}{17}, \frac{8}{17}\right)$ (b) $\left(-\frac{\sqrt{5}}{3}, \frac{2}{3}\right)$ (c) $(16, -63)$

    (d) $(-7, -9)$ (e) $(20, -21)$
12. (2.1) Find the function named that satisfies the given conditions.

    (a) $\sin 270°$
    (b) $\sin \alpha$, if $\tan \alpha = \frac{15}{8}$, and α in QIII
    (c) $\cot \beta$, if $\sin \beta = -1$
    (d) $\sec \gamma$, if $\sin \gamma = -\sqrt{3}/4$, γ in QIV
    (e) $\cos \delta$, if $\cot \delta = 0$, and $180° < \delta < 360°$

13. (2.2) Convert to degrees or radian measure.

(a) $135°$ (b) $-\frac{\pi}{3}$ (c) $\frac{5\pi}{3}$

(d) $-1450°$ (e) $\left(\frac{\pi}{2}\right)^{\circ}$

14. (2.2) Express each function in terms of its reference angle.

(a) $\sin 142°$ (b) $\cos\left(-\frac{4\pi}{3}\right)$ (c) $\tan 1100°$

15. (2.3) Express each of the other five functions in terms of $\sin\theta$.

16. (2.3) Prove the identities.

(a) $\frac{\sin\theta}{\csc\theta} = \sin^2\theta$

(b) $\cot^2\theta(1 + \tan^2\theta) = \csc^2\theta$

(c) $\frac{1 + \cot^2\theta}{1 + \tan^2\theta} = \cot^2\theta$

17. (2.4) Find the value or angle. Round the answer to the nearest ten-thousandth or the nearest minute. All angles are the smallest positive value less than $3\pi/2$.

(a) $A = \sin 94°32'$ (b) $\tan\beta = 1.6904$
(c) $C = \sec(-47.1°)$ (d) $\delta = \arcsin(-0.6253)$
(e) $E = \csc 303.55°$

18. (2.4) Find the index of refraction of fused silica if light enters at 39.75° and is refracted to 26°.

19. (2.4) The index of refraction of table salt is 1.5443. If light enters at 18°, to what angle is it refracted?

20. (2.4)

(a) Find the volume of a segment of a cylinder with radius 12 cm and height 35 cm if $\theta = 20°$.
(b) Find the exact volume of a segment of a cylinder with $\theta = \pi/3$, radius 10 in., and height 6 in.
(c) Find the volume of a 4.72-m-high segment with width 1.73 m and radius 2.41 m.

graphs of the trigonometric functions

three

3.1 THE SINE WAVE

The definition of the trigonometric functions, derivations of certain properties, the ratios on the right triangle, and applications to physical problems have provided a certain feeling for these functions. A more complete understanding might best be acquired by looking at the graphs of the functions. The graph of a function often displays its behavior as only a picture can. We may begin by plotting some points to determine the general shape. Once this is done, however, sketching should be possible with few if any actual calculations.

The Sine Function

A sufficient number of points for $y = \sin x$ can be generated by the special values we have seen before. These familiar values are shown in Table 3.1. These values can then be plotted as shown in Figure 3.1. The x axis in the following figures is marked off in multiples of π so that the radian measures, $\pi/3$, $\pi/2$, $2\pi/3$, π, . . . , can be located easily. Moreover, both scales will be real numbers. If a length of 3 units from the y axis is used for π on the x axis, then the scale will be nearly equivalent. So if 4 squares are used for 1 unit on the y axis, then 12 squares will be

TABLE 3.1 VALUES OF y = sin x

x	$\sin x$	x	$\sin x$
0	0	π	0
$\frac{\pi}{6}$	$\frac{1}{2} = 0.5$	$\frac{7\pi}{6}$	$-\frac{1}{2} = -0.5$
$\frac{\pi}{4}$	$\frac{\sqrt{2}}{2} = 0.71$	$\frac{5\pi}{4}$	$-\frac{\sqrt{2}}{2} = -0.71$
$\frac{\pi}{3}$	$\frac{\sqrt{3}}{2} = 0.87$	$\frac{4\pi}{3}$	$-\frac{\sqrt{3}}{2} = -0.87$
$\frac{\pi}{2}$	1	$\frac{3\pi}{2}$	-1
$\frac{2\pi}{3}$	$\frac{\sqrt{3}}{2} = 0.87$	$\frac{5\pi}{3}$	$-\frac{\sqrt{3}}{2} = -0.87$
$\frac{3\pi}{4}$	$\frac{\sqrt{2}}{2} = 0.71$	$\frac{7\pi}{4}$	$-\frac{\sqrt{2}}{2} = -0.71$
$\frac{5\pi}{6}$	$\frac{1}{2} = 0.5$	$\frac{11\pi}{6}$	$-\frac{1}{2} = -0.5$
π	0	2π	0

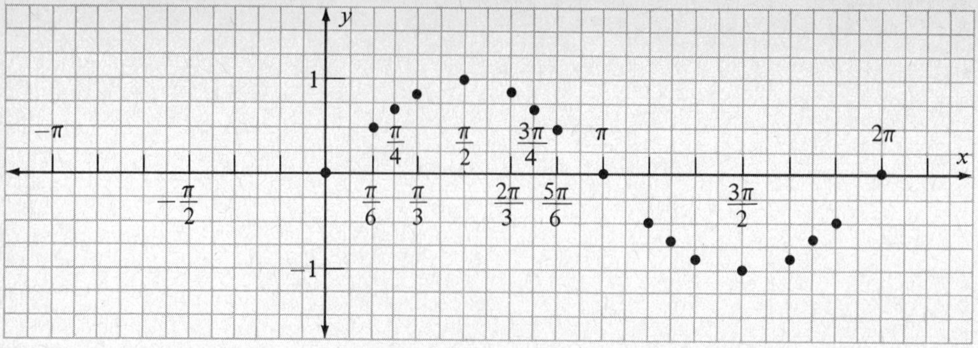

Figure 3.1 Special values for $y = \sin x$.

used for π on the x axis. With this convention the scales will be approximately the same and the graphs will not be distorted.

The points of the graph of $y = \sin x$ can be connected by a smooth curve that we call a **sine wave.** Working with coterminal angles, it was seen that $\sin \theta = \sin(\theta + 2\pi) = \sin(\theta + 4\pi) = \ldots$. That is, $\sin \theta = \sin(\theta + 2k\pi)$ for any integer k. A function for which $f(x) = f(x + p)$ for all x in its domain is called **periodic,** with **period** p.

$$\sin \theta = \sin(\theta + 2k\pi) \qquad \text{for any integer } k$$

The sine is periodic with period 2π.

Because the function is periodic, this period of the curve from 0 to 2π simply repeats itself in both directions as shown in Figure 3.2. The shape and form is distinctive and could be reproduced arbitrarily far. The initial period of the sine wave from 0 to 2π will be called the **base cycle.** Each trigonometric function will have a base cycle. If we concen-

Note 4 squares for 1 unit on the y axis and 12 squares for π on the x axis.

Figure 3.2 The graph of the sine wave, $y = \sin x$.

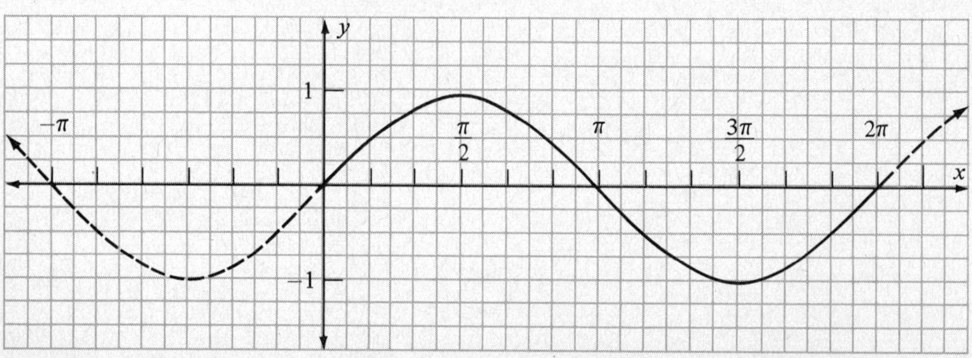

trate on sketching the base cycle of these functions, then that cycle may be repeated as necessary to fulfill our requirements. Because of its distinctive shape, a few key points on the base cycle are all that is necessary.

We know that one cycle of the sine starts at zero, ends at 2π, and rises and falls one unit above and below the x axis. This **frames** the base cycle of $y = \sin x$. See Figure 3.3.

We observe that it begins and ends with $y = 0$ and is zero at the midpoint of the base cycle as shown in Figure 3.4.

Further, as we see in Figure 3.5, it is at its extremes at the "quarter points" between zeros.

These five **key points** on the frame of the graph should be enough to complete a sketch of the sine wave over the base cycle as shown in Figure 3.6. This cycle may then be repeated as needed.

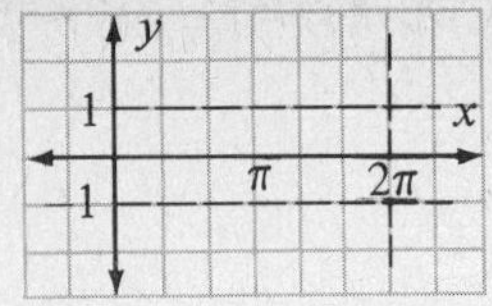

Figure 3.3

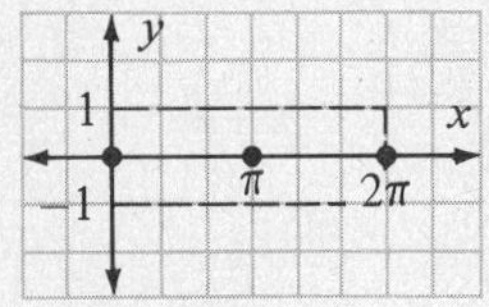

Figure 3.4

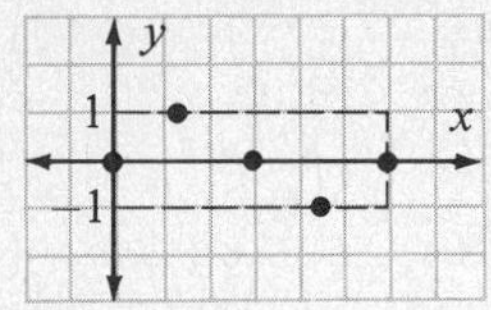

Figure 3.5

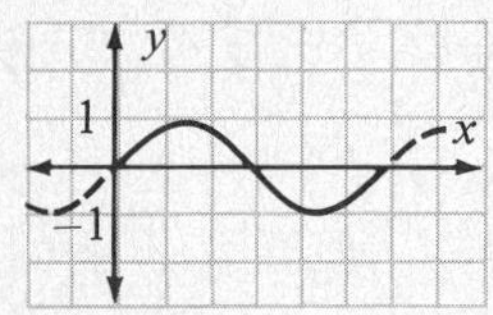

Figure 3.6

Amplitude

Not all sine waves, however, are completely alike. The length of the base cycle or its height may be changed, or it may be moved away from the origin. Let us observe the differences if $y = \sin x$, $y = 2 \sin x$, and $y = \frac{1}{3} \sin x$ are plotted on the same axis as shown in Figure 3.7. As we have seen before, $y = \sin x$ completes its cycle in 2π units, rising and falling one unit above and below the x axis. $y = 2 \sin x$ rises and falls two full units, and $y = \frac{1}{3} \sin x$ varies by one-third unit, whereas both maintain the same period as $y = \sin x$. It would be reasonable to expect $y = a \sin x$ to reach a units above and below the x axis. This is in fact the case, and $|a|$ is called the **amplitude** of the sine wave. Now when "framing" a sine wave,

Figure 3.7 Comparison of sine waves.

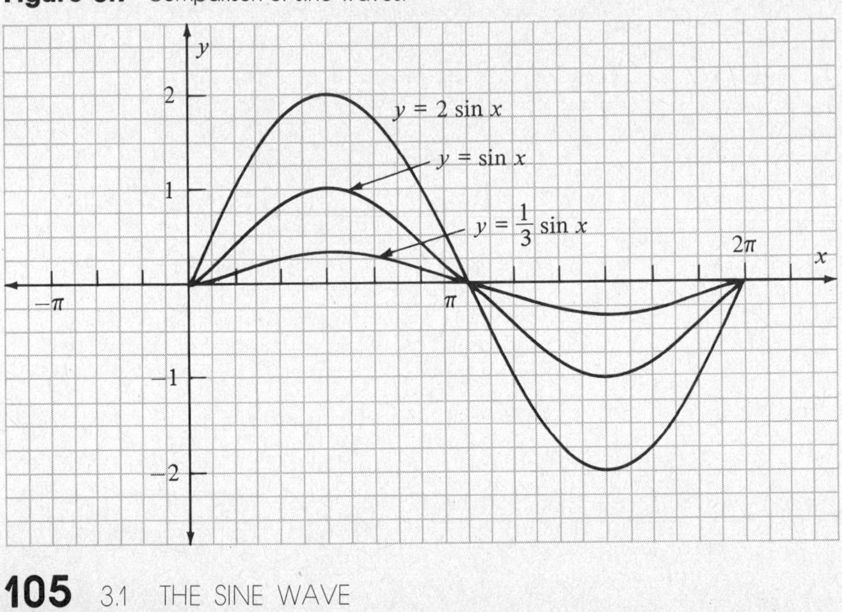

The curves $y = 2 \sin x$ and $y = \frac{1}{3} \sin x$ can be plotted by a table of values similar to Table 3.1.

the amplitude must be observed and the "frame" of the base cycle adjusted in height.

EXAMPLE

1. Find the amplitude, then frame and sketch the base cycle of $y = \frac{5}{2} \sin x$.

   By inspection, $a = \frac{5}{2}$ or 2.5.

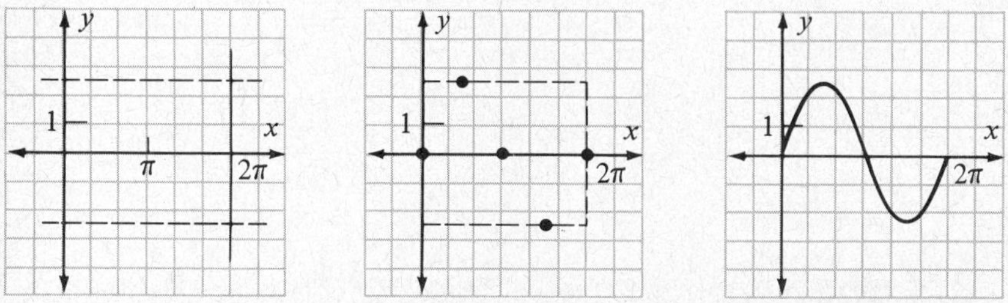

Notice here that one square is used for one unit on the y axis and three squares are used for π on the x axis.

The coefficient of sin x may be negative. The graph of $y = -f(x)$ is the reflection of the graph of $y = f(x)$ in the x axis. Thus, if $y = \sin x$ shows an initial increase, $y = -\sin x$ will begin by decreasing as shown in Figure 3.8.

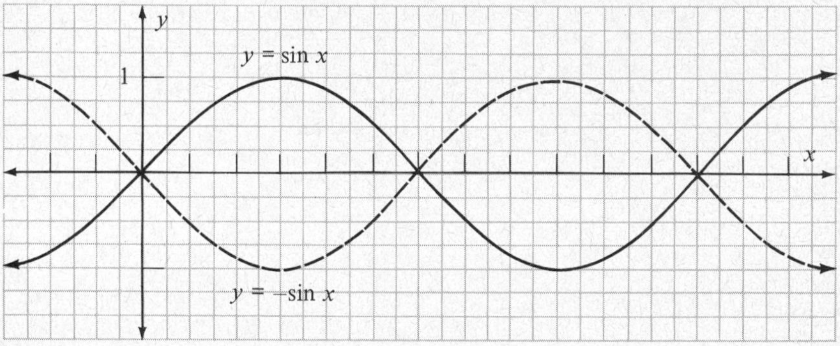

Figure 3.8 Graphs of $y = \sin x$ and $y = -\sin x$.

EXAMPLE

2. Sketch $y = -2 \sin x$.

   By inspection, $|a| = 2$, $a < 0$.

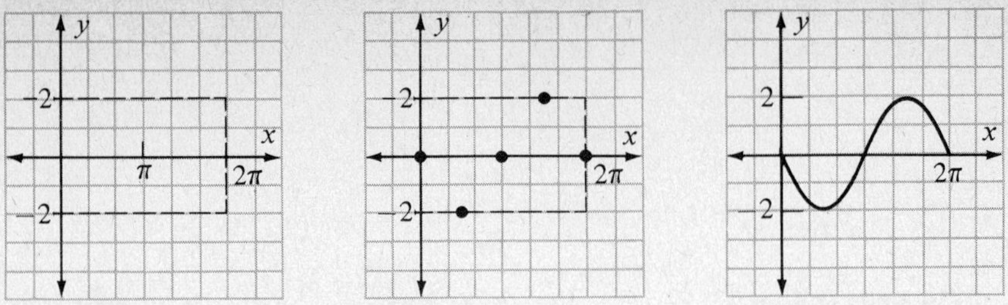

Essentially the only difference with $a < 0$ is that two of the key points are plotted differently.

Frequency

Another difference can be observed by plotting $y = \sin x$, $y = \sin 2x$, and $y = \sin 3x$ on the same set of axes, as shown in Figure 3.9.

Again these curves could be plotted with a table of values similar to Table 3.1.

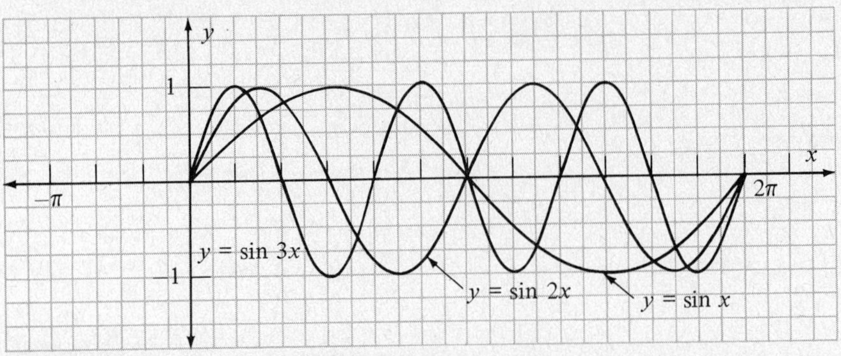

Figure 3.9 Comparison of sine waves.

As we have seen before, $y = \sin x$ completes one cycle in 2π, but $y = \sin 2x$ completes *two* cycles and $y = \sin 3x$ completes *three* cycles of the wave in the same interval. It would be reasonable to expect $y = \sin bx$ to complete b cycles of the wave in 2π units. This is true, and b is called the **frequency** of the sine wave and $2\pi/b$ is the length of one cycle or the **period** of the function.

EXAMPLES

3. Determine the amplitude, frequency, and period, then frame and sketch one cycle of $y = \sin \frac{3}{2}x$.

   By inspection, $|a| = 1$ is the amplitude, $b = \frac{3}{2}$ is the frequency, and

$$p = \frac{2\pi}{\frac{3}{2}} = \frac{2}{3} \cdot 2\pi = \frac{4\pi}{3}$$

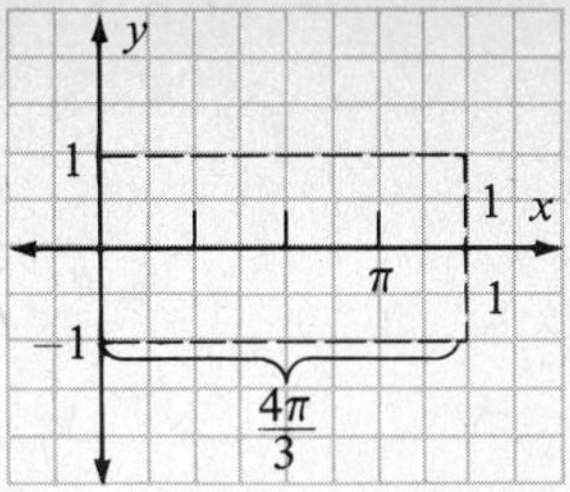

(a)

We use the amplitude and frequency to *frame* one cycle.

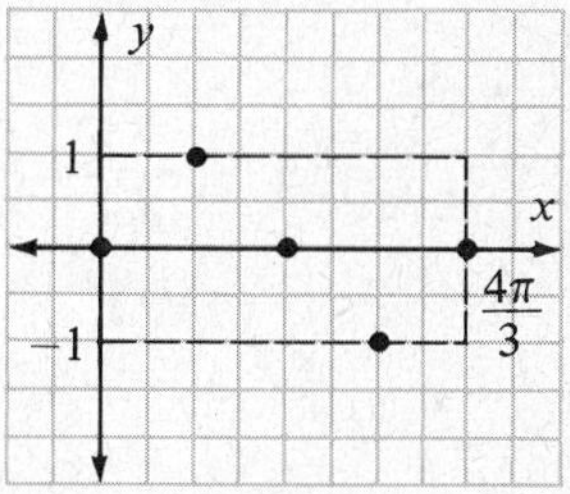

(b)

Then we *plot* the five "key points" for sine.

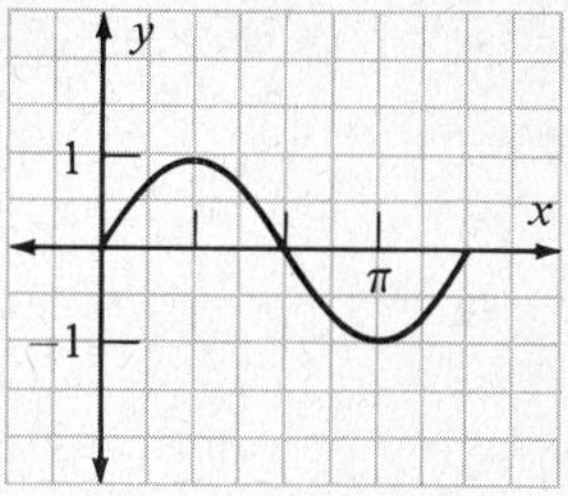

(c)

And then we *sketch* one cycle of the sine (and extend it as needed).

4. Find the amplitude and period, then frame and sketch the graph of $y = 2 \sin 3x$.
By inspection, $|a| = 2$ and $b = 3$, so $p = 2\pi/3$.

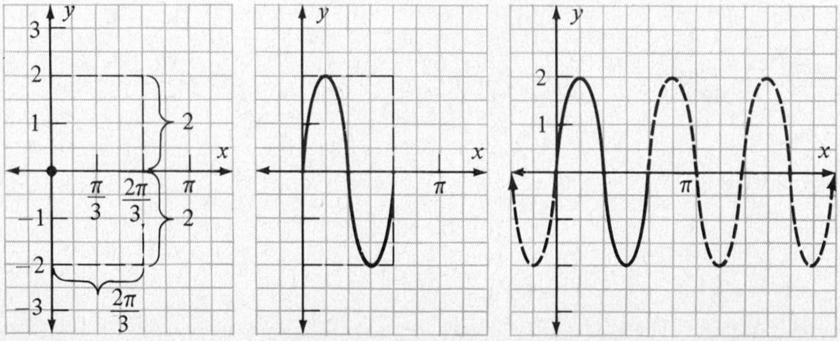

Frequency and amplitude can be thought of as changing the scale on the x and y axes. Given the function $y = f(x)$, it is clear that in $y = af(x)$ the y value is amplified $|a|$ times, stretching or shrinking the graph of $y = \sin x$ in the vertical direction. $y = f(bx)$ accomplishes a similar feat in the horizontal or x direction. There is, however, an important difference. If $|a| > 1$, the graph is *stretched* vertically; whereas if $|b| > 1$, the graph is *shrunk* horizontally.

Translation

A third change is $y - k = f(x - h)$, where $y = f(x)$ is **translated** h units horizontally and k units vertically. That is, the entire graph is moved over (or back) and up (or down) to a new position. h is called the **horizontal displacement** and k is the **vertical displacement.** See Appendix A.2 for a discussion of translations. In the case of the sine wave, it would be sufficient to move the initial point of the base cycle and proceed to construct the curve in the new position. The wave will be framed and sketched as before, but it will not necessarily begin at the origin. Thus, $y - k = a \sin b(x - h)$ will have an initial point at (h, k), amplitude $|a|$, and frequency b. Translation of sine waves is best shown by example.

EXAMPLES

5. Sketch $y - 3 = \sin(x - \pi/2)$.

Comparing with $y - k = a \sin b(x - h)$, we find that (h, k) is $(\pi/2, 3)$. The amplitude is 1, the frequency is 1, so the period is 2π.

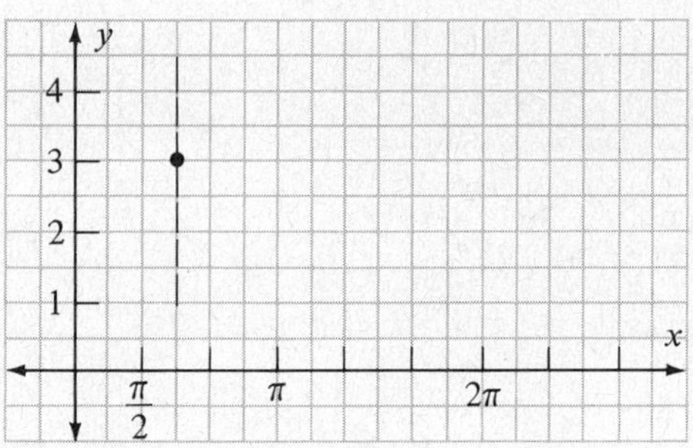

1. Plot initial point $(\pi/2, 3)$.

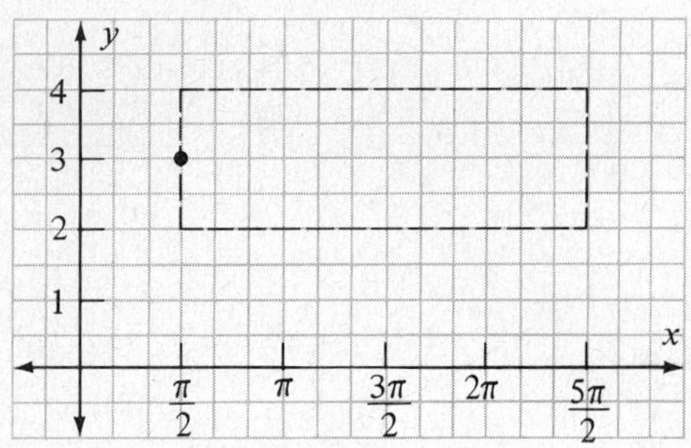

2. Frame the base cycle.

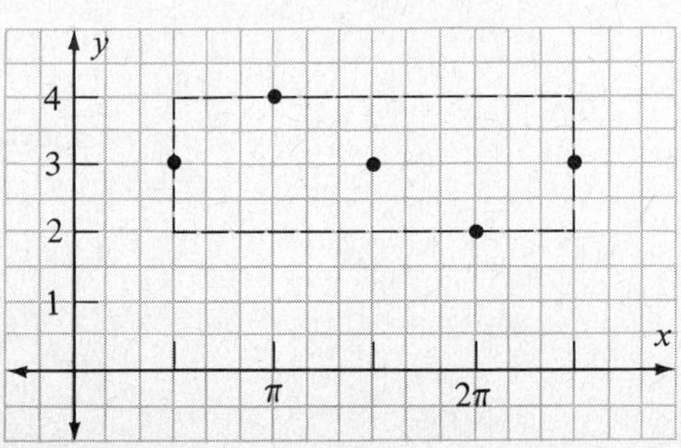

3. Plot the five key points.

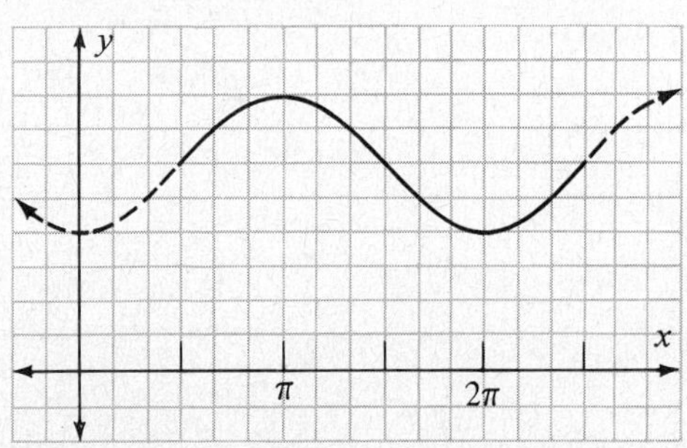

4. Sketch the base cycle.

6. Sketch $y - 1 = \frac{3}{2} \sin 2(x + \pi)$.

$$y - k = a \sin b(x - h)$$

$(h, k) = (-\pi, 1)$, $|a| = \frac{3}{2}$, $b = 2$, so $p = \pi$.

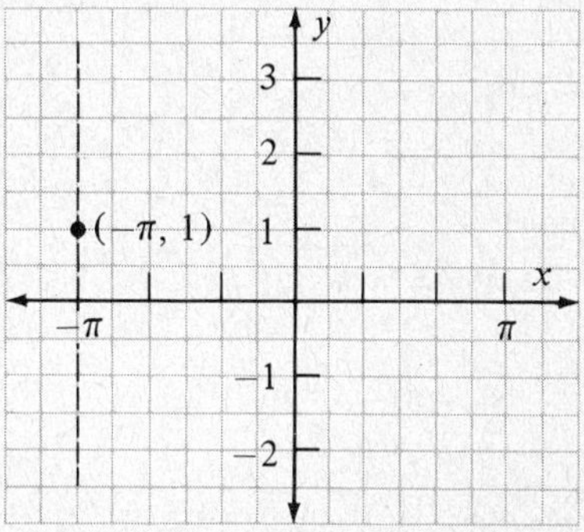

1. Plot initial point.

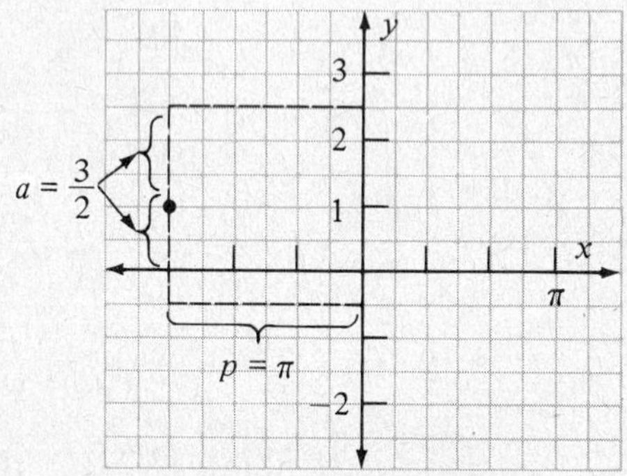

2. Frame base cycle.

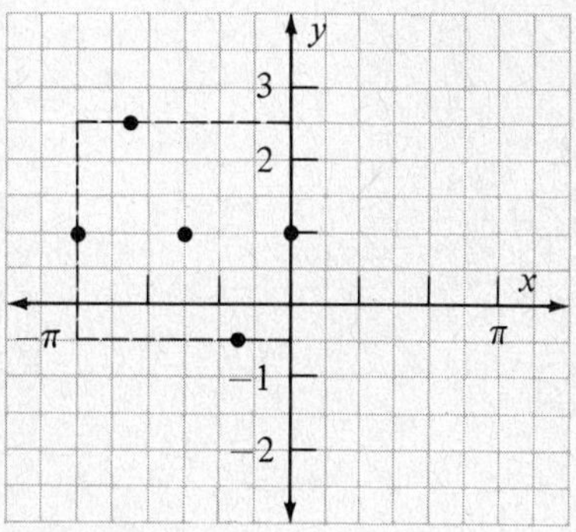

3. Plot key points.

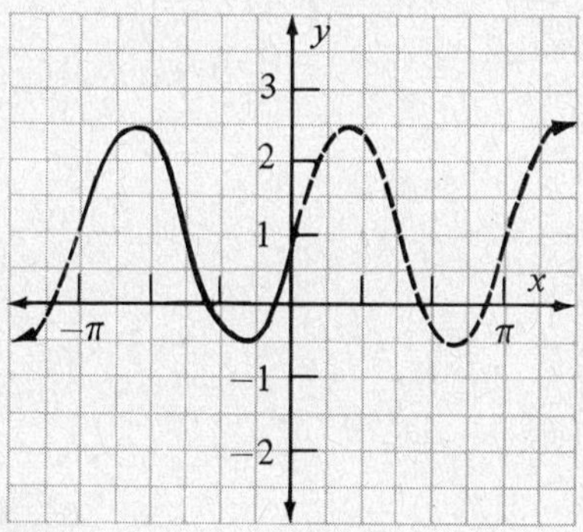

4. Sketch and repeat.

7. Sketch

$$y + 3 = -\frac{1}{2} \sin \frac{2}{3}\left(x - \frac{\pi}{3}\right) : -\pi \leq x \leq 2\pi$$

$$(h, k) = \left(\frac{\pi}{3}, -3\right) \qquad |a| = \frac{1}{2} \qquad b = \frac{2}{3}$$

so $p = 3\pi$, but $a < 0$, so the graph opens down first.

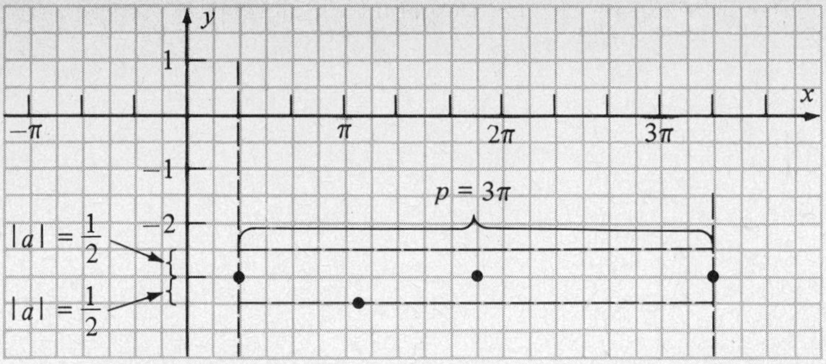

1. Plot initial point ($\pi/3$, -3), frame, and plot key points.

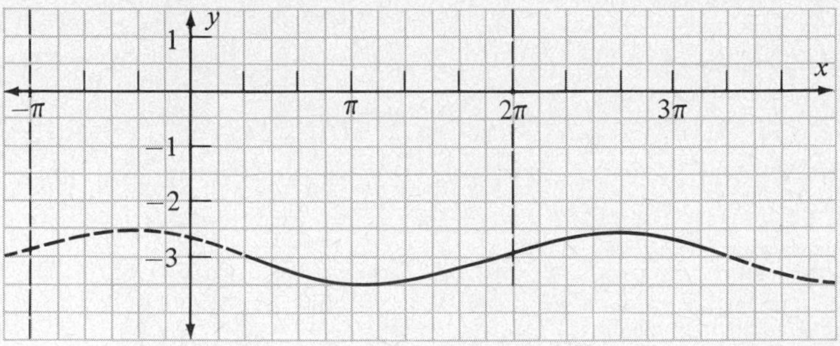

2. Sketch base cycle, establish limits $-\pi \le x \le 2\pi$, and repeat cycles as necessary.

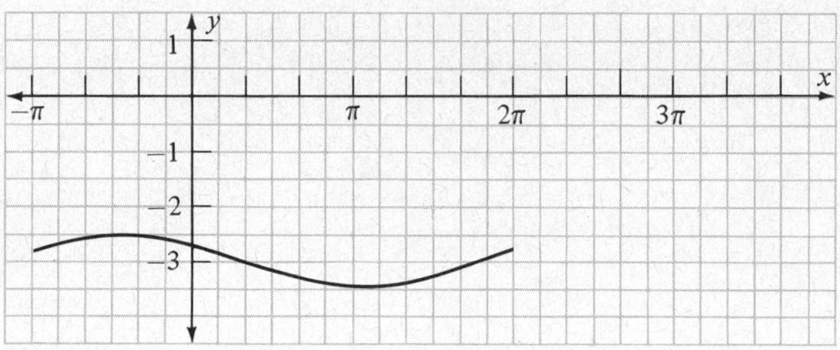

3. $y + 3 = -\frac{1}{2} \sin \frac{2}{3}(x - \pi/3)$: $-\pi \le x \le 2\pi$.

Summary

The "framing method" of sketching the sine assumes that the function is in the form

$$\boldsymbol{y - k = a \sin b(x - h)}$$

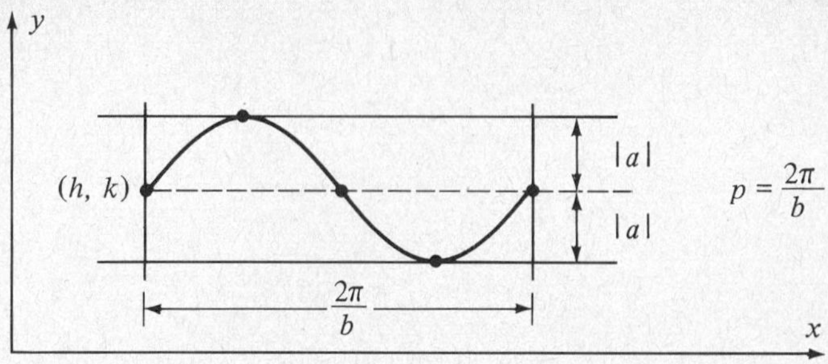

1. Plot the initial point (h, k) of the base cycle.
2. Use the period p to find the terminal point of the base cycle.
3. Use the amplitude a to complete the frame.
4. Plot the five key points.
5. Sketch the base cycle and extend by repeating as necessary.

Sketching the Graph of the Sine Wave: **The Framing Method**

1. Mark left end.
2. Mark right end.
3. Mark top and bottom.
4. Plot key points.
5. Sketch.

exercise set 3.1

A. exercises

1. Construct a table of values similar to Table 3.1 for $y = \cos x$.

See Table 3.1 on 103 and its discussion.

2. Construct a table of values similar to Table 3.1 for $y = \tan x$.
3. Use the table constructed in Exercise 1 to plot and sketch the base cycle of $y = \cos x$.
4. Use the table constructed in Exercise 2 to plot and sketch the base cycle of $y = \tan x$.

Do you notice anything unusual about the graph of $y = \tan x$?

Sketch the base cycle of the functions given in Exercises 5–20.

See Examples 1–4.

5. $y = -\sin x$
6. $y = \sin x$
7. $y = 3 \sin x$
8. $y = -3 \sin x$
9. $y = \sin 2x$
10. $y = \sin 3x$
11. $y = \sin \frac{1}{2}x$
12. $y = \sin \frac{1}{3}x$
13. $y = 2 \sin 3x$
14. $y = 3 \sin 2x$
15. $y = \frac{1}{2} \sin 3x$
16. $y = -\frac{3}{2} \sin 2x$

17. $y = \frac{9}{2} \sin \frac{2}{3}x$

18. $y = \frac{4}{3} \sin \frac{3}{2}x$

19. $y = -\frac{5}{2} \sin \frac{3}{4}x$

20. $y = -\frac{3}{2} \sin \frac{4}{3}x$

B. exercises

In Exercises 21–30, identify the initial point of the base cycle, amplitude, frequency, and period.

See Examples 5–7.

21. $y - 2 = 3 \sin 4(x - \pi)$

22. $y - 1 = 2 \sin 3(x - \pi)$

23. $y - 3 = 2 \sin \frac{1}{2}\left(x + \frac{\pi}{2}\right)$

24. $y + \frac{1}{2} = 3 \sin \frac{1}{3}\left(x - \frac{\pi}{6}\right)$

25. $y + \frac{3}{2} = -\frac{1}{2} \sin 2\left(x + \frac{\pi}{3}\right)$

26. $y + \frac{5}{2} = -\frac{3}{2} \sin 3\left(x + \frac{\pi}{4}\right)$

27. $y - \frac{1}{3} = 3 \sin \frac{2}{3}\left(x + \frac{4\pi}{3}\right)$

28. $y - \frac{2}{3} = 2 \sin \frac{3}{2}\left(x + \frac{3\pi}{4}\right)$

29. $y + 1 = \frac{5}{2} \sin \frac{3}{4}\left(x - \frac{5\pi}{6}\right)$

30. $y + 2 = -\frac{5}{2} \sin \frac{4}{3}\left(x - \frac{3\pi}{2}\right)$

Because of the tilt of the earth's rotation away from the plane of its orbit about the sun, the length of the day at a given latitude varies throughout the year. The phenomenon is periodic, with period equal to the yearly orbit about the sun. Table 3.2 lists the times of sunset and sunrise at 20-day intervals at latitude 38°N. The graph of times of sunset through the year is, in fact, a sine wave.

TABLE 3.2 SUNRISE AND SUNSET AT 38°N LATITUDE, 1979.

DAY	DATE	SUNRISE	SUNSET	DAY	DATE	SUNRISE	SUNSET
20	1/20	7:15	17:08	200	7/19	4:51	19:21
40	2/9	6:58	17:31	220	8/8	5:08	19:03
60	3/1	6:33	17:53	240	8/28	5:25	18:37
80	3/21	6:03	18:12	260	9/17	5:43	18:06
100	4/10	5:33	18:31	280	10/7	6:00	17:35
120	4/30	5:05	18:50	300	10/27	6:20	17:08
140	5/20	4:46	19:07	320	11/16	6:42	16:48
160	6/9	4:36	19:22	340	12/6	7:02	16:40
180	6/29	4:39	19:27	360	12/26	7:15	16:45

Sacramento, CA, Colorado Springs, CO, St. Louis, MO, Cincinnati, OH, and Washington, DC, are all at this approximate latitude.

31. Plot the sunset data from Table 3.2 using 1 square = 10 days on the x axis and 1 square = 20 min on the y axis. Sketch a sine wave through these points and estimate the amplitude and displacement (from 0 day, 0 min). Finally, write an equation with this data that approximates the sunrise time given the day of the year at 38°N.

32. Repeat Exercise 31 for the sunrise data.

Use the methods of this section to sketch the graphs of the functions given in Exercises 33–39. Be careful to apply the restrictions.

It is suggested that you use 4 squares per unit and 12 squares per π on your graph paper for Exercises 33–39.

33. $y = 3 \sin \frac{3}{2} x;\ \pi \leq x \leq \frac{4\pi}{3}$

34. $y = 2 \sin \left(x + \frac{\pi}{3}\right);\ -\frac{\pi}{6} \leq x \leq \frac{2\pi}{3}$

35. $y + 1 = 2 \sin 3 \left(x + \frac{5\pi}{6}\right);\ -\frac{5\pi}{6} \leq x \leq -\frac{\pi}{2}$

36. $y + 1 = -2 \sin \left(x + \frac{5\pi}{6}\right);\ -\frac{5\pi}{6} \leq x \leq -\frac{\pi}{3}$

37. $y + 1 = -2 \sin \left(x + \frac{\pi}{2}\right);\ -\frac{\pi}{2} \leq x \leq \frac{4\pi}{3}$

38. $y - \frac{3}{4} = -\sin 3 \left(x - \frac{\pi}{3}\right);\ \frac{\pi}{3} \leq x \leq \frac{\pi}{2}$

39. $y + 2 = \frac{1}{8} \sin \frac{3}{2} \left(x + \frac{\pi}{3}\right);\ -\frac{\pi}{3} \leq x \leq \frac{\pi}{3}$

C. exercises

40. **palatable plotting** Set up your coordinate axes with $-\pi \leq x \leq 3\pi/2$ and $-4 \leq y \leq 3$ and graph Exercises 33–39 on the same axes and add the following "finishing touches."

(a) $y = \frac{18}{\pi} x + 4;\ -2 < y < 1$

(b) $$\begin{cases} y + 1 > -2 \sin \left(x + \frac{\pi}{2}\right) \\ y + 2 < \frac{1}{8} \sin \frac{3}{2} \left(x + \frac{\pi}{3}\right) \\ y + \frac{9}{4} \geq \frac{1}{8} \sin \frac{3}{2} \left(x + \frac{\pi}{3}\right) \end{cases}$$

3.2 COSINE AND TANGENT

Having dealt exclusively with the graph of the sine in the first section, the examination of the other functions should now be easier. In each case, however, plotting the graph from a table of values is probably a good beginning to observe the general behavior of the curve.

Cosine

Using a table of values similar to Table 3.1, the graph of $y = \cos x$ is shown in Figure 3.10. It is remarkably similar to $y = \sin x$.

In Figure 3.11, both $y = \sin x$ and $y = \cos x$ are shown on the same

> **HISTORICAL ASIDE**
>
> "Trigonometry," said Augustus DeMorgan, "is the science of continually undulating magnitude. . . . " The oft-quoted DeMorgan was an important nineteenth-century mathematician. His wit, insight, and candor were notorious. When asked when he was born or his age, he would reply with a riddle: "I was x years old in the year x^2." When was the cagey DeMorgan born?

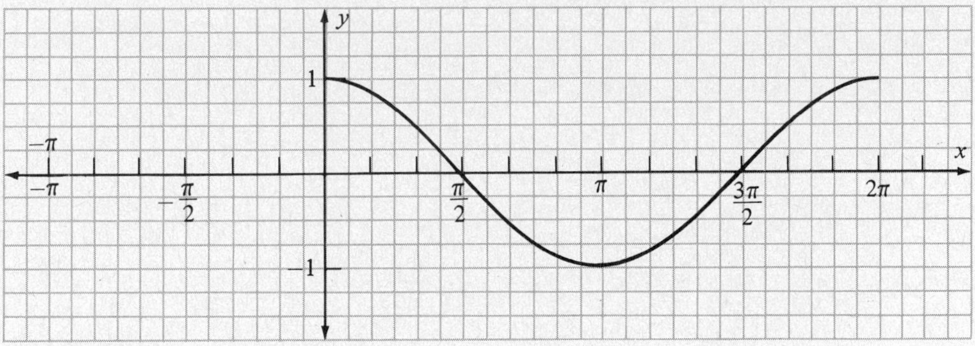

Figure 3.10 Graph of y = cos x.

axes. We can see that $y = \cos x$ is horizontally displaced $\pi/2$ units from $y = \sin x$ but otherwise identical. We say that the two waves are out of **phase** $\pi/2$ units. The cosine wave could be moved to coincide with the sine by

$$y = \cos\left(x - \frac{\pi}{2}\right) = \sin x$$

Figure 3.11 Graphs of y = sin x and y = cos x.

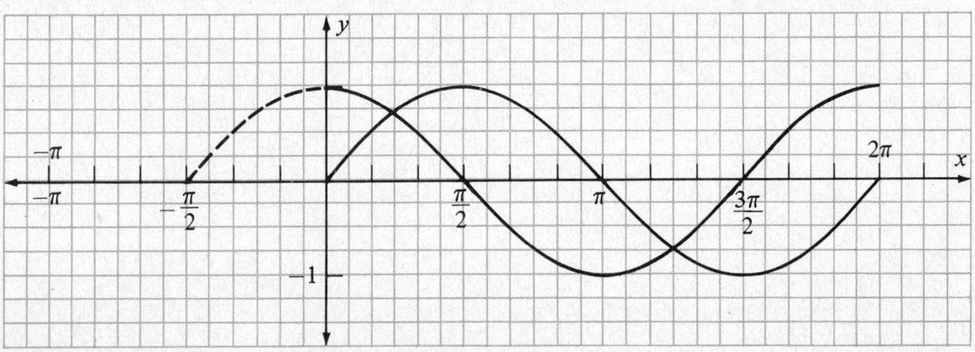

or the sine moved to coincide with the cosine:

$$y = \sin\left(x + \frac{\pi}{2}\right) = \cos x$$

The cosine then should be "framed" as the sine was and in fact, except for the placement of the key points, the method is identical.

EXAMPLES

Sketch each of the following.

1. $y = 3 \cos 2x$

$(h, k) = (0, 0)$, $|a| = 3$, $b = 2$, and $p = 2\pi/2 = \pi$.

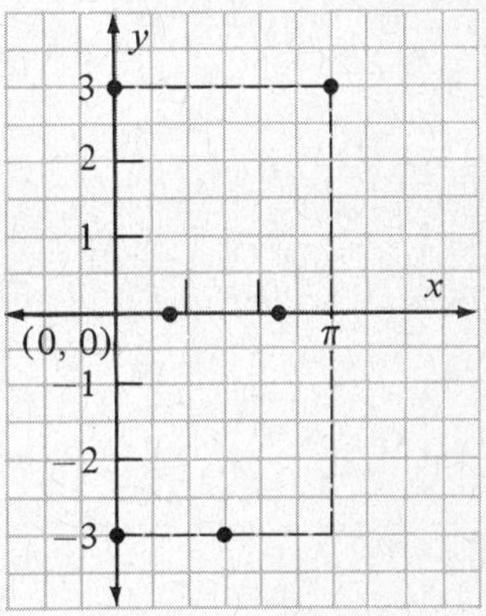

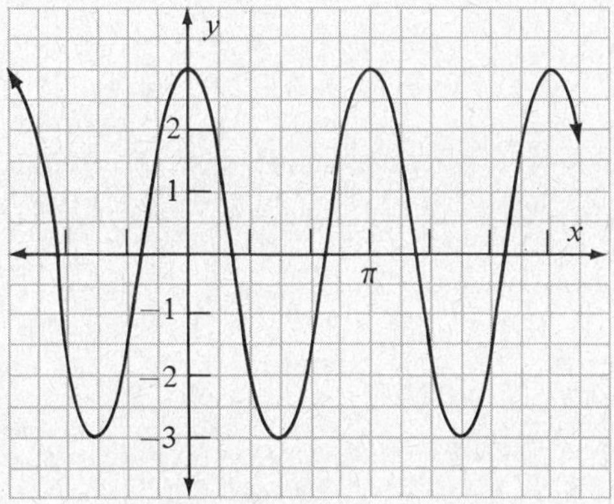

For cosine the point (h, k) is *not* on the graph. The end points are at the top, the midpoint is at the bottom, and the quarter points are on the axis of the curve.

2. $y - 1 = 2 \cos \frac{3}{2}\left(x + \frac{\pi}{3}\right)$

$(h, k) = \left(-\frac{\pi}{3}, 1\right)$ $\quad |a| = 2 \quad b = \frac{3}{2} \quad$ and $\quad p = \frac{2\pi}{\frac{3}{2}} = \frac{4\pi}{3}$

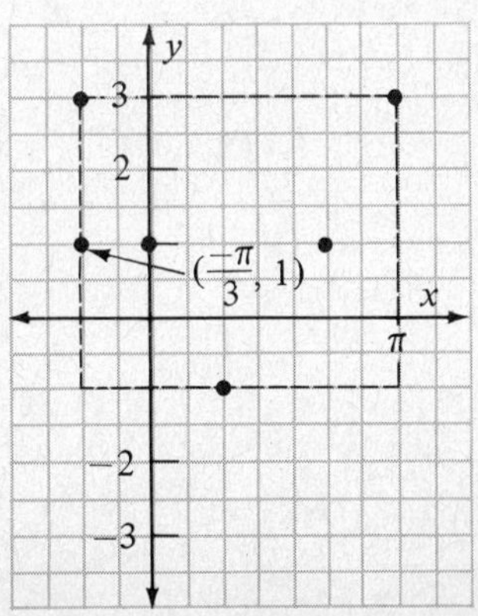

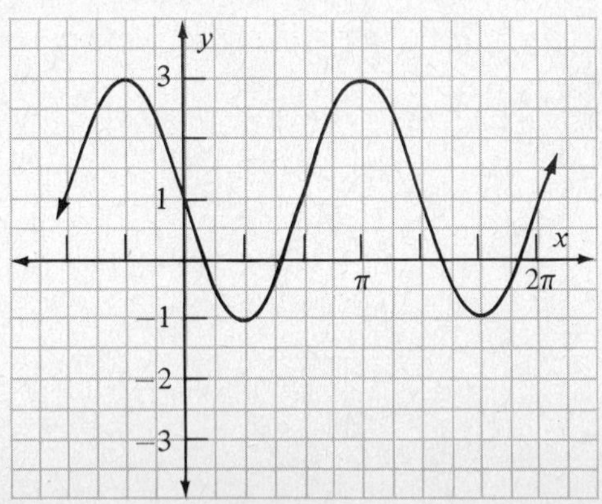

3. $y + 2 = -\frac{1}{2} \cos \frac{2}{3}(x + \pi); -\pi \leq x \leq \pi$

$(h, k) = (-\pi, -2), |a| = \frac{1}{2}, a < 0, b = \frac{2}{3},$ and $p = \frac{2\pi}{\frac{2}{3}} = 3\pi$

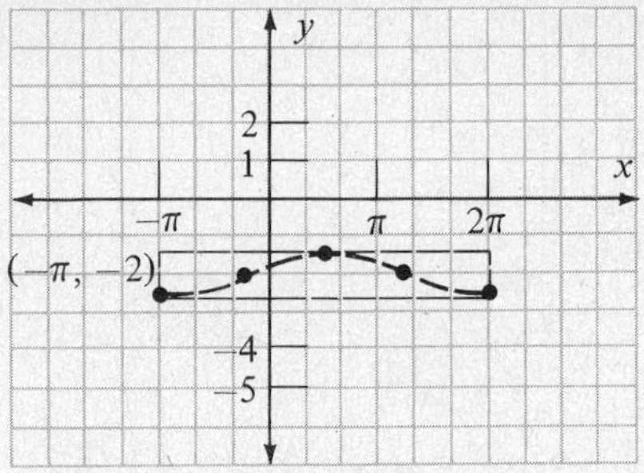

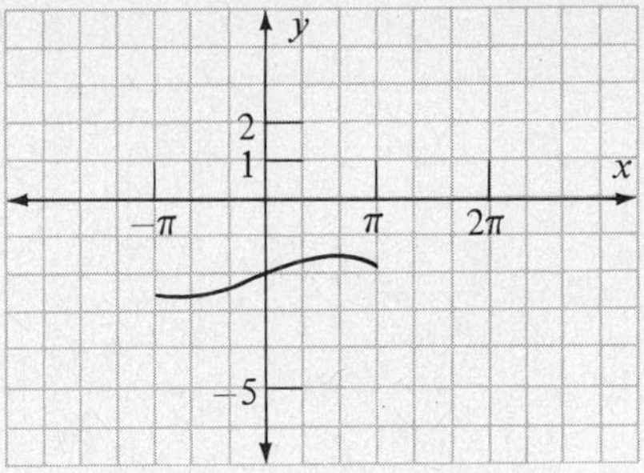

Tangent

In Figure 3.12 the values of tangent have been plotted similar to Figure 3.10. Immediately, $y = \tan x$ appears different because it is not the same kind of wave and it completes two cycles of its values in 2π units. The dashed vertical lines are called **asymptotes.** The $\tan x$ is undefined at those values of x and the values of $\tan x$ increase (or decrease) without

Figure 3.12 Graphing y = tan x.

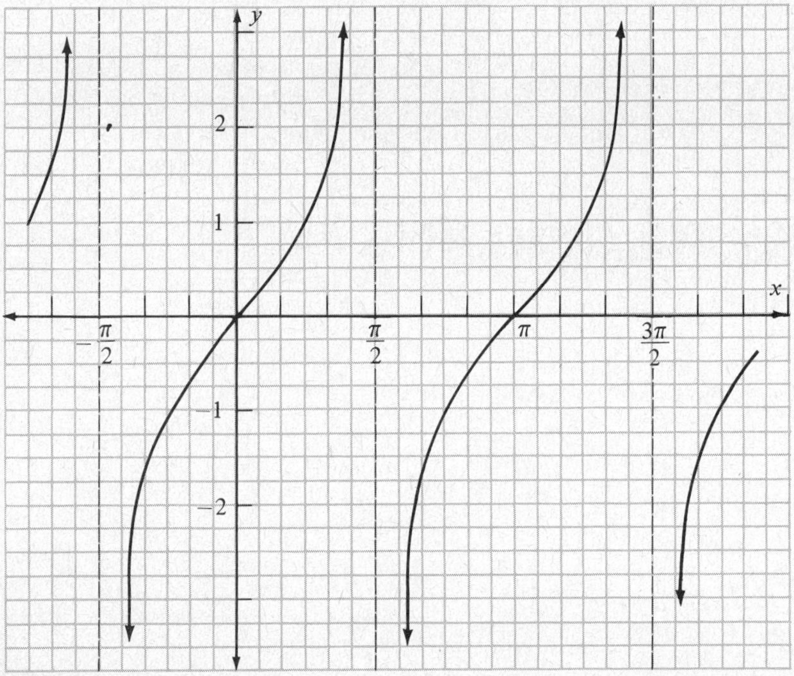

bound as x nears these values. Because a cycle is completed in π units, the period for tangent is π/b, where b is the frequency. The tangent does not actually have an amplitude in the sense of extreme values. But notice in Figure 3.13 that $y = \tan x$ is 1 at the quarter points, so $y = a \tan x$ will be a at its quarter points of each cycle. The initial point of the base cycle of $y = \tan x$ is actually in the middle of the cycle with asymptotes equally spaced on either side. With these observations in mind, let us frame and sketch

$$y - k = a \tan b(x - h)$$

as in Figure 3.14.

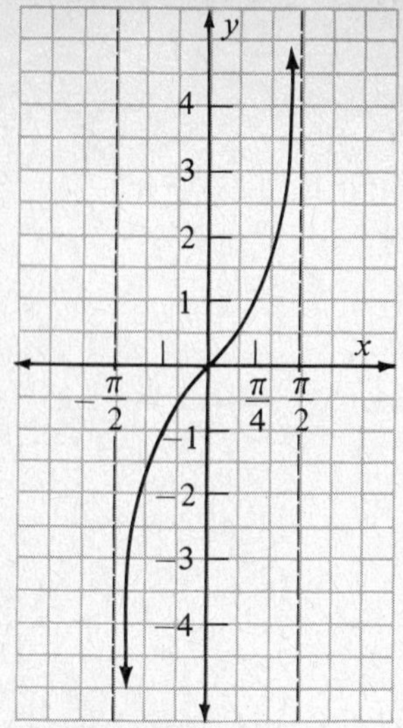

Figure 3.13 Base cycle of y = tan x.

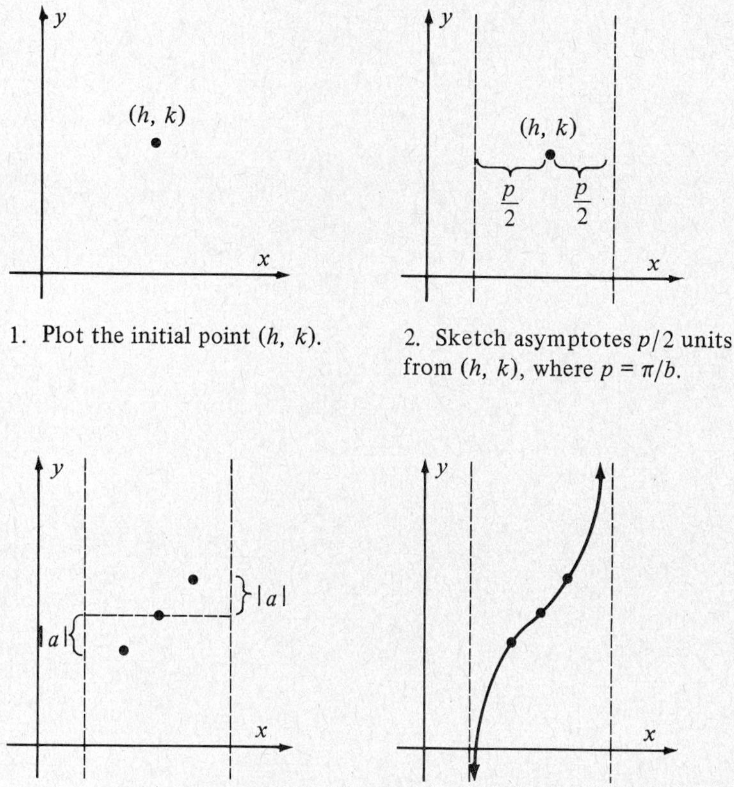

1. Plot the initial point (h, k).
2. Sketch asymptotes $p/2$ units from (h, k), where $p = \pi/b$.
3. Plot quarter points.
4. Sketch base cycle.

Figure 3.14 Sketching the base cycle of $y - k = a \tan b(x - h)$.

EXAMPLES

Sketch each of the following.

4. $y = \frac{3}{2} \tan \frac{1}{2}x$

$$(h, k) = (0, 0),\ |a| = \tfrac{3}{2},\ b = \tfrac{1}{2},\ \text{and } p = \frac{\pi}{\frac{1}{2}} = 2\pi$$

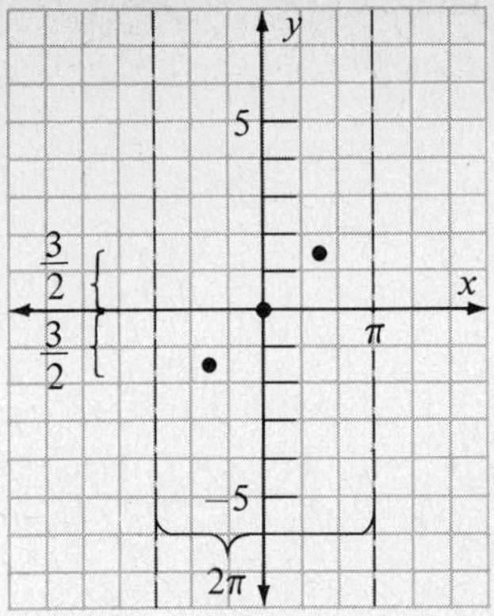

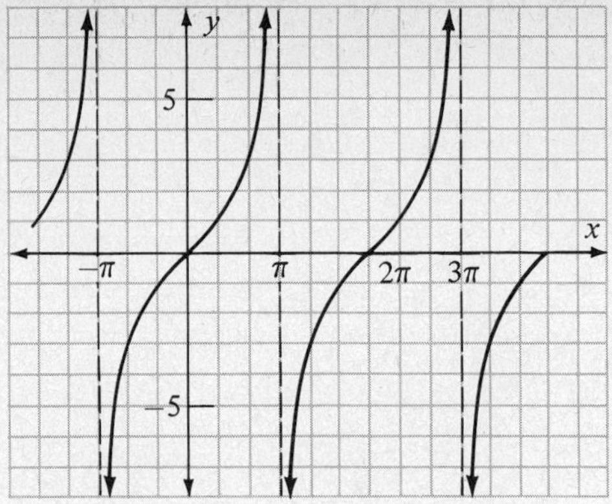

5. $y - 2 = -\frac{1}{2} \tan \frac{2}{3}x;\ -\pi \le x \le 2\pi$

$$(h, k) = (0, 2),\ |a| = \tfrac{1}{2},\ a < 0,\ b = \tfrac{2}{3},\ \text{and } p = \frac{\pi}{\frac{2}{3}} = \frac{3\pi}{2}$$

With $a < 0$, notice that the curve decreases instead of increasing as it does when $a > 0$.

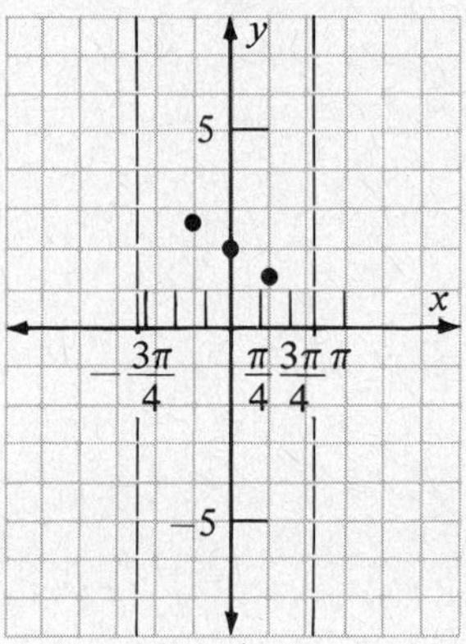

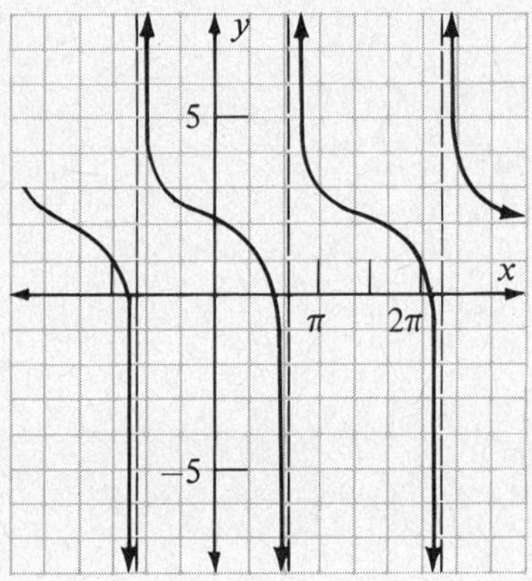

6. $y + 1 = 2 \tan 3(x - \pi/2);\ 0 < x < \pi$

$$(h, k) = (\pi/2, -1),\ |a| = 2,\ b = 3,\ \text{and } p = \pi/3.$$

The restriction $0 < x < \pi$ gives three full cycles of the curve.

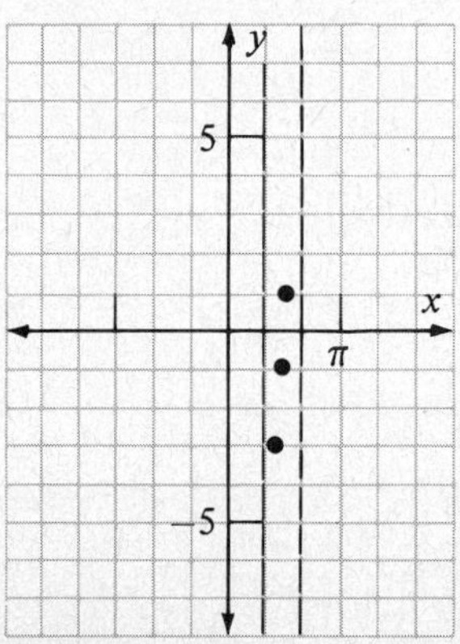

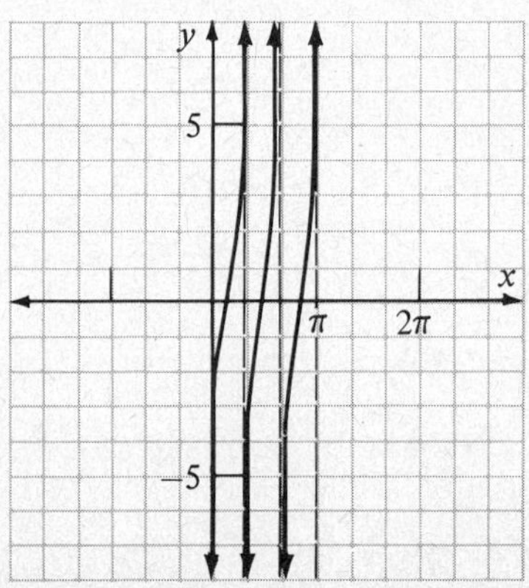

Standard Forms

By inspection of

$$\mathbf{y - k = a \sin b(x - h)}$$

$$\mathbf{y - k = a \cos b(x - h)}$$

$$\mathbf{y - k = a \tan b(x - h)}$$

(h, k), a, b, and hence p may be determined, and this data enables the framing and sketching of the curve. Each of these equations is called the **standard form** of the curve. It is unlikely that every trigonometric function will be found conveniently in standard form. More often the equation will have to be rewritten in this form in order to gather the data necessary to sketch a graph. Consider the following examples and their simplification to standard form. In each step, observe what must be done to proceed toward the standard form.

EXAMPLES

Rewrite in standard form and identify the initial point of the base cycle, the amplitude, and the period.

7. $2y = \cos(3x + 2\pi) + 2$ — Isolate cosine.

$2y - 2 = \cos(3x + 2\pi)$ — Divide by the coefficient of y.

$y - 1 = \frac{1}{2}\cos(3x + 2\pi)$ — Factor out the coefficient of x.

$$\mathbf{y - 1 = \frac{1}{2}\cos 3\left(x + \frac{2\pi}{3}\right)}$$

By inspection,

$(h, k) = \left(-\frac{2\pi}{3}, 1\right)$ — initial point

$|a| = \frac{1}{2}$ — amplitude

$\frac{2\pi}{b} = \frac{2\pi}{3}$ — period

8. $5 \tan 4\left(\frac{x}{3} - \frac{8\pi}{3}\right) - 2y = 3$

$2y + 3 = 5 \tan 4\left(\frac{x}{3} - \frac{8\pi}{3}\right)$ — Isolate

$$y + \frac{3}{2} = \frac{5}{2} \tan 4\left(\frac{x}{3} - \frac{8\pi}{3}\right) \qquad \text{Divide}$$

$$\mathbf{y + \frac{3}{2} = \frac{5}{2} \tan \frac{4}{3} (x - 8\pi)} \qquad \text{Factor}$$

By inspection,

$$(h, k) = \left(8\pi, -\frac{3}{2}\right) \qquad \text{initial point}$$

$$|a| = \frac{5}{2} \qquad \text{amplitude}$$

$$p = \frac{\pi}{b} = \frac{\pi}{\frac{4}{3}} = \frac{3\pi}{4} \qquad \text{period}$$

Inequalities

Sine and cosine divide the plane into two half-planes just like a line, a parabola, and certain other curves. If $y = \sin x$ is all the points *on* the sine wave, then $y > \sin x$ or $y < \sin x$ is the set of points *not on* the wave. The graph of $y > \sin x$ is the set of points *above* $y = \sin x$, and $y < \sin x$ is *below*. To graph an inequality with a trigonometric function, first graph the allied equality. Then decide whether it is the region above or below the graph of the equality.

EXAMPLES

9. $y + 1 > 2 \cos \frac{3}{2}\left(x - \frac{2\pi}{3}\right); x > 0$

First sketch the equation

$$y + 1 = 2 \cos \frac{3}{2}\left(x - \frac{2\pi}{3}\right)$$

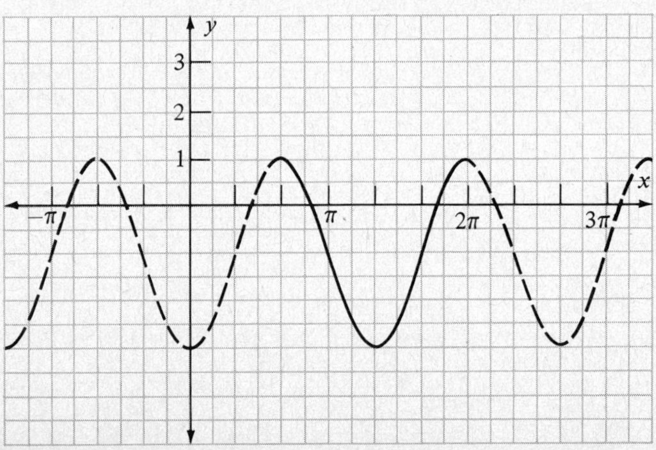

Now take the points above the curve and to the right of the y axis $(x > 0)$.

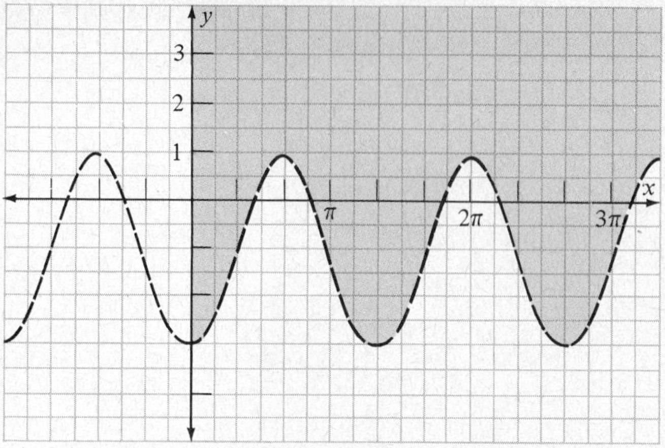

The point (0, 0) is above the curve and satisfies the inequality, so the graph of the inequality is the set of points above the curve.

10. $\begin{cases} y \geq \sin x \\ y - 2 \leq \cos x \\ 0 \leq x \leq 3\pi \end{cases}$

The brace symbol { takes the place of the word "and" between inequalities and indicates the intersection of the graphs.

First lightly sketch the individual graphs of the allied equations.

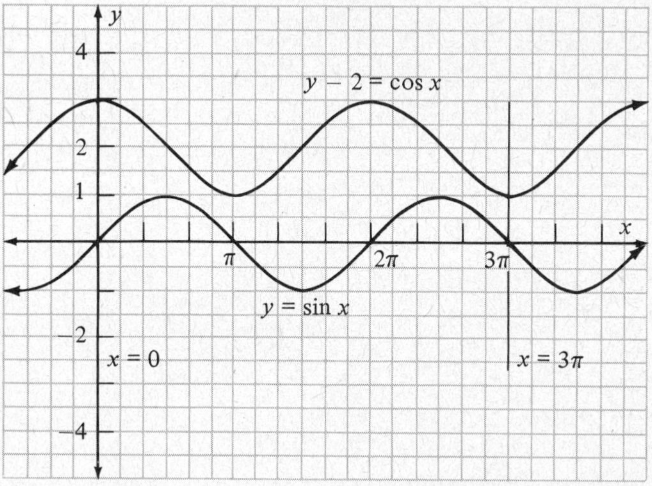

Then shade the intersection of the inequalities and erase the construction lines.

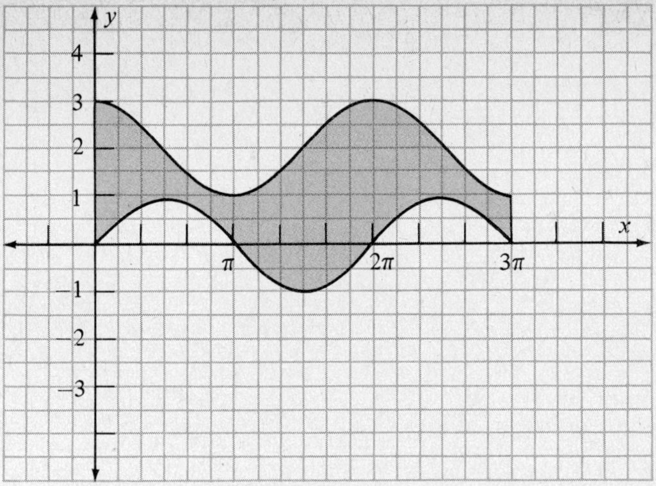

Check the intersection region by substituting the point $(2\pi,\ 1)$ into

$$\begin{cases} 1 \geq \sin 2\pi \\ 1 - 2 \leq \cos 2\pi \\ 0 \leq 2\pi \leq 3\pi \end{cases}$$

and see that it checks.

$$\begin{cases} 1 \geq 0 \\ -1 \leq 1 \\ 0 \leq 2\pi \leq 3\pi \end{cases}$$

11. $\begin{cases} y > \dfrac{1}{2} \tan x \\ -\dfrac{\pi}{2} < x < \pi \end{cases}$

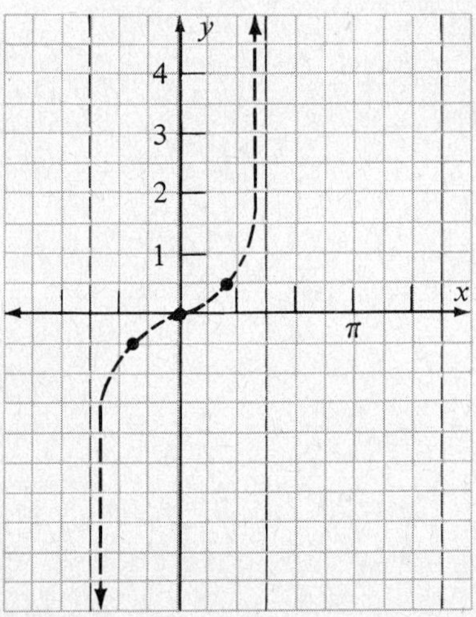

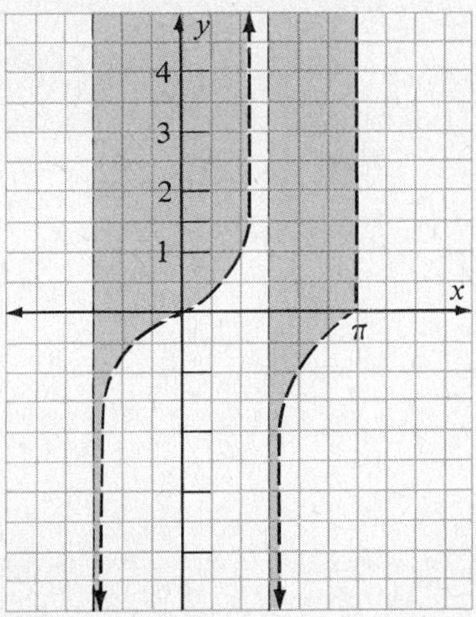

exercise set 3.2

A. exercises

In Exercises 1–10, sketch each pair of equations on the same coordinate axes.

See Examples 1–6.

1. $y = 2 \sin x$, $y = 2 \cos x$
2. $y = \sin 2x$, $y = \cos 2x$
3. $y = \tan x$, $y = 2 \tan x$
4. $y = \tan x$, $y = \tan 2x$
5. $y - 1 = -\cos x$, $y + 1 = -\cos x$
6. $y = \cos\left(x - \frac{\pi}{2}\right)$, $y = \cos\left(x + \frac{\pi}{2}\right)$
7. $y = \cos 2x$, $y = -\tan\left(x - \frac{\pi}{2}\right)$
8. $y = -\frac{1}{2} \cos x$, $y = \tan\left(x - \frac{\pi}{2}\right)$
9. $y = \cos x$, $y = \cos(x + \pi)$
10. $y = \frac{1}{2} \tan x$, $y = -\frac{1}{2} \tan x$

Rewrite each equation in Exercises 11–20 in standard form and identify h, k, a, b, and p.

See Examples 7 and 8.

11. $y = 2 + 3 \sin(4x - 2\pi)$
12. $1 + 2 \sin(3x - \pi) = y$
13. $2y + 3 = 4 \cos\left(x - \frac{\pi}{6}\right)$
14. $3y - 2 = 6 \cos\left(x + \frac{\pi}{4}\right)$
15. $3y - 6 = 2 \tan \frac{1}{3}(6x - \pi)$
16. $4y - 8 = 3 \tan \frac{1}{2}(3x + \pi)$
17. $5 = 6y + 3 \sin \frac{2}{3}(3x + 4\pi)$
18. $7 = 8y + 6 \cos \frac{3}{4}(2x - 3\pi)$
19. $\frac{3}{2} - \frac{4}{3}y = \frac{2}{9} \tan \frac{3}{2}(4x - 3\pi)$
20. $\frac{5}{3} - \frac{3}{4}y = \frac{3}{8} \tan \frac{2}{3}(6x + 2\pi)$

Sketch the graphs of the systems of inequalities in Exercises 21–30.

See Examples 9–11.

21. $\begin{cases} y \geq \sin x \\ y - 3 \leq \sin x \end{cases}$
22. $\begin{cases} y \leq \cos x \\ y + 2 \geq \cos x \end{cases}$
23. $\begin{cases} y + 1 \geq \cos 2x \\ y - 2 \leq 2 \cos x \end{cases}$
24. $\begin{cases} y - 1 \leq \sin x \\ y + 2 \geq 2 \sin 2x \end{cases}$
25. $\begin{cases} y \geq \tan x \\ y - 2 \leq \frac{1}{2} \tan x \\ -\frac{\pi}{2} \leq x \leq \frac{\pi}{2} \end{cases}$
26. $\begin{cases} y \leq \frac{1}{2} \tan x \\ y + 2 \geq \tan x \\ -\frac{\pi}{2} \leq x \leq \frac{\pi}{2} \end{cases}$

27. $\begin{cases} y > 2 \tan\left(x - \dfrac{\pi}{4}\right) \\ y - 2 < \sin 2x \\ 0 < x < \dfrac{\pi}{2} \end{cases}$

28. $\begin{cases} y - 2 < 2 \tan x \\ y + 2 > 2 \sin\left(x - \dfrac{\pi}{2}\right) \\ -2\pi < x < 2\pi \end{cases}$

29. $\begin{cases} y \geq 2 \cos x \\ y < 2 \sin x \\ y > -1 \end{cases}$

30. $\begin{cases} y < \tan x \\ y > \tan 2x \\ y < 1 \end{cases}$

B. exercises

The following is used in Exercises 31–35.

Certain periodic motion can be described by the sine or cosine function. Actually, any object that moves such that its position on a straight line is given at any time t by $d = a \sin bt$ is called a **simple harmonic motion.**

A weight suspended on a spring will produce simple harmonic motion when pulled a units down and released. Its vertical displacement d from rest is given by

$$d = a \cos \sqrt{\frac{k}{m}}\, t$$

where m is the mass, t is the time after release, and k is a constant for a given spring.

31. Determine the frequency of the oscillating motion in cycles per second.

32. What is the period of the motion?

Let k be π^2 cm/s^2, the mass be 10 g, and the amplitude equal 5 cm.

33. What is the total vertical movement of the weight?

34. What is the period of this oscillation?

35. How long does each cycle take?

A mass suspended by a string of length l is a simple pendulum when pushed to the side and allowed to swing to and fro. Its horizontal displacement d from rest position is given by *Refers to Exercises 36–40.*

$$d = a \cos \sqrt{\frac{g}{l}}\, t$$

where g is the gravitational constant.

36. What is the frequency of the simple harmonic motion in oscillations per sec?

37. Find the period of the oscillation.

Let the length of the string be 327 cm, the amplitude 25 cm, and use 981 cm/s² for g.

38. How far does the pendulum swing side to side?

39. What is the period of this motion?

40. What is the total time for one swing?

Graph the equations in Exercises 41–50 carefully, noting the restrictions on each.

41. $y - \frac{3}{2} = -\frac{1}{2}\sin\frac{1}{2}\left(x + \frac{\pi}{2}\right);\ -\frac{\pi}{2} \le x \le \frac{5\pi}{2}$

42. $y - \frac{15}{2} = -\frac{1}{2}\cos\frac{3}{2}\left(x + \frac{\pi}{2}\right);\ -\frac{\pi}{2} \le x \le \frac{\pi}{2}$

43. $2y - 13 = 2\cos\frac{1}{4}(6x + 7\pi);\ -7\pi \le 6x \le -3\pi$

44. $2\tan x + 9 = 2y;\ |x| \le \frac{\pi}{2},\ 3 \le 2y \le 15$

45. $y = 2\left[1 - 2\cos\left(\frac{x}{2} - \frac{3\pi}{4}\right)\right];\ 3\pi \le 2x \le 5\pi$

46. $y + 3\sin\left(\frac{6x - 7\pi}{6}\right);\ 10\pi \le 6x \le 13\pi$

47. $15 - 2y = \cos 3\left(\frac{6x + \pi}{2}\right);\ -7\pi \le 6x \le -3\pi$

48. $2y - 6 = 3\tan\frac{1}{6}(6x + 5\pi);\ -4\pi \le 3x \le \pi$

49. $4y - 29 = \sin\left(\frac{6x + \pi}{2}\right);\ |x| \le \frac{\pi}{6}$

50. $4y = 29 + \cos\left(\frac{9x + \pi}{3}\right);\ -2\pi \le 6x \le \pi$

C. exercises

51. **palatable plotting** Set up your coordinate axes with $-2\pi \le x \le 3\pi$ and $-5 \le y \le 10$ and graph Exercises 41–50 on the same axes, then add the following finishing touches.

Suggestion: Use 2 squares per unit on graph paper.

(a) $4y \leq \sin \frac{2}{3}(2x - \pi) - 11$

(b) $\begin{cases} 2y < \frac{1}{2} \sin \frac{1}{2}(6x + \pi) + \frac{29}{2} \\ 2y > \frac{1}{2} \sin \frac{1}{3}(9x + \pi) + \frac{29}{2} \\ -\pi < 6x < \pi \end{cases}$

3.3 RECIPROCAL FUNCTIONS

We are able to graph sine, cosine, and tangent functions, and it is known that

$$\cot x = \frac{1}{\tan x} \qquad \sec x = \frac{1}{\cos x} \qquad \csc x = \frac{1}{\sin x}$$

Thus these identities should provide the means to graph cotangent, secant, and cosecant.

Cotangent

The points of $y = \cot x$ can be obtained by taking the reciprocals of the y values, the ordinates, of $y = \tan x$. Figure 3.15 shows $y = \tan x$ and $y = \cot x$ on the same axes. Note that at $x = 0$, tangent is zero and has no reciprocal, so cotangent is undefined. As x increases, the reciprocal or the cotangent decreases until it is again undefined at $x = \pi$, because the

Figure 3.15 Graphs of tangent and cotangent.

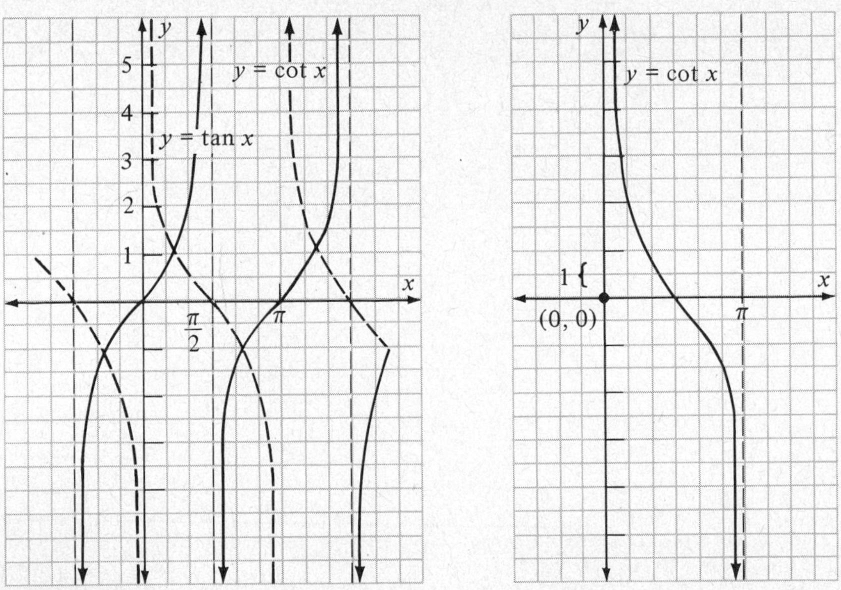

tangent is again zero. It is seen that this interval $0 \leq x \leq \pi$ is taken as the base cycle of $y = \cot x$. The cotangent has the same period as tangent and behaves similarly. The shape is familiar but decreases instead of increasing between asymptotes, like the tangent. Like the cosine, the initial point of the base cycle is *not* on the graph of $y = \cot x$.

In general, the graph of $y - k = a \cot b(x - h)$ can be framed and sketched as shown in Figure 3.16.

$$\mathbf{y - k = a \cot b(x - h)}$$

1. Initial point (h, k) on the asymptote.
2. Period $p = \pi/b$ gives terminal point, which is on asymptote.
3. $|a|$ gives quarter points.

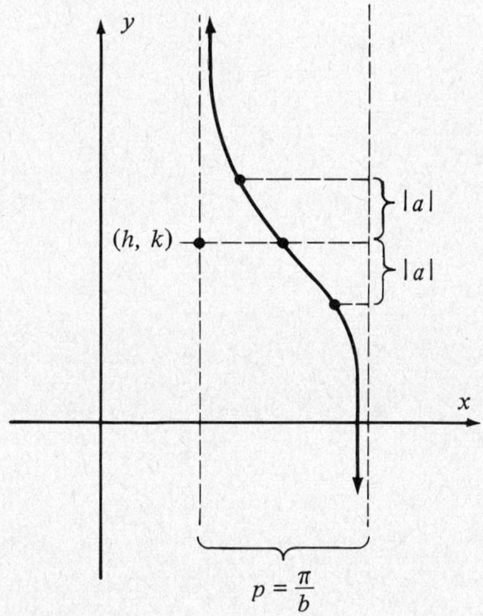

Figure 3.16 Graphing cotangent.

EXAMPLES

1. $y = 2 \cot \frac{1}{2}x$

$(h, k) = (0, 0)$

$|a| = 2$

$$b = \frac{1}{2}$$

$$p = \frac{\pi}{\frac{1}{2}} = 2\pi$$

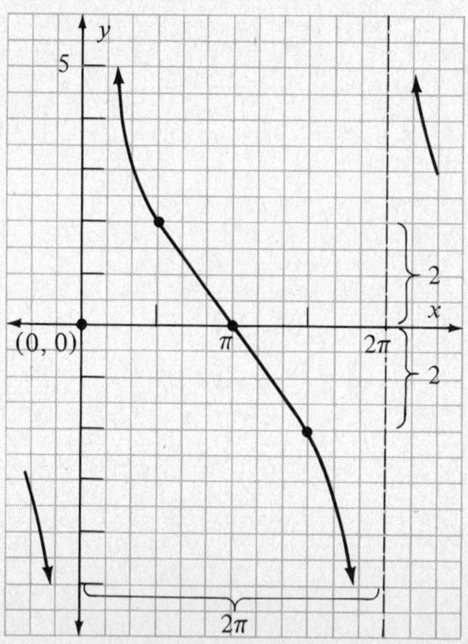

2. $y + 2 = \frac{1}{2} \cot \frac{2}{3}(x - \pi)$

$(h, k) = (\pi, -2)$

$|a| = \dfrac{1}{2}$

$b = \dfrac{2}{3}$

$p = \dfrac{\pi}{\frac{2}{3}} = \dfrac{3\pi}{2}$

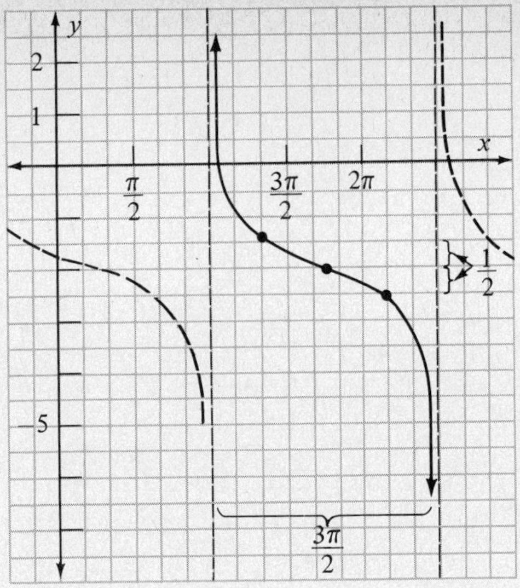

Secant and Cosecant

Using the graphs of sine and cosine, the graphs of secant and cosecant can also be obtained by reciprocals. See Figures 3.17 and 3.18. The periods of secant and cosecant are 2π, as for sine and cosine. Further, the key points of sine and cosine are asymptotes and minimums and maximums for secant and cosecant. These same points can be used to sketch these functions, so the analysis will be the same for sine/cosecant and cosine/secant even though the behavior of cosecant and secant are somewhat different. This is illustrated for cosecant in Figure 3.19. Secant would be very similar, and you should outline the procedure yourself.

Figure 3.17 Sine and cosecant.

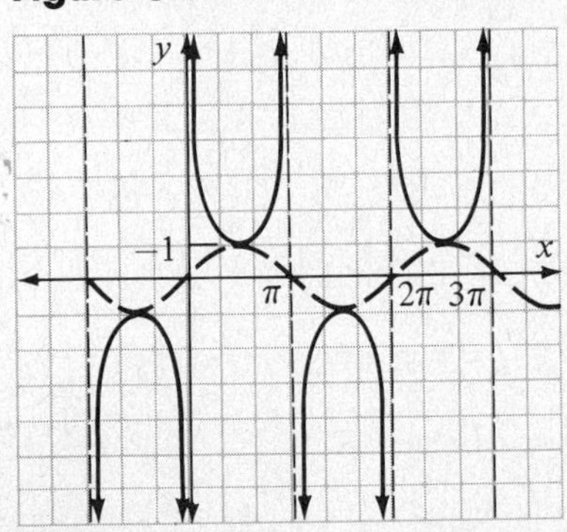

Figure 3.18 Cosine and secant.

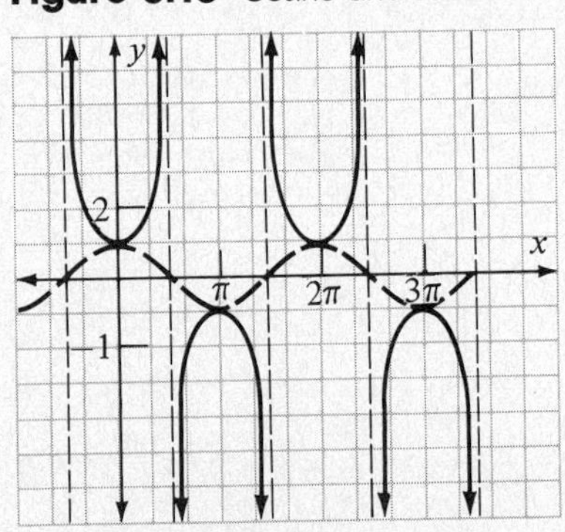

$y - k = a \csc b(y - h)$

1. Plot key points for $y - k = a \sin b(x - h)$.
2. End points and midpoint determine asymptotes.
3. Quarter points determine maximum and minimum values between asymptotes.

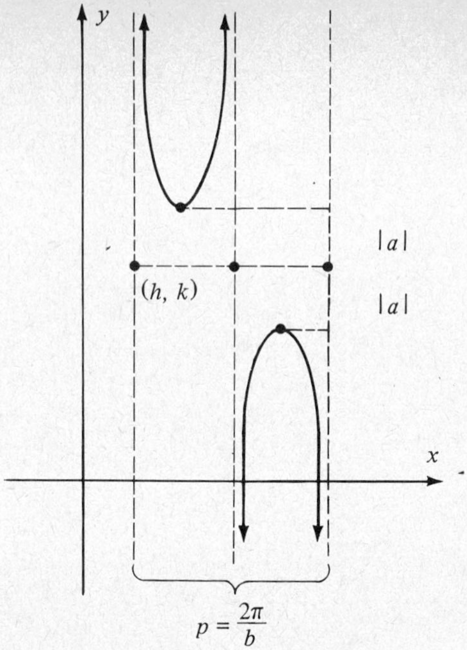

Figure 3.19 Sketching cosecant.

EXAMPLES

Sketch the functions.

3. $y = 4 \csc \frac{1}{3}x$

$$(h, k) = (0, 0)$$

$$|a| = 4$$

$$b = \frac{1}{3}$$

$$p = \frac{2\pi}{\frac{1}{3}} = 6\pi$$

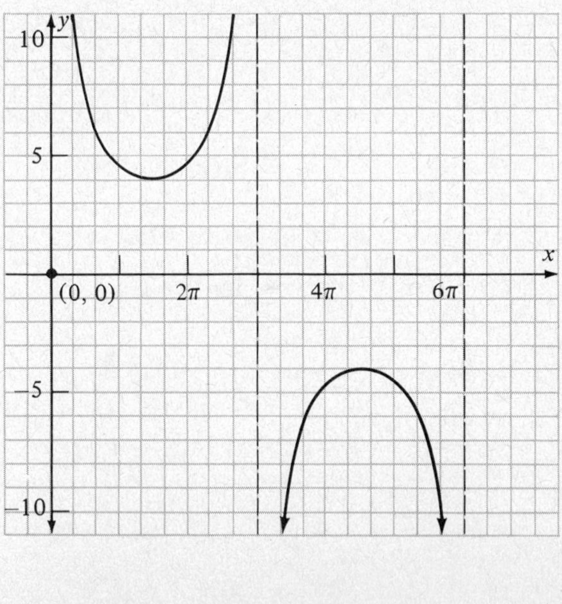

4. $y > 2 \sec 2x$

$$(h, k) = (0, 0)$$
$$|a| = 2$$
$$b = 2$$
$$p = \frac{2\pi}{2} = \pi$$

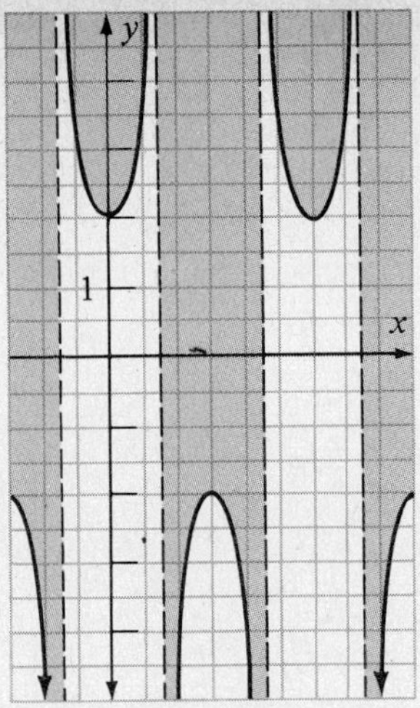

The inequality will be all points above $y = 2 \sec 2x$. Mark the key points for cosine. Sketch the asymptotes, then the secant curve through the quarter points. Notice the unusual graph produced by the secant inequality.

5. $y + 1 = \csc \frac{2}{3}\left(x + \frac{\pi}{2}\right)$

$$(h, k) = \left(-\frac{\pi}{2}, -1\right)$$
$$|a| = 1$$
$$b = \frac{2}{3}$$
$$p = \frac{2\pi}{\frac{2}{3}} = 3\pi$$

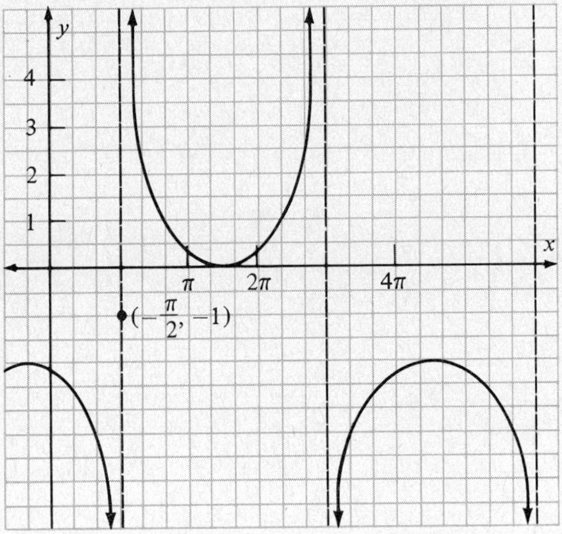

In summary, cotangent, secant, and cosecant are sketched by making use of an understanding of the graphs of their reciprocal functions. This means preparing to graph cosine or sine, but sketching the distinctive secant or cosecant behavior on that framework. However, this illustrates an important concept of graphing. Once the behavior of a function is well known, only a skeleton of a framework is necessary to produce a decent sketch. There is certainly no need to calculate numerous points in order to sketch the trigonometric functions.

exercise set 3.3

A. exercises

In Exercises 1–10, sketch each pair of equations on the same coordinate axes.

1. $y = \sin x$
   $y = \csc x$
2. $y = \cos x$
   $y = \sec x$

See Figures 3.15, 3.17, and 3.18.

3. $y = \tan x$
   $y = \cot x$
4. $y = \tan x$
   $y = -\cot x$
5. $y = 2 \sin x$
   $y = 2 \csc x$
6. $y = \cos 2x$
   $y = \sec 2x$

See Examples 1 and 3.

7. $y = -\tan\left(x - \frac{\pi}{2}\right)$
   $y = \cot x$
8. $y = \csc x$
   $y = \sec\left(x - \frac{\pi}{2}\right)$
9. $y - 2 = \sin \frac{3}{2}x$
   $y - 2 = \csc \frac{3}{2}x$
10. $y = \cos \frac{2}{3}(x - \pi)$
    $y = \sec \frac{2}{3}(x - \pi)$

Rewrite each equation in Exercises 11–20 in standard form and identify h, k, a, b, and p.

11. $3y = \csc \frac{1}{3}(2x - \pi)$
12. $\frac{1}{2}y = \cot \frac{1}{9}(3x - \pi)$

See Examples 2 and 5.

13. $3y = 2 \sec\left(\frac{3x}{2} + \frac{3\pi}{2}\right)$
14. $3y = 2 \cos\left(\frac{3x}{2} - \frac{3\pi}{2}\right)$
15. $y = 4 - \tan \frac{1}{4}(9x - \pi)$
16. $y = 4 - \cot 9\left(\frac{x}{4} + \frac{\pi}{4}\right)$
17. $3y + 20 + \sin \frac{1}{4}(3x + 3\pi) = 0$
18. $3y + 18 + \sin \frac{1}{4}(3x + 3\pi) = 0$
19. $y = \sec \frac{9}{8}(\frac{8}{9}x + \pi) + 2$
20. $y = \sec(3x + \pi) + 6$

Sketch the intersection of each system of inequalities given in Exercises 21–30.

21. $\begin{cases} y \geq \tan x \\ 0 \leq x \leq 2\pi \end{cases}$
22. $\begin{cases} y \leq \cot x \\ 0 \leq x \leq 2\pi \end{cases}$

See Example 4.

23. $\begin{cases} y < \csc x \\ 0 < x < 3\pi \end{cases}$
24. $\begin{cases} y > \sec x \\ -\frac{3\pi}{2} < x < \frac{3\pi}{2} \end{cases}$

25. $\begin{cases} y \geq \cot x \\ y \leq \csc x \end{cases}$

26. $\begin{cases} y \geq \tan x \\ y \leq \sec x \end{cases}$

27. $\begin{cases} y \geq \tan x \\ y \geq \cot x \end{cases}$

28. $\begin{cases} y \geq \sin x \\ y \geq \csc x \end{cases}$

29. $\begin{cases} y < \sec x \\ y > \csc\left(x - \frac{\pi}{2}\right) \end{cases}$

30. $\begin{cases} y > \cot x \\ y > \tan\left(x - \frac{\pi}{2}\right) \end{cases}$

B. exercises

An object is H units high and d units from an observer h units tall. The distance to the object is a function of the angle of elevation to the top of the object as given by

$$d = (H - h) \cot \theta$$

31. Sketch d as a function of θ for $H = 105$ and $h = 5$.

Note the implicit restriction on θ.

32. Rewrite the equation so that H is a function θ with the distance known.

An electrical circuit consists of a coil and a condenser. The charge Q on the condenser t seconds after the circuit is closed is given by

Assume negligible resistance.

$$Q = Q_0 \sin \pi \left(\frac{t}{\sqrt{MC}} + \frac{1}{2}\right)$$

where Q_0 is the initial charge (in coulombs), M is the inductance of the coil (in henrys), and C is the capacitance of the condenser (in farads).

33. What is the frequency of oscillation of the current?

34. What is the maximum charge on the condenser?

35. What is the phase shift (horizontal displacement)?

36. Find the frequency if $M = 0.25$ and $C = 5.0 \times 10^{-4}$.

Water waves in their simplest form are a simple harmonic motion and can be represented by

$$y = a \sin \frac{2\pi}{T}\left(t - \frac{x}{5.12T}\right)$$

where t is the time and x is the distance from the source.

37. What is the meaning of T?

38. If the equation of the wave is

$$y = 4 \sin \frac{\pi}{3}\left(t - \frac{x}{30.72}\right)$$

what is the period of the wave (in seconds)?

39. For the wave in Exercise 38, what is the distance from crest to trough (in feet)?

40. Write the equation of Exercise 38 in the form $y = a \sin b(x - k)$, where t is a constant. That is, write y as a function of x.

Recall that $\sin(-x) = -\sin x$.

Graph the equations in Exercises 41–50, carefully noting the restrictions on each.

41. $y = \frac{1}{3} \csc \frac{2}{3}\left(x - \frac{\pi}{2}\right); \; -\pi \leq x \leq 2\pi$

Suggestion: Use 3 squares for 1 unit on the x axis. Notice that then each square on the x axis is $\pi/9$ unit.

42. $y = 2 \cot \frac{1}{3}\left(x - \frac{\pi}{3}\right); \; -\pi \leq x \leq \pi$

43. $y = \frac{2}{3} \sec \frac{3}{2}(x - \pi); \; -\pi \leq x \leq \pi$

44. $y = \frac{2}{3} \cos \frac{3}{2}(x + \pi); \; -\pi \leq x \leq -\frac{\pi}{3}$

45. $y - 4 = -\tan \frac{9}{4}\left(x - \frac{\pi}{9}\right); \; -\frac{\pi}{9} \leq x \leq \frac{\pi}{3}, \; -20 \leq 3y \leq 27$

46. $y - 4 = -\cot \frac{9}{4}(x + \pi); \; -\pi \leq x \leq -\frac{5\pi}{9}, \; -20 \leq 3y \leq 27$

47. $y + \frac{20}{3} = -\frac{1}{3} \sin \frac{3}{4}(x + \pi); \; -\pi \leq x \leq \frac{\pi}{3}$

48. $y + 6 = -\frac{1}{3} \sin \frac{3}{4}(x + \pi); \; 3\pi \geq -3x \geq \pi$

49. $y - 2 = \sec\left(x + \frac{8\pi}{9}\right); \; 8\pi \geq -9x \geq 5\pi$

50. $y - 6 = \sec 3\left(x + \frac{\pi}{3}\right); \; 6\pi \geq -9x \geq 5\pi$

C. exercises

51. **palatable plotting** Set up your coordinate axes with $-2\pi \le x \le \pi$ and $-7 \le y \le 10$ and graph Exercises 44–50 on the same axes, then add the following touches.

Suggestion: Use 3 squares per unit, 9 squares per π.

(a) $3x + \pi = 0;\ 19 \ge -3y \ge 2$

(b) $y - 9 \le \frac{1}{9}\cos\frac{9}{4}\left(x + \frac{\pi}{3}\right);\ y \ge 6,\ -\frac{5\pi}{9} < x < -\frac{\pi}{9}$

(c) $y - 4 \le -\tan\frac{9}{4}\left(x - \frac{\pi}{9}\right);\ y \ge 6,\ -\frac{\pi}{9} \le x \le 0$

(d) $y - 4 \le -\cot\frac{9}{4}(x + \pi);\ y \ge 6,\ -\frac{2\pi}{3} \le x \le -\frac{5\pi}{9}$

3.4 INVERSE TRIGONOMETRIC FUNCTIONS

Problems arose previously where the value of a trigonometric function of an angle was known and the measure of the angle unknown. From Chapter 2, recall that

$$\sin \theta_m \approx 0.14811653$$

was equivalent to

$$\theta_m \approx \text{the angle whose sine is } 0.14811653$$

and the notation

$$\theta_m \approx \arcsin 0.14811653$$

or

$$\sin^{-1}(0.14811653)$$

was introduced to deal with the problem. The angle at that time was found on a calculator using the [INV], [ARC], or [2nd] keys.

$$\theta_m \approx 8.518°$$

However, there are infinitely many angles with that value of sine. 8.518° was reasonable for the application, index of refraction, and in Chapter 1 all angles were acute. At that time we expressed certain reservations about the arc or inverse functions, and it is now time to deal with them.

The Graphs of the Inverse Trigonometric Functions

Recall that the graph of the inverse relation is symmetric to the line $y = x$. See Appendix A.2 for a discussion of inverse functions. Graphing $y = \sin x$ and $x = \sin y$ (or $y = \sin^{-1} x$) on the same axes as in Figure 3.20, it can be seen that $x = \sin y$ is a relation that is not a function. However, restricting the range of the inverse relation, we can define an **inverse function** for sine and for each of the other trigonometric functions. This restriction is quite arbitrary, because there are many intervals that are appropriate. Convention, however, has us choose the interval with the smallest possible angles as in Figure 3.21. So the following defines an inverse for $y = \sin x$ that is a function.

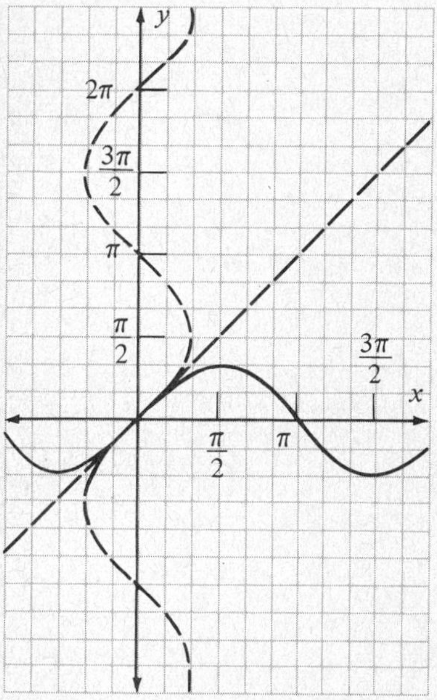

Figure 3.20 Graph of y = sin x and x = sin y.

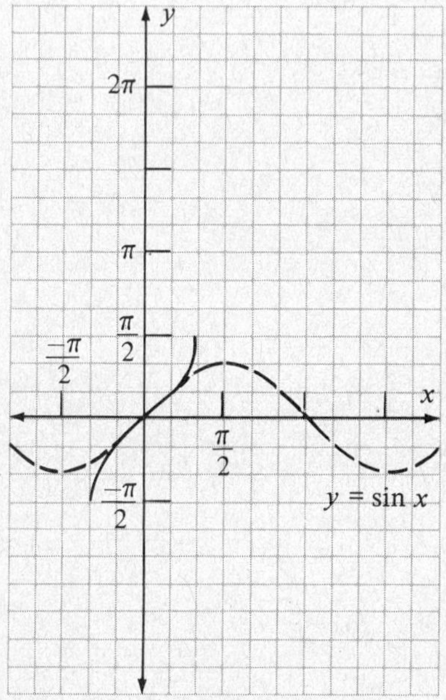

Figure 3.21 Graph of the function y = Sin⁻¹ x.

$y = \mathrm{Sin}^{-1}\, x$ is equivalent to

$$x = \sin y : -\frac{\pi}{2} \le y \le \frac{\pi}{2}$$

This notation, using a capital letter, is preferred to a completely new symbol, but can be confusing.

The other functions are defined below with the same arbitrary restrictions to create functions. These values within the restrictions are sometimes called the **principal values** of the inverse functions. Note that these are the values that your calculator displays for the inverse keys.

INVERSE TRIGONOMETRIC FUNCTIONS

$$\mathbf{y = Sin^{-1}\,x} \leftrightarrow x = \sin y \quad : \quad -\frac{\pi}{2} \le y \le \frac{\pi}{2}$$

$$\mathbf{y = Cos^{-1}\,x} \leftrightarrow x = \cos y \quad : \quad 0 \le y \le \pi$$

$$\mathbf{y = Tan^{-1}\,x} \leftrightarrow x = \tan y \quad : \quad -\frac{\pi}{2} < y < \frac{\pi}{2}$$

$$y = \text{Cot}^{-1}\,x \leftrightarrow x = \cot y \quad : \quad -\frac{\pi}{2} \le y \le \frac{\pi}{2},\ y \ne 0$$

$$y = \text{Sec}^{-1}\,x \leftrightarrow x = \sec y \quad : \quad 0 \le y \le \pi,\ y \ne \frac{\pi}{2}$$

$$y = \text{Csc}^{-1}\,x \leftrightarrow x = \csc y \quad : \quad -\frac{\pi}{2} \le y \le \frac{\pi}{2},\ y \ne 0$$

In order to fix these restrictions in your mind, look closely at the graphs in Figure 3.22. The range of Arcsin, Arccos, and Arctan are chosen so that the graph is a smooth unbroken curve within the restrictions. The Arccsc, Arcsec, and Arccot are taken with the same range as the corresponding reciprocal function. Notice that Arccot could be defined over $0 < y < \pi$, but the Arctan is used to compute Arccot, especially if a calculator is employed. The restrictions are arbitrary, as we mentioned, but reasonable, as you can see.

Graphing the Inverse Functions

To graph an inverse relation, first graph the corresponding trigonometric function, then reflect it in the line $y = x$. To graph the inverse function, the restrictions must be applied to the relation.

EXAMPLES

Sketch the graphs.

1. $y = 2 \sin^{-1} \frac{2}{3}x$

$\frac{1}{2}y = \sin^{-1} \frac{2}{3}x$

is the inverse of

$$\tfrac{2}{3}y = \sin \tfrac{1}{2}x \quad \text{or } y = \tfrac{3}{2} \sin \tfrac{1}{2}x$$

First sketch $y = \frac{3}{2} \sin \frac{1}{2}x$ ($a = \frac{3}{2}$ and $p = 4\pi$), shown as the dashed curve below. Then sketch the reflection in $y = x$, the solid curve.

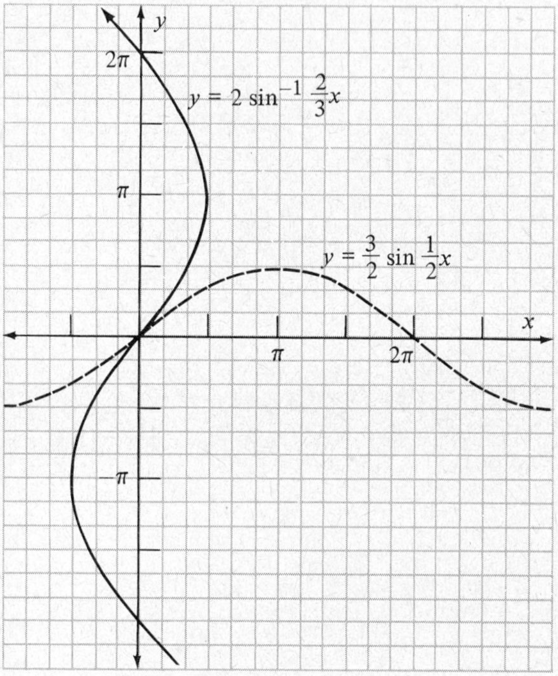

2. $2x = \text{Tan}^{-1} (y + 2)$

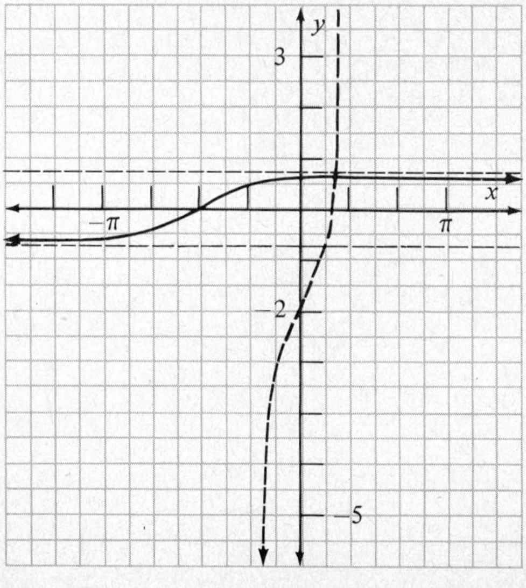

$$2x = \text{Tan}^{-1}(y + 2)$$

is the inverse of

$$y + 2 = \tan 2x \quad : \quad -\frac{\pi}{2} < 2x < \frac{\pi}{2}$$

First sketch $y + 2 = \tan 2x$ with $-\pi/4 < x < \pi/4$. Then sketch the reflection in $y = x$ as shown above.

Figure 3.22 Graphs of the inverse trigonometric functions.

Graphs of the Inverse Trigonometric Functions

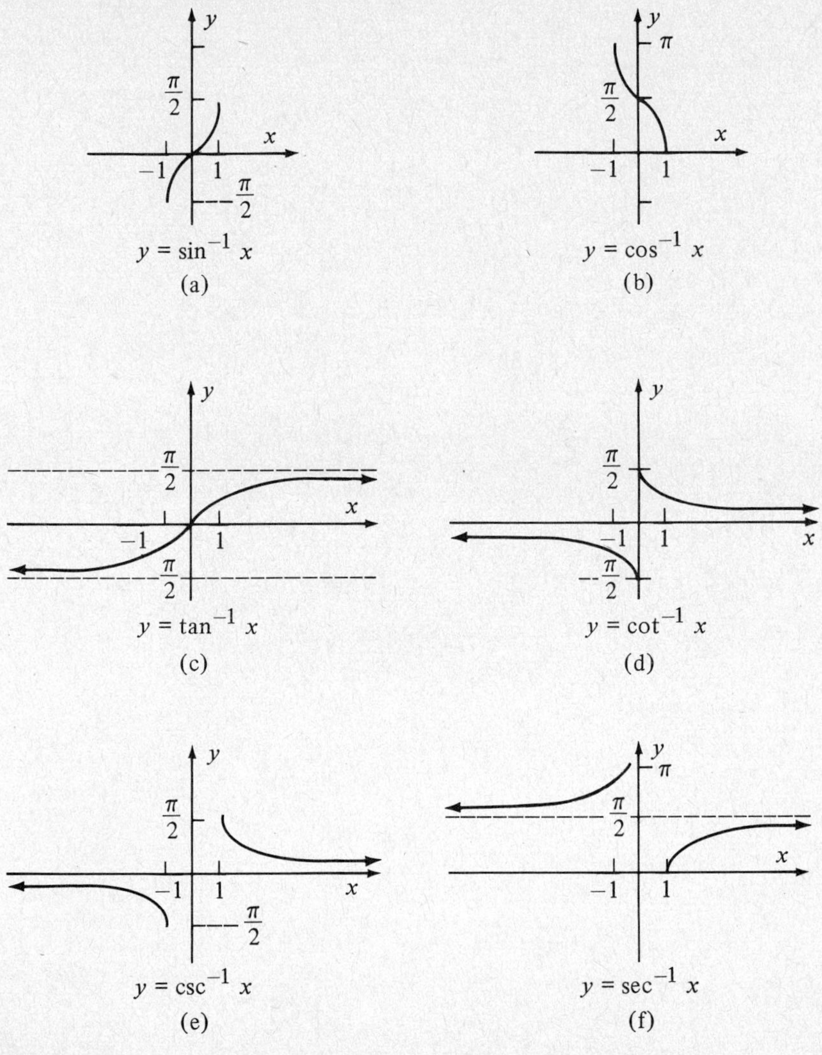

Functional Values

Although values of inverse functions have been evaluated in Chapters 1 and 2, the concept has been refined somewhat. The examples that follow use exact values because most of the previous attention has been on calculators.

EXAMPLES

Find the exact value of each.

3. arcsin $\frac{1}{2}$

Let $\theta = \arcsin \frac{1}{2}$, then $\sin \theta = \frac{1}{2}$. $\sin \theta$ is $\frac{1}{2}$ at $\theta = \pi/6$ and $5\pi/6$ and all coterminal angles.

$$\mathbf{arcsin}\ \frac{1}{2} = \frac{\pi}{6} \pm 2\pi k \quad \text{or} \quad \frac{5\pi}{6} \pm 2\pi k$$

Degree measure could be used here; $30° \pm 360k°$ or $150° \pm 360k°$.

4. Arcsin $\frac{1}{2}$

Using Example 3, $\sin \theta = \frac{1}{2}$ *but* $-\pi/2 \le \theta \le \pi/2$, so

$$\mathbf{Arcsin}\ \frac{1}{2} = \frac{\pi}{6}$$

5. $\text{Cos}^{-1}\left(-\frac{\sqrt{3}}{2}\right)$

Let $\theta = \text{Arccos}\ (-\sqrt{3}/2)$; then $\cos \theta = -\sqrt{3}/2$ with $0 \le \theta \le \pi$. A sketch might be helpful.

$$\cos \theta = \frac{-\sqrt{3}}{2} = \frac{x}{r}$$

$$\theta = \frac{5\pi}{6}$$

$$\mathbf{Cos}^{-1}\left(-\frac{\sqrt{3}}{2}\right) = \frac{5\pi}{6}$$

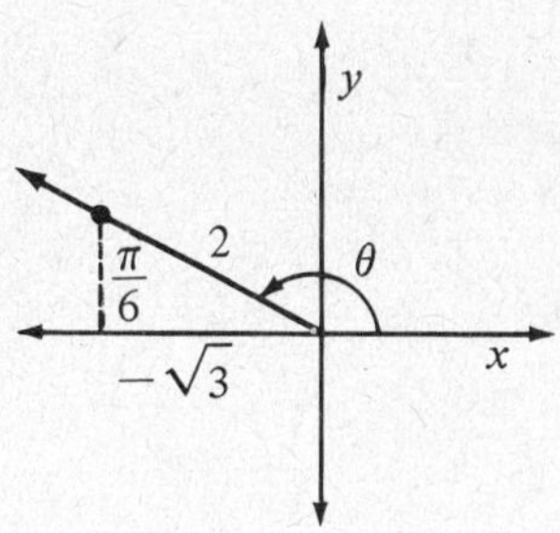

6. $\sin[\text{Tan}^{-1}(-1)]$

This can be written as $\sin \theta$, where $\theta = \text{Tan}^{-1}(-1)$ or $\tan \theta = -1$ with $-\pi/2 \le \theta \le \pi/2$.

From the sketch it is apparent that

$$\sin \theta = -\frac{1}{\sqrt{2}}$$

$$\mathbf{sin[Tan^{-1}(-1)]} = -\frac{1}{\sqrt{2}} \quad \text{or} \quad -\frac{\sqrt{2}}{2}$$

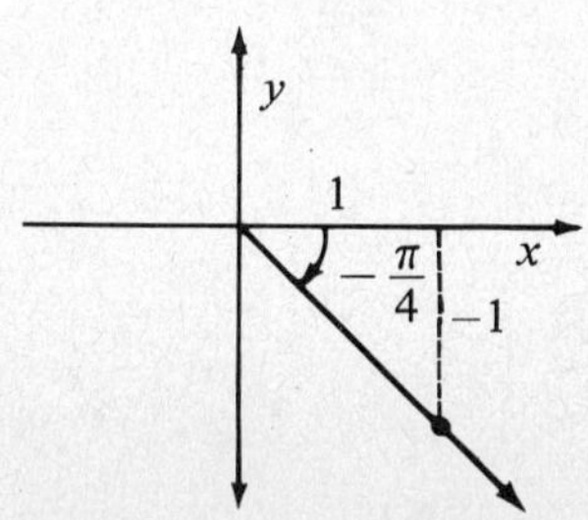

7. $\tan(\text{Arccos } \frac{3}{4})$
$\tan\theta$ where $\theta = \text{Arccos } \frac{3}{4}$ or $\cos\theta = \frac{3}{4}$ where $0 \le \theta \le \pi$.

$$3^2 + y^2 = 4^2$$
$$y^2 = 16 - 9$$
$$y = \sqrt{7}$$

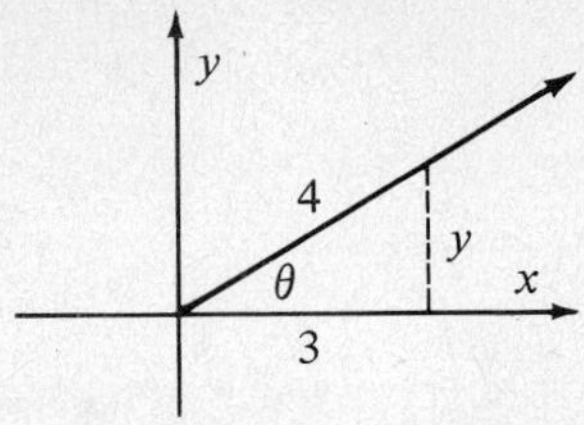

$$\mathbf{\tan\left(\text{Arccos } \frac{3}{4}\right) = \frac{\sqrt{7}}{3}}$$

8. $\sin\left[\text{Arccos } \frac{1}{2} + \text{Arcsin } (-1)\right]$

$$\sin\left[\frac{\pi}{3} + \left(-\frac{\pi}{2}\right)\right]$$
$$\sin\left(-\frac{\pi}{6}\right) = -\frac{1}{2}$$

$$\mathbf{\sin\left[\text{Arccos } \frac{1}{2} + \text{Arcsin } (-1)\right] = -\frac{1}{2}}$$

9. $\sin(\text{Cos}^{-1} x)$
$\sin\theta$, where $x = \cos\theta$, $0 \le \theta \le \pi$. Consider the sketch below.

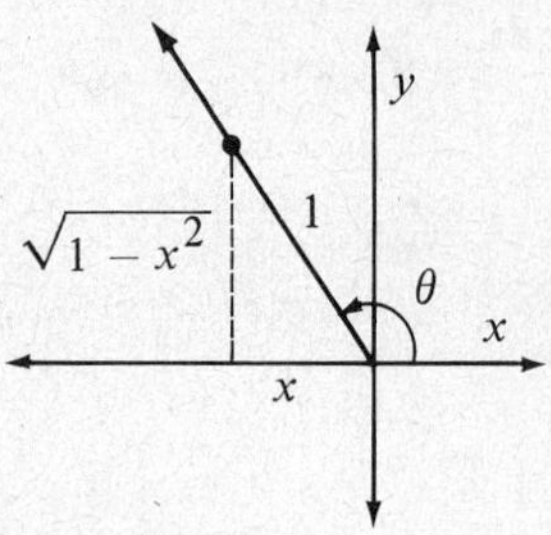

if $x < 0$

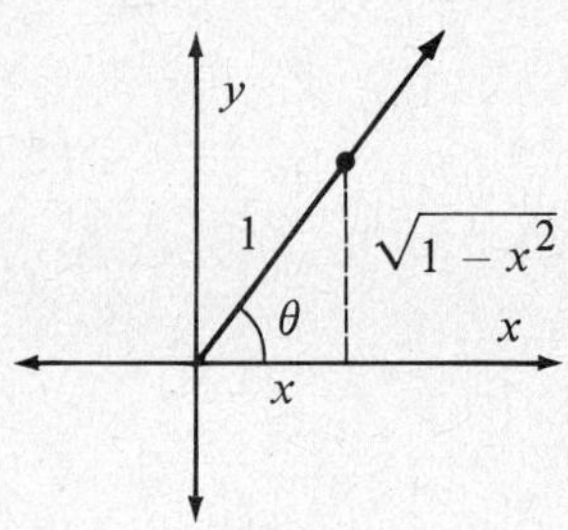

if $x > 0$

$\sin\theta = \dfrac{\sqrt{1 - x^2}}{1}$ $\quad$ $\sin\theta \ge 0$ because $0 \le \theta \le \pi$.

$$\mathbf{\sin(\text{Cos}^{-1} x) = \sqrt{1 - x^2}}$$

Alternative Solution

$$\sin^2\theta + \cos^2\theta = 1$$
$$\sin^2\theta = 1 - \cos^2\theta$$
$$\sin\theta = \sqrt{1 - \cos^2\theta}$$
$$\sin(\text{Cos}^{-1} x) = \sqrt{1 - [\cos(\text{Cos}^{-1} x)]^2}$$
$$\mathbf{\sin(\text{Cos}^{-1} x) = \sqrt{1 - x^2}}$$

At times it is important to be able to rewrite an expression using the inverse relations to solve for the angle. Recall the formula in Exercise Set 3.3, Exercises 31–32, where the distance d to an object is a function of the angle of elevation θ.

$$\boldsymbol{d = (H - h) \cot \theta}$$

where H is the height of the top of the object and h is the eye-level height of the observer.

EXAMPLES

10. Rewrite the expression above, solving for θ.

$$\frac{d}{H - h} = \cot \theta \qquad \text{(isolate cotangent)}$$

$$\theta = \text{Arccot}\,\frac{d}{H - h} \qquad \text{(use inverse)}$$

or

$$\boldsymbol{\theta = \textbf{Arctan}\,\frac{H - h}{d}} \qquad \text{(or reciprocal)}$$

11. Find θ if $H = 1200$, $h = 6$, $d = 199$ (all in feet) in Example 10.

$$\theta = \text{Tan}^{-1}\,\frac{1200 - 6}{199} = \text{Tan}^{-1}\,6$$

$$\theta \approx 80.537678°$$

or

$$\boldsymbol{\theta \approx 80°32'}$$

exercise set 3.4

A. exercises

Sketch the graphs of the inverse relations and functions in Exercises 1–10.

See Examples 1 and 2.

1. $2y = \arcsin 3x$
2. $y = 3 \arccos 2x$
3. $y = \tan^{-1} \frac{1}{2}x$
4. $\frac{1}{2}y = \cot^{-1} x$
5. $y = \text{Sin}^{-1}\, 2x$
6. $y = 2\, \text{Cos}^{-1}\, x$
7. $y = \text{Arccos}(x + 1)$
8. $y + 1 = \text{Arcsin}\, x$
9. $\frac{1}{2}y = \text{Tan}^{-1}(x - 3)$
10. $2y = \text{Tan}^{-1}(x + 3)$

Find the exact values of the inverse relations and functions is Exercises 11–30.

11. $\arccos 0$ 12. $\arcsin(-1)$ 13. $\tan^{-1} 0$ See Examples 3–5.

14. $\sec^{-1} 1$ 15. $\text{Arcsin } 0$ 16. $\text{Arccos } 1$

17. $\text{Sin}^{-1} \dfrac{1}{\sqrt{2}}$ 18. $\text{Cos}^{-1} \dfrac{\sqrt{3}}{2}$ 19. $\text{Tan}^{-1} 1$

20. $\text{Tan}^{-1} \sqrt{3}$ 21. $\text{Arccos}\left(-\dfrac{\sqrt{2}}{2}\right)$ 22. $\text{Arccsc}\,(-\sqrt{2})$

23. $\text{Sec}^{-1} 2$ 24. $\text{Cot}^{-1}(-1)$ 25. $\text{Arccos } \frac{1}{2}$

26. $\text{Arcsin}\left(-\dfrac{1}{2}\right)$ 27. $\text{Csc}^{-1}(-2)$ 28. $\text{Sec}^{-1} \dfrac{2}{\sqrt{3}}$

29. $\text{Tan}^{-1} \dfrac{1}{\sqrt{3}}$ 30. $\text{Sec}^{-1}(-1)$

B. exercises

Find the exact value of the functions in Exercises 31–44.

31. $A = \cos(\text{Arctan } \sqrt{3})$ 32. $C = \cos\,[\text{Arcsin }(-\frac{1}{2})]$ See Examples 6 and 7.

33. $E = \tan\left[\text{Arccos}\left(-\dfrac{1}{\sqrt{2}}\right)\right]$ 34. $H = \cot\left(\text{Arccos} \dfrac{\sqrt{3}}{2}\right)$

35. $I = \cos(\text{Tan}^{-1} \frac{3}{4})$ 36. $L = \tan\,(\text{Sin}^{-1} \frac{7}{25})$ See Exercise 62.

37. $M = \cot\,[\text{Cos}^{-1}(-\frac{5}{13})]$ 38. $N = \sec(\text{Cos}^{-1} \frac{24}{25})$

39. $P = \cos\,(\text{Arcsin } \frac{1}{5})$ 40. $R = \tan[\text{Arccos }(-\frac{1}{3})]$

41. $S = \tan\left[\text{Arcsin} \dfrac{1}{2} + \text{Arccos}\left(-\dfrac{\sqrt{3}}{2}\right)\right]$ See Example 8.

42. $T = \sin(\text{Arctan } 0 + \text{Arctan } 1)$

43. $U = \cos\,[\text{Tan}^{-1}(-1) + \pi]$

44. $Y = \tan\,[\pi + \text{Sin}^{-1}(-\frac{1}{2})]$

In Exercises 45–50, express each strictly in terms of x.

45. $\tan(\text{Cos}^{-1} x)$ 46. $\sin(\text{Tan}^{-1} x)$ See Example 9.

47. $\sin\left(\operatorname{Tan}^{-1}\dfrac{x}{\sqrt{1-x^2}}\right)$

48. $\tan\left(\operatorname{Sin}^{-1}\dfrac{1}{x}\right)$

49. $\cos(\operatorname{Arcsin}\sqrt{1-4x^2})$

50. $\cos\left(\operatorname{Arcsin}\dfrac{\sqrt{4x^2-1}}{2x}\right)$

The top of an object is H units high and d units from an h unit tall observer. The angle of elevation to the top of the object is given by

See Exercises 31 and 32 in Section 3.3.

$$\theta = \textbf{Arccot}\,\frac{d}{H-h} \quad \textbf{or} \quad \textbf{Arctan}\,\frac{H-h}{d}$$

51. calculator problem If a 6-ft viewer stands back 10 ft from an object hung 8 ft from the floor, what is the angle of elevation?

See Example 11.

52. calculator problem If a 5-ft 6-in person views an object 100 ft in the air and stands 75 ft back from directly beneath the object, what is the angle of elevation?

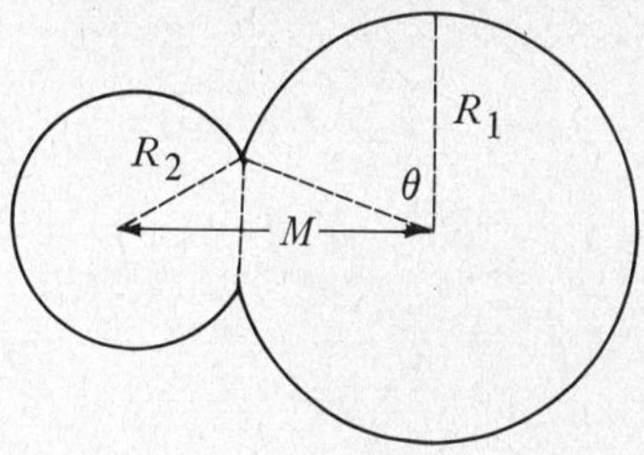

Two intersecting parallel circular cylinders have radii R_1 and R_2, *where* $R_1 > R_2$. *If the larger cylinder has an offset of M, then the offset angle is given by*

$$\cos\theta = \frac{R_1^2 + M^2 - R_2^2}{2MR_1}$$

53. Prove the formula for $M = R_1 + R_2$.

54. calculator problem Find θ if cylinders of 10 ft and 7.5 ft are offset 6 ft.

See Example 11.

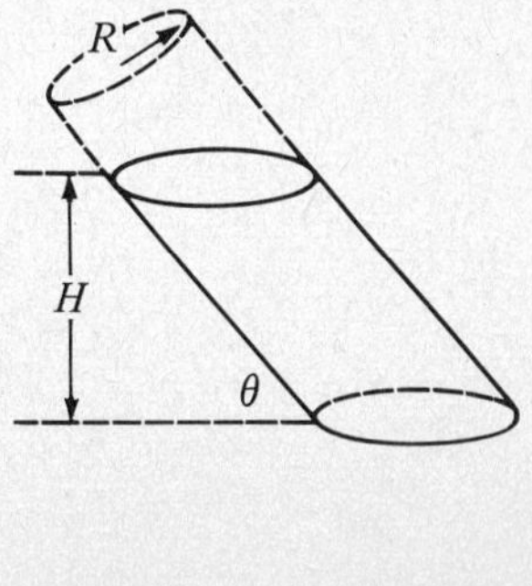

If a circular hole with radius R is drilled at an angle θ through material H units thick, the shape is called an oblique cylinder. The volume is given by

$$V = \pi R^2 H \csc \theta$$

55. Rewrite the volume formula, solving for θ.

56. **calculator problem** An 11-cm-radius hole is drilled through a 6-cm-thick plate. If 7380 cm^3 of waste is created, at what angle was the hole drilled?

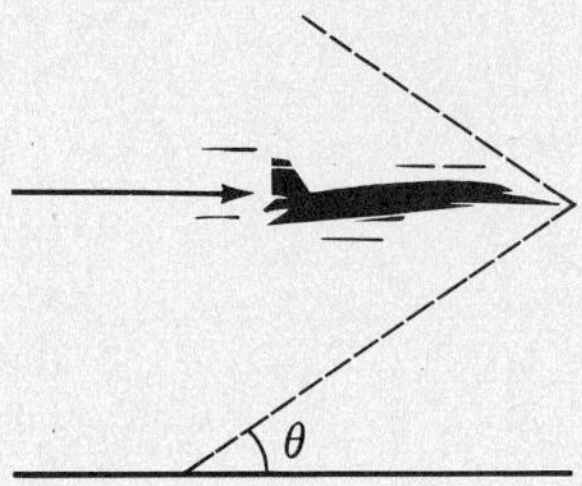

Figure 3.23 Supersonic sound waves

Supersonic aircraft leave a conical wake of sound waves. See Figure 3.23. The SST dispute of a couple years ago centered on the effects of those waves as the plane swept across the country. The angle that the waves strike the ground depends on the mach number of the plane.

$$m \sin \theta = 1 \quad : \quad m \geq 1$$

The mach number m is the ratio of the planes speed to the speed of sound (approximately 741.82 mi/h).

HISTORICAL ASIDE

The mach number of supersonic speed is named after Ernst Mach (1838–1916), an Austrian physicist. "There is no problem in all mathematics that cannot be solved by direct counting," Mach said in 1898. "But," he continued, "with the present implements of mathematics many operations can be performed in a few minutes which without mathematical methods would take a lifetime." Recall that that was in 1898! What would Mach think about present-day computers and calculators?

57. Rewrite the equation, solving for θ.

58. What is the angle of the shock waves when a plane "breaks the sound barrier"?

By way of contrast, the speed record for a piston engine craft is 469.549 mi/h set by a Grumman F8F Bearcat in 1969. Strictly subsonic!

59. **calculator problem** A USAF lockheed SR-71 established a new speed record of 2193.16 mi/h on July 28, 1976. Find the angle at which the sound waves hit the ground in its wake.

60. **calculator problem** The record speed on the Los Angles-to New York run is 1214.65 mi/h as set by a USAF B58 Hustler in 1962. Find the angle at which its sound waves struck the ground.

The total LA-NY flight time was 2 h and 58.71 s.

C. exercises

61. **brain teaser** An old rule of thumb says that one should stand back nine times the height of a painting to view it properly. Pre-

sumedly at this distance the viewer is able to take in the whole work with minimum eye movement or angle of vision. Suppose that a painting is centered at eye level and you stand at the "proper distance." What angle of vision is necessary for a work of art x units tall? (Hint: Begin with Exercises 51–52.)

The angle of vision is the angle subtended by the top and bottom of the object viewed.

62. **trig talk** Use Exercises 31–44 to fill in the blanks with the capital letter that corresponds to the value below the blank. When all the blanks are filled, the Lord Kelvin quotation may be read.

"___ ___ ___ ___ ___ ___ ___ ___ ___ ___ ___

$-\frac{5}{12}$ $\frac{1}{2}$ $\frac{\sqrt{2}}{2}$ $\sqrt{3}$ -1 $-\frac{5}{12}$ $\frac{1}{2}$ $\frac{\sqrt{2}}{2}$ $\frac{4}{5}$ $\frac{\sqrt{3}}{2}$ 0

___ ___   ___ ___ ___

$\frac{4}{5}$ 0 $\frac{\sqrt{2}}{2}$ $\sqrt{3}$ -1

O ___ ___ ___   ___ ___ ___ ___

$\frac{25}{24}$ $\frac{7}{24}$ $-\frac{\sqrt{3}}{3}$ $\frac{\sqrt{2}}{2}$ $-2\sqrt{2}$ $-\frac{\sqrt{2}}{2}$ -1

___ ___ ___ ___ ___ ___ ___ ___ ___ ___ ___."

$-\frac{5}{12}$ -1 $\frac{\sqrt{2}}{2}$ $\frac{1}{2}$ $\frac{2\sqrt{6}}{5}$ $\sqrt{3}$ $-\frac{\sqrt{3}}{3}$ 0 $\frac{4}{5}$ $\frac{\sqrt{3}}{2}$ 0

(Lord Kelvin)

3.5 COMBINING FUNCTIONS

Often a complex function is made up of a number of much simpler ones. Even a musical instrument playing a single note produces a complex tone made up of a pure tone augmented by harmonics giving distinctive overtones that identify the instrument. The complex tone can be broken into components each of which is itself a pure tone.

Addition of Ordinates

Consider two functions $y = f(x)$ and $y = g(x)$ combined to form a new function $y = f(x) + g(x)$. Graphing such a function is best illustrated by example.

EXAMPLES

1. Sketch $y = f(x) + g(x)$, given $f(x)$ and $g(x)$.

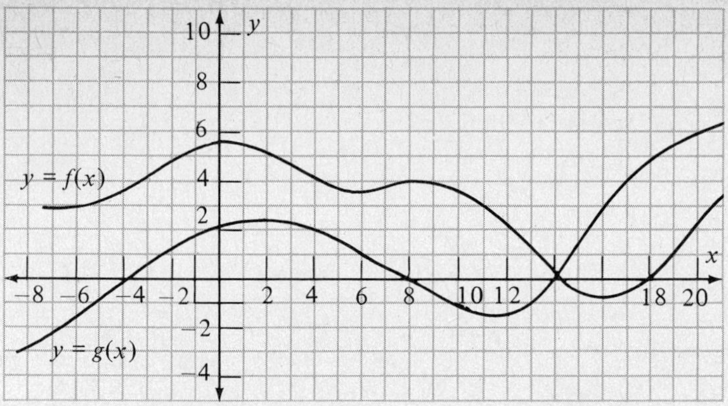

Given the graphs of f and g in this case, usually you would first sketch these graphs.

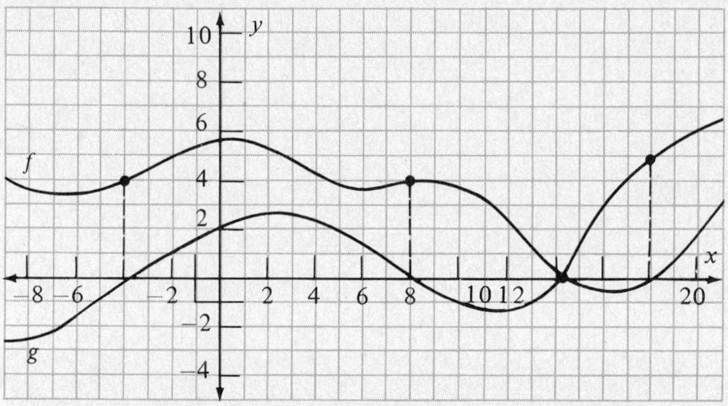

Second, if both (x) and $g(x)$ are zero for any value of x, then $f(x) + g(x)$ is zero. If one of the two is zero, $f(x) + g(x)$ is the nonzero value at that value of x.

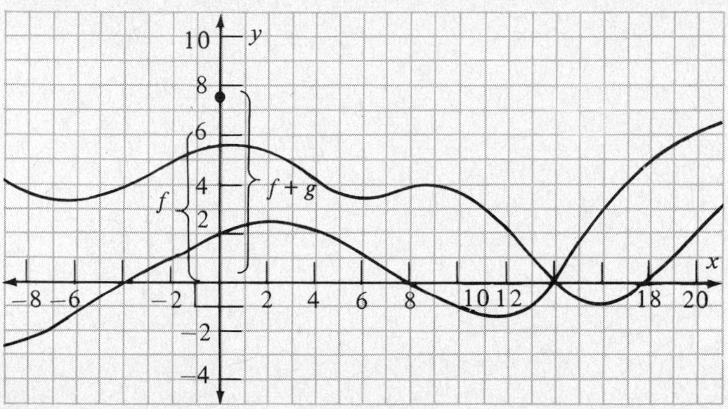

Third, use your eye or a ruler to measure the y coordinate of g and then add that distance to the y coordinate of f. The result will be a y coordinate of $f + g$.

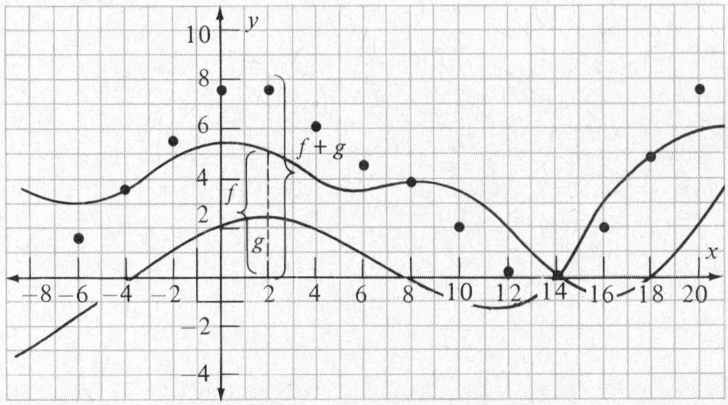

Then move across at regular intervals on the x axis and add y coordinates to locate sufficiently many points on $f + g$. Beware that one or both may be negative.

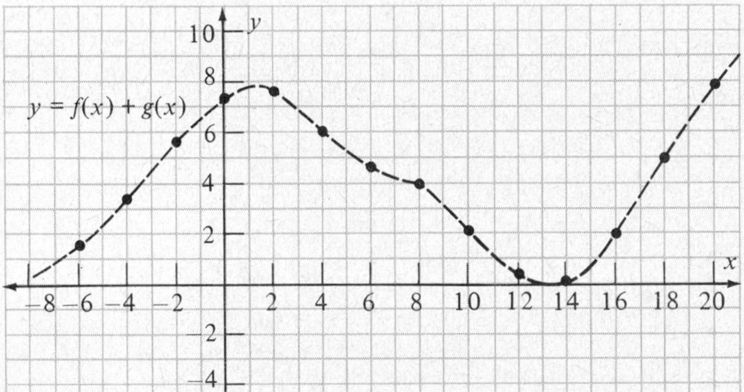

Finally, trace a smooth curve through the points to obtain a sketch of $f + g$.

2. Sketch $y = 2 \sin x + \sin 2x$

(a) Sketch $y = 2 \sin x$ and $y = \sin 2x$.

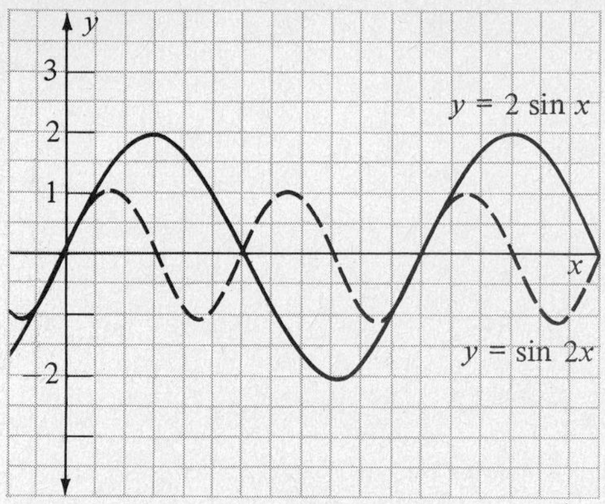

(b) Look at the zeros of each.

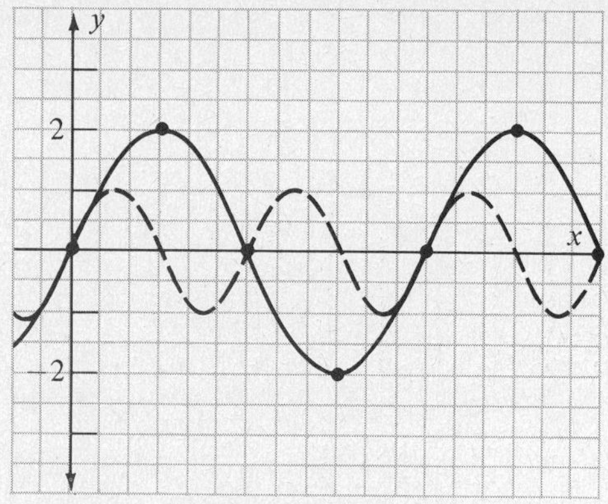

(c) Add y coordinates for a number of points.

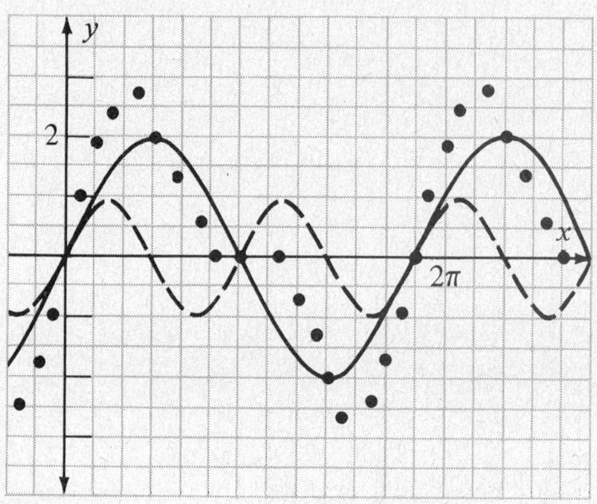

(d) Trace a smooth curve through these points.

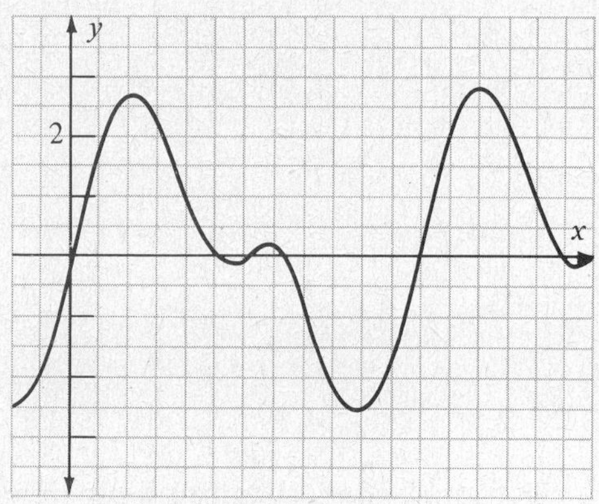

In the remainder of the examples, one graph will be used to show each of the functions (dashed) and the composite (solid) with many key points of construction. The key points should allow you to verify the accuracy of the sum.

EXAMPLES

3. Sketch $y = \dfrac{x}{3} + \sin x$.

Note that such a graph might quite accurately describe the business cycle of a seasonal product that fluctuates but continues to grow

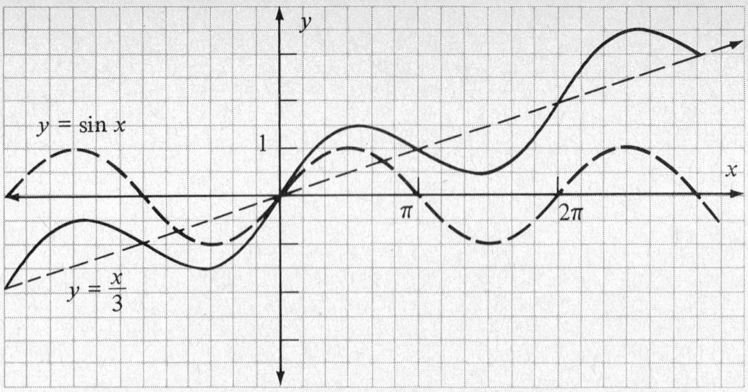

season to season. (Rototillers? Lawn mowers? Snow tires? Pollution???)

4. Sketch $y = \frac{1}{2}x \sin 2x$.
Because this is a product, concentrate on points where y is either 1 or 0.

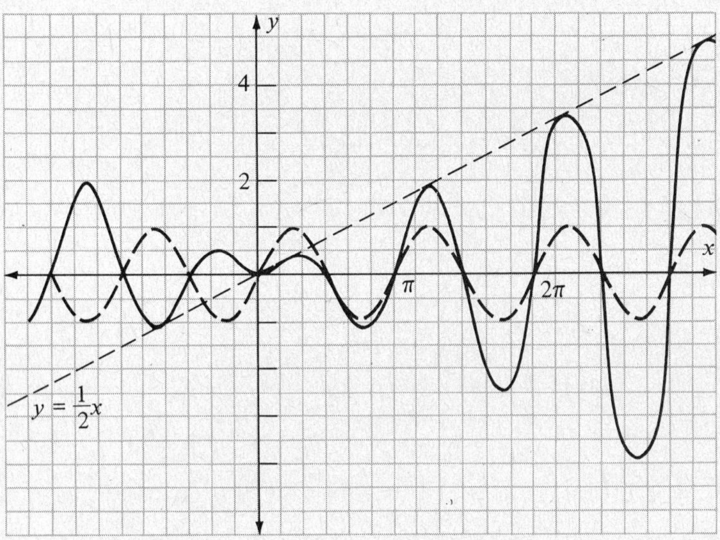

Notice this is the *product of y coordinates* $y = (\frac{1}{2}x)(\sin 2x)$.

5. Sketch $y = 2^{-x} \sin 4x$.
This last curve is called a **damped wave.** $y = 2^{-x}$ is called the **damping function.**

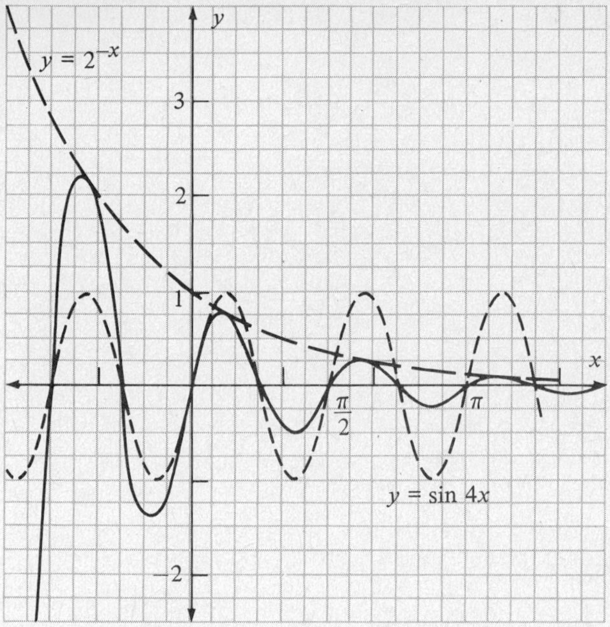

An AM radio station transmits a signal that is basically of the form

$$y = A \sin bt \sin Bt$$

where $2\pi/B$ is the fixed frequency of the so-called carrier signal of the station, which is between 550 and 1600 kc/s. $2\pi/b$ is the frequency of the audio signal, which varies from about 100 to 10,000 cycle/s. **AM** stands for **amplitude modulated,** and $A \sin bt$ may be thought of as the variable (modulated) amplitude of $\sin Bt$. Figure 3.24 shows an amplitude-modulated carrier wave with fixed frequency and variable amplitude.

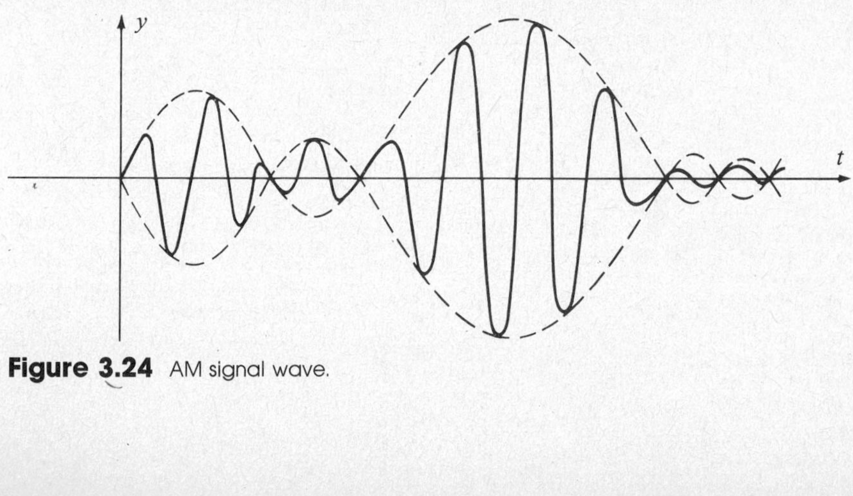

Figure 3.24 AM signal wave.

Most ratio interference is in the amplitude of the signal. An FM station offers a signal that avoids this kind of interference. It transmits a wave that is basically of the form

$$y = A \sin(Bt + a \sin bt)$$

where $2\pi/B$ is the frequency of the carrier wave $y = A \sin Bt$, which is modulated by the audio signal $y = a \sin bt$. **FM** stands for **frequency modulated,** and the amplitude is constant so that interference may be gated or filtered out. Figure 3.25 shows a frequency-modulated carrier wave with fixed amplitude and variable frequency.

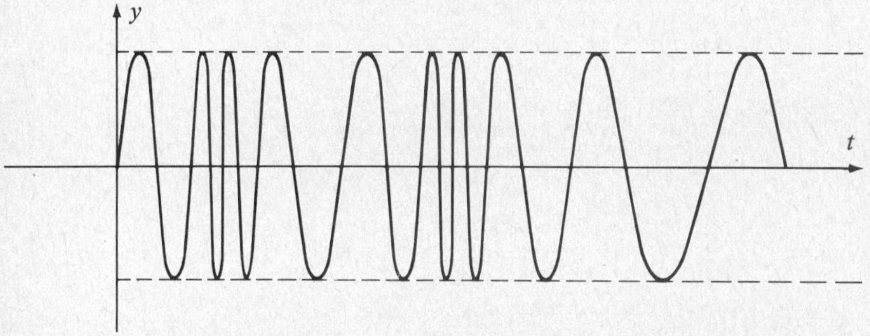

Figure 3.25 FM signal wave.

Whereas a pure tone such as that made by a tuning fork is described by a sine wave of the form

$$y = A \sin Bt = A \sin 2\pi ft$$

the tone of most sound is made up of many pure tones. The tone of a musical instrument playing the same note as the tuning fork would be given by

$$y = A_1 \sin 2\pi ft + A_2 \sin 4\pi ft + A_3 \sin 6\pi ft + \cdots + A_n \sin 2\pi(nf)t$$

where f is the **fundamental frequency** and $2f, 3f, 4f, \ldots, nf$ are **harmonics** giving **overtones** to the note. It is these overtones that give a distinct sound to a particular instrument. Figure 3.26 shows such a note broken down into 12 components, a fundamental note and 11 overtones.

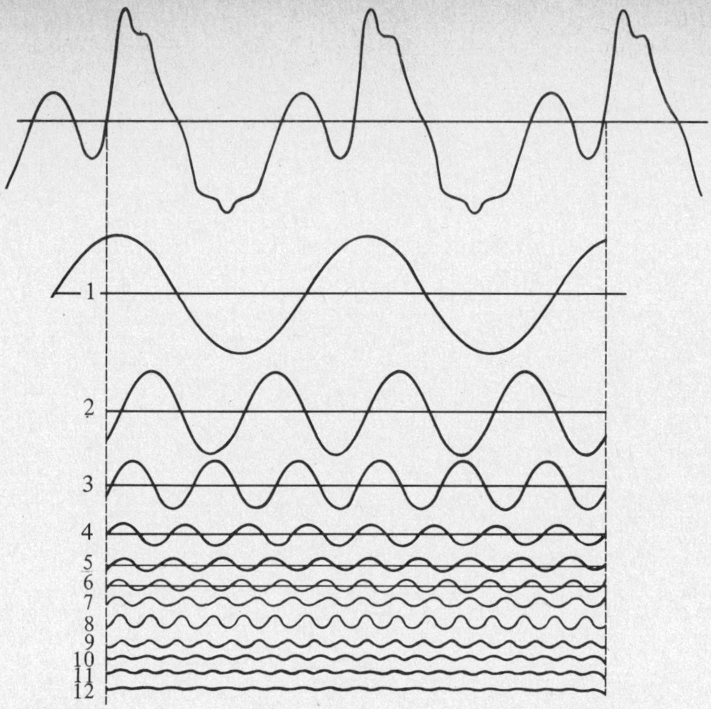

Figure 3.26 Note with overtones. From Dayton C. Miller, The Science of Musical Sounds, courtesy of Case Institute of Technology.

Note that each wave shares the same horizontal axis, and they would be composed by adding the ordinates as in Examples 1–3.

EXAMPLES

Sketch each of the following functions.

6. $y = 3 \sin x + 2 \sin 2x$

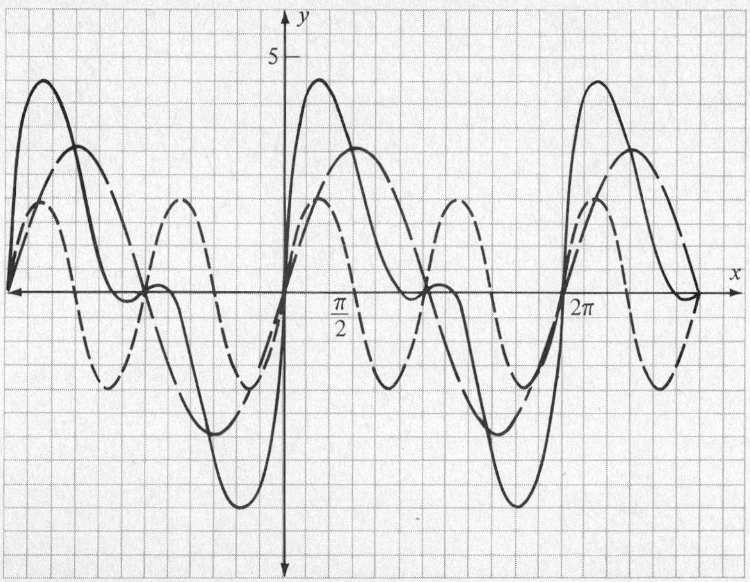

7. $y = 3 \sin x + 2 \sin 2x + \sin 3x$
Make use of Example 6, because this is just $\sin 3x$ added to the previous wave.

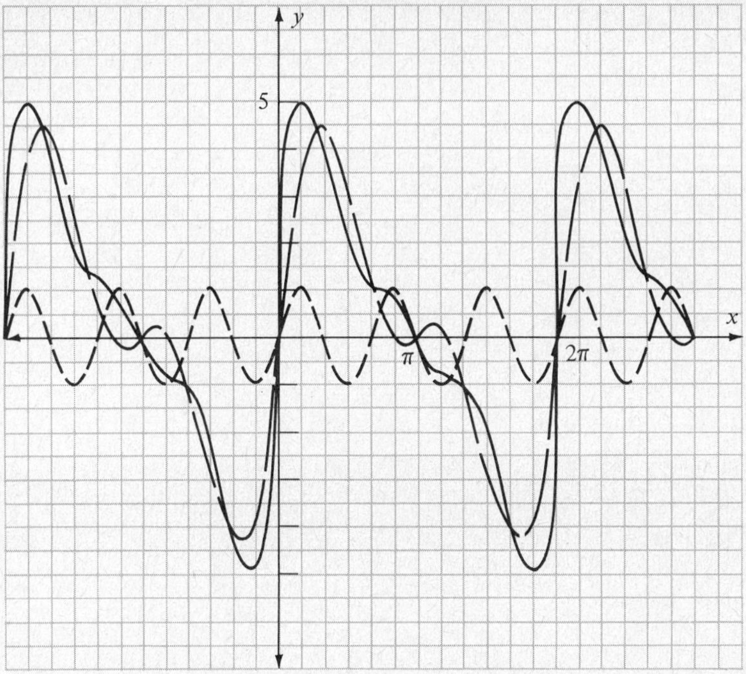

exercise set 3.5

A. exercises

Sketch each of the functions in Exercises 1–24 by composition of ordinates.

1. $y = \cos x + 3 \sin x$
2. $y = 2 \cos x + 3 \sin x$

See Examples 1 and 2.

3. $y = \sin x + \cos x$
4. $y = \sin x - \cos x$
5. $y = \cos x + \sin 2x$
6. $y = \sin x + \sin 2x$
7. $y = 2 \sin x + \sin 2x$
8. $y = \frac{1}{2} \cos x + \sin \frac{1}{2}x$
9. $y = \sin x + \cos 2x$
10. $y = \sin 2x - \cos x$
11. $y = 2 \sin x - \cos 2x$
12. $y = 2 \cos x + \sin 3x$
13. $y = 2 \cos x - \frac{1}{2} \sin 2x$
14. $y = \sin 3x + 2 \cos 2x$
15. $y = \cos\left(x + \frac{\pi}{4}\right) - 2 \sin 2x$

16. $y = 2\cos 2\left(x + \frac{\pi}{4}\right) - 3\sin\left(x - \frac{\pi}{4}\right)$

17. $y = \frac{1}{3}x + \sin x$

18. $y = \frac{x}{2} - \cos x$

See Example 3.

19. $y = \frac{1}{3}e^x + \sin 3x$

20. $y = \frac{1}{2}x^2 + \cos 2x$

21. $y = x \sin x$

22. $y = x \cos 2x$

See Example 4.

23. $y = 2^{-x} \sin x$

24. $y = \frac{\cos x}{2^x}$

See Example 5.

B. exercises

Two light waves interfere destructively if the amplitude of the composite wave is everywhere zero.

25. Show that $y = \sin(x + \pi/2) + \cos(x + \pi)$ illustrates destructive interference.

26. Determine if the following is destructive.

$$y = \cos x + \cos\left(x + \frac{2\pi}{3}\right) + \cos\left(x + \frac{4\pi}{3}\right)$$

The wave displayed on an oscilloscope for the sound of a tuning fork vibrating at f c/s and amplitude 4 is given by

$$\boldsymbol{y = 4\sin 2\pi f x}$$

If an instrument sounds the same tone, the first overtone created would be

$$\boldsymbol{y = 2\sin 4\pi f x}$$

27. Sketch the sound wave

$$y = 4\sin 2\pi f x + 2\sin 4\pi f x$$

28. Add the second overtone

$$y = \sin 8\pi f x$$

to the wave in Exercise 27.

29. Add a third overtone to the wave in Exercise 28,

$$y = \tfrac{1}{2}\sin 16\pi f x$$

Once a tone is created by a tuning fork, for instance, the amplitude begins to decay. That is, the loudness diminishes steadily until it is inaudible. This damped wave can be described by

$$\boldsymbol{y = Ae^{-rt}\sin 2\pi f t}$$

To simplify the sketch, let $f = 1$. Actually,

$$f = \frac{1}{2L}\sqrt{\frac{T}{m}}$$

for a vibrating string of length L, tension T, and mass m per unit length. The D string on a violin, for example, is 33 cm long, has a mass of 0.0115 g/cm, and tension of 4.32×10^6 dyn/cm.

$$f = \frac{1}{2 \times 33}\sqrt{\frac{4.32 \times 10^6}{0.0115}}$$

$$\approx 294$$

thus the frequency of D on the musical scale is 294 c/s.

30. If the tone is D ($f = 294$) and the damping rate r is 0.01, sketch the damped sound wave.

31. The current I in a given circuit is composed of power from two sources.

$$I = 6 \sin 60\pi t + 3 \cos 120\pi t$$

where t is in seconds. Sketch the graph of this current.

32. Sketch the graph of the current (see Exercise 31)

$$y = 2 \sin 80\pi t + 4 \cos 120\pi t$$

C. exercises

GARFIELD/Jim Davis

33. **brain teaser** Suppose that each "BOING" in the cartoon is a part of a sine wave and that the graph in Figure 3.27 represents what is happening to Garfield in the cartoon. Write an equation for his path.

Figure 3.27

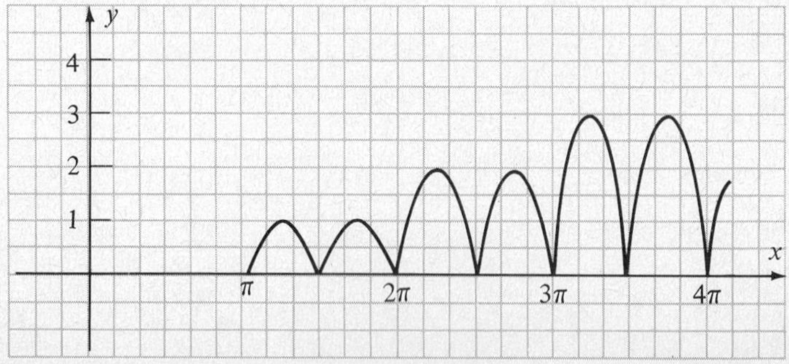

3.6 CHAPTER REVIEW

Graphing the trigonometric functions provides an insightful view of their behavior. The graphs supply a fuller understanding of the functions themselves. The method of framing emphasizes sketching a base cycle relying on an initial point, the amplitude, and the frequency. This approach highlights the similarities with other graphing experience, such as the line and the parabola. We observe a broad range of applications that make these functions invaluable to the analysis of many physical problems. Combining these functions (as well as certain algebraic ones) indicates how other, even more significant, problems can be analyzed trigonometrically.

Fill in the symbol, word, or words that make the statements in Exercises 1–10 complete and correct.

1. (3.1) The graph of $y = a \sin bx$ has an ________ of a and a ________ b.
2. (3.1) The graph of $y = a \sin bx$ has a ________ of $2\pi/b$.
3. (3.1) The graph of $y - k = a \sin b(x - h)$ has a horizontal displacement of ________ and a vertical displacement of ________.
4. (3.2) The graph of $y = a \tan bx$ has a period of ________.
5. (3.2) The graph of tangent has an ________ at those values of x for which it is undefined.
6. (3.2) The graph of ________ is the set of points above the graph of $y - k = a \cos b(x - h)$.
7. (3.3) The graph of $y = a \sec bx$ has a period of ________ units.
8. (3.4) $y = \text{Sin}^{-1} x$ is equivalent to ________, where ________.
9. (3.4) $\sin(\text{Arctan } x)$ means the sine of the ________ whose ________ is x.
10. (3.5) The graph of $y = a \sin bx + A \cos Bx$ can be sketched by ________ the ________ of $y = a \sin x$ and $y = A \cos Bx$.

Rewrite each of the following in standard form and identify the initial point of the base cycle, the amplitude, and the period.

11. (3.1) $5y = 2 \sin (3x - 2\pi)$
12. (3.2) $3 \tan 2(x + \pi) - 2y + 4 = 0$
13. (3.3) $4y + 8 = 12 \sec \frac{2}{3}(\frac{9}{4}x - \pi)$

Sketch each of the following for the interval $0 < x < 2\pi$.

14. (3.1) (a) $y = 2 \sin \frac{3}{2}x$
    (b) $y + 2 = 3 \sin 2(x - \pi)$
15. (3.2) (a) $y - 1 = \dfrac{1}{2} \cos \dfrac{2}{3}\left(x - \dfrac{\pi}{2}\right)$
    (b) $y + 1 = 2 \tan \dfrac{1}{2} x$

16. (3.3) (a) $y = \frac{1}{2} \cot \left(x - \frac{\pi}{2}\right)$

(b) $y = 2 \sec \frac{3}{2} x$

17. (3.3) (a) $\begin{cases} y \leq \sin x \\ y + 2 \geq \cos x \end{cases}$

(b) $\begin{cases} y > \tan x \\ y < \sec x \end{cases}$

18. (3.4) (a) $y = \text{arcin } 2x$

(b) $y = 3 \text{ Arccos } \frac{1}{2}x$

19. (3.5) (a) $y = 2 \cos x + \sin 2x$

(b) $y = x + \cos 3x$

Determine the exact value of each.

20. (3.4) $\sin\left(\text{Arccos} \frac{\sqrt{2}}{2}\right)$

21. (3.4) $\sec\left[\text{Arctan}(-\frac{8}{15})\right]$

22. (3.4) $\cos\left(\text{Arctan} \frac{1}{\sqrt{3}} - \text{Arccsc } 2\right)$

23. (3.4) $\tan(\text{Arcsin} \sqrt{x})$

For a circular tube of radius r, the distance h to which capillary action will elevate (or depress) a liquid of density p and surface tension T (against air) is

$$h = \frac{2T \cos \alpha}{rpg}$$

where g is gravity and α is the "angle of contact" between the liquid surface and the tube surface.

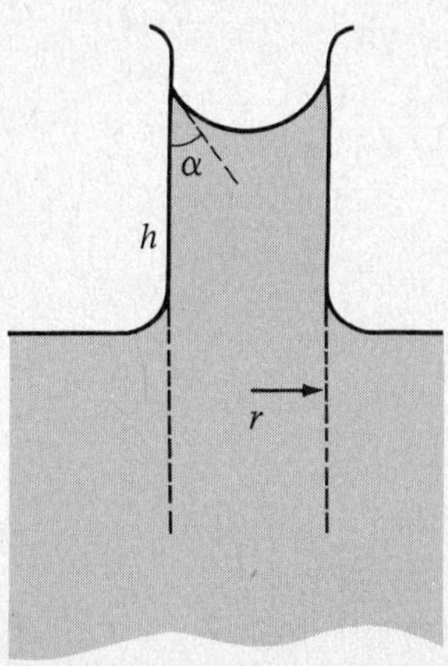

24. (3.4) Rewrite the expression, solving for α.

25. (3.4) The density of aceton is 0.8 g/ml and the surface tension is 23.7 dyn/cm at 20°C. Find the angle of contact (to the nearest degree) in a 0.5-cm-radius tube that has a 0.0989-cm elevation. (Use $g = 981.274$ cm/s^2.)

identities and equations four

4.1 TRIGONOMETRIC IDENTITIES

The history of trigonometry is tied to measurement and manipulation. Surveying and the construction of the pyramids, astronomy and the measurement of heavenly sizes and distances, and the necessary calculation of elaborate tables mark early trigonometry. Although the manipulation aspects are still important, their place has been minimized by the computer and more recently by the hand-held calculator. Our analysis of uniform circular motion as seen in sound waves, periodic phenomena such as the pendulum, and the displacement of a piston illustrate a more modern aspect of the subject. The very character of the functions helps to explain these phenomena and becomes of primary interest. It becomes necessary to be able to simplify or otherwise change complex trigonometric expressions. This is an important part of applying the calculus to trigonometry.

HISTORICAL ASIDE
You might expect the first mathematics book in the New World to have been published in Philadelphia, Boston, or New York. Actually, it was a small commercial arithmetic by Juan Diez, published in 1556 in Mexico City.

As we saw in Chapter 2, an identity is an equation that is true for all values of the variable for which it is defined. The definitions of the functions were used to develop a set of fundamental identities. These identities were in turn used to prove other identities. This process is continued in this chapter, and you will see how identities have other applications, including the solution of trigonometric equations. Let us review the fundamental identities before we begin to develop others.

FUNDAMENTAL IDENTITIES

$$\sin\theta = \frac{1}{\csc\theta}$$

$$\cos\theta = \frac{1}{\sec\theta}$$

$$\tan\theta = \frac{1}{\cot\theta} = \frac{\sin\theta}{\cos\theta}$$

$$\cos^2\theta + \sin^2\theta = 1$$

$$1 + \tan^2\theta = \sec^2\theta$$

$$\cot^2\theta + 1 = \csc^2\theta$$

Proving Identities

In proving identities in Chapter 2, we saw that there are many ways to proceed. This flexibility is at once a boon and a blessing. You may start anywhere and do anything, but some tactics lead seemingly nowhere, whereas others go directly to your goal. Let us try to provide some guidelines for your decisions.

SUGGESTED STRATEGIES FOR PROVING IDENTITIES

I. Choose the more complicated side of the identity to simplify.
II. Take your clues from the other member, your goal.
   1. Express in terms of functions in the other member.
   2. If there are powers, remember the Pythagorean identities.
   3. If there are powers, factoring may help.
   4. If there are products, expand and simplify.
   5. If there are fractions, combine when they are separate.
   6. If there are fractions, separate when they are combined (especially where the denominator is relatively simple).
   7. If there are fractions, multiplying by the conjugate may help.
III. When in doubt, change everything to sines and cosines.

This choice may be your most important decision. Keep your eye on where you are going.

In this list, notice that the form of the identity should suggest possibilities. Rarely, if ever, will there be *the* thing to do, just something to do. Start, begin, get into it!

Now let us try to illustrate these hints with examples. However, remember there is no one best way to achieve a proof. The suggestions above are just that—*suggestions.* Follow your instincts.

EXAMPLES

Prove the identities.

1. $\sec\theta = \sec\theta\sin^2\theta + \cos\theta$

Choose the right side.

You have powers and products, look to factor or use Pythagorean identities. Notice that you have secant on the left, so factor out $\sec\theta$.

$\sec\theta = \sec\theta\sin^2\theta + \cos\theta$

$= \sec\theta\sin^2\theta + \cos\theta(\cos\theta\sec\theta)$ $\cos\theta\sec\theta = 1$

$= \sec\theta\sin^2\theta + \cos^2\theta\sec\theta$

$= \sec\theta(\sin^2\theta + \cos^2\theta)$ Factor.

$= \sec\theta(1)$ $\sin^2\theta + \cos^2\theta = 1$

$\boldsymbol{\sec\theta = \sec\theta}$

2. $\sec^2\alpha + \csc^2\alpha = \sec^2\alpha\csc^2\alpha$

The big question is which side should be attacked. It might be easier to get a sum from the product than vice versa.

$\sec^2\alpha + \csc^2\alpha = \sec^2\alpha\csc^2\alpha$ Power, so use Pythagorean.

$= \sec^2\alpha(1 + \cot^2\alpha)$

$$= \sec^2 \alpha + \frac{1}{\cos^2 \alpha} \frac{\cos^2 \alpha}{\sin^2 \alpha} \qquad \text{Use sine and cosine.}$$

$$= \sec^2 \alpha + \frac{1}{\sin^2 \alpha}$$

$$\mathbf{\sec^2 \alpha + \csc^2 \alpha = \sec^2 \alpha + \csc^2 \alpha}$$

3. $\dfrac{\tan \beta - \cot \beta}{\sec \beta \csc \beta} = \sin^2 \beta - \cos^2 \beta$

No question here: The left is more involved. Try separating the fraction.

$$\sin^2 \beta - \cos^2 \beta = \frac{\tan \beta - \cot \beta}{\sec \beta \csc \beta}$$

$$= \frac{\tan \beta}{\sec \beta \csc \beta} - \frac{\cot \beta}{\sec \beta \csc \beta}$$

$$= \tan \beta \cos \beta \sin \beta - \cot \beta \cos \beta \sin \beta$$

Use reciprocals.

$$= \frac{\sin \beta}{\cos \beta} \cos \beta \sin \beta - \frac{\cos \beta}{\sin \beta} \cos \beta \sin \beta$$

Use sine and cosine.

$$\mathbf{\sin^2 \beta - \cos^2 \beta = \sin^2 \beta - \cos^2 \beta}$$

4. $\cos \theta - \sin \theta = \dfrac{\cos^2 \theta - \sin^2 \theta}{\cos \theta + \sin \theta}$

Work on the fraction. But don't be too hasty with a conjugate; factor first.

$$\cos \theta - \sin \theta = \frac{\cos^2 \theta - \sin^2 \theta}{\cos \theta + \sin \theta}$$

$$= \frac{(\cos \theta - \sin \theta)(\cos \theta + \sin \theta)}{\cos \theta + \sin \theta}$$

$$\mathbf{\cos \theta - \sin \theta = \cos \theta - \sin \theta}$$

5. $\dfrac{\csc \theta - \sec \theta}{\csc \theta + \sec \theta} = \dfrac{\cos \theta - \sin \theta}{\cos \theta + \sin \theta}$

Notice that $\csc \theta$ and $\sec \theta$ are reciprocals of $\cos \theta$ and $\sin \theta$. Multiply by $\sin \theta \cos \theta$.

$$\frac{\cos \theta - \sin \theta}{\cos \theta + \sin \theta} = \frac{\csc \theta - \sec \theta}{\csc \theta + \sec \theta}$$

$$= \left(\frac{\csc \theta - \sec \theta}{\csc \theta + \sec \theta}\right) \frac{\sin \theta \cos \theta}{\sin \theta \cos \theta}$$

$$= \frac{\csc\theta\sin\theta\cos\theta - \sec\theta\sin\theta\cos\theta}{\csc\theta\sin\theta\cos\theta + \sec\theta\sin\theta\cos\theta}$$

$$= \frac{(\csc\theta\sin\theta)\cos\theta - (\sec\theta\cos\theta)\sin\theta}{(\csc\theta\sin\theta)\cos\theta + (\sec\theta\cos\theta)\sin\theta}$$

Don't forget that $\sin\theta = 1/\csc\theta$ means $\sin\theta\csc\theta = 1$.

$$\boldsymbol{\frac{\cos\theta - \sin\theta}{\cos\theta + \sin\theta} = \frac{\cos\theta - \sin\theta}{\cos\theta + \sin\theta}}$$

In the last example, you may think the choice of $\sin\theta\cos\theta$ was a "miracle" as in the Harris cartoon. "Then a miracle occurs"; unfortu-

Courtesy of Sidney Harris, from *What's So Funny About Science?* (William Kaufman, Inc., Los Altos, Ca. 1978).

nately, it is not a valid step. But if you thought the last solution was a miracle, remember that these are examples and could be worked any number of ways. Use the examples as you use the suggestions, as *one* model of many that might be shown. Trust your instincts.

EXAMPLES

6. Try Example 5 again.

$$\frac{\cos\theta - \sin\theta}{\cos\theta + \sin\theta} = \frac{\csc\theta - \sec\theta}{\csc\theta + \sec\theta}$$

$$= \frac{(1/\sin\theta) - (1/\cos\theta)}{(1/\sin\theta) + (1/\cos\theta)}$$

$$= \frac{(\cos\theta - \sin\theta)/(\sin\theta\cos\theta)}{(\cos\theta + \sin\theta)/(\sin\theta\cos\theta)}$$

Now you can appreciate the choice of $\sin\theta\cos\theta$ in Example 5.

$$\boldsymbol{\frac{\cos\theta - \sin\theta}{\cos\theta + \sin\theta} = \frac{\cos\theta - \sin\theta}{\cos\theta + \sin\theta}}$$

7. $\dfrac{1 + \sin x}{\cos x} = \dfrac{\cos x}{1 - \sin x}$

Of the two, the more complicated denominator makes the right side more complex.

$$\frac{1 + \sin x}{\cos x} = \frac{\cos x}{1 - \sin x}$$

$$= \frac{\cos x}{1 - \sin x}\frac{1 + \sin x}{1 + \sin x} \qquad \text{Try the conjugate!}$$

$$= \frac{\cos x(1 + \sin x)}{1 - \sin^2 x} \qquad \text{Pythagorean.}$$

$$= \frac{\cos x(1 + \sin x)}{\cos^2 x} \qquad \text{Reduce.}$$

$$\boldsymbol{\frac{1 + \sin x}{\cos x} = \frac{1 + \sin x}{\cos x}}$$

It may be helpful to review the examples of Section 2.3 before attempting the exercises.

exercise set 4.1

A. exercises

Prove the identities in Exercises 1–40.

See Examples 1 and 2.

1. $\cos\theta + \tan\theta \sin\theta = \sec\theta$
2. $\sin\theta + \cot\theta \cos\theta = \csc\theta$
3. $1 - 2\sin^2\lambda = 2\cos^2\lambda - 1$
4. $(\cos\lambda - \sin\lambda)^2 = 1 - 2\cos\lambda \sin\lambda$
5. $\cot x + \tan x = \cot x \sec^2 x$
6. $\cos^2 x \csc x + \sin x = \csc x$
7. $(1 - \cos\alpha)(\csc\alpha + \cot\alpha) = \sin\alpha$
8. $(\sec\alpha - \tan\alpha)(\csc\alpha + 1) = \cot\alpha$
9. $\cot^2\beta + \cos^2\beta + \sin^2\beta = \csc^2\beta$
10. $\sin^2\beta + \tan^2\beta + \cos^2\beta = \sec^2\beta$

B. exercises

See Examples 3–7.

11. $\dfrac{\tan A - \cot A}{\tan A + \cot A} = 2\sin^2 A - 1$
12. $\dfrac{1 - \tan^2 B}{1 + \tan^2 B} = 1 - 2\sin^2 B$
13. $\dfrac{\sin x}{\csc x - \cot x} = 1 + \cos x$
14. $\dfrac{1}{\csc x - \sin x} = \tan x \sec x$
15. $\dfrac{\cot^2\alpha - 1}{1 + \cot^2\alpha} = 2\cos^2\alpha - 1$
16. $\dfrac{1 - \tan^2\alpha}{\tan^2\alpha + 1} = \cos^2\alpha - \sin^2\alpha$
17. $\dfrac{1 + \csc\beta}{\csc\beta - 1} = \dfrac{1 + \sin\beta}{1 - \sin\beta}$
18. $\dfrac{\sin\beta + \cos\beta}{\cos\beta - \sin\beta} = \dfrac{\cot\beta + 1}{\cot\beta - 1}$
19. $\dfrac{\sin A}{1 + \cos A} = \dfrac{1 - \cos A}{\tan A \cos A}$
20. $\dfrac{1 + \cot A}{\sin A} = \dfrac{1 - \cot^2 A}{\sin A - \cos A}$
21. $2\sec^2\theta = \dfrac{1}{1 + \sin\theta} + \dfrac{1}{1 - \sin\theta}$
22. $2\sec^2\theta = \dfrac{1}{1 - \cos\theta} + \dfrac{1}{1 + \cos\theta}$

23. $\dfrac{\cos\lambda}{1-\csc\lambda}+\dfrac{\cos\lambda}{1+\csc\lambda}=-2\sin\lambda\tan\lambda$

24. $\dfrac{\cos\lambda}{\csc\lambda+1}+\dfrac{\cos\lambda}{\csc\lambda-1}=2\tan\lambda$

25. $\dfrac{1}{\cot x}+\dfrac{\cos x}{1+\sin x}=\sec x$

26. $\dfrac{\sin x}{1+\cos x}+\dfrac{1+\cos x}{\sin x}=2\csc x$

27. $\dfrac{\sec\alpha}{\cos\alpha}-\dfrac{\tan\alpha}{\cot\alpha}=1$

28. $\dfrac{\cos\alpha}{\sec\alpha}+\dfrac{\sin\alpha}{\csc\alpha}=1$

29. $\dfrac{\csc\beta}{\sin\beta}-\dfrac{\cot\beta}{\tan\beta}=1$

30. $\dfrac{\sec\beta}{\sin\beta}-\dfrac{\sin\beta}{\cos\beta}=\cot\beta$

31. $\dfrac{\sin^4\alpha-\cos^4\alpha}{\sin^2\alpha-\cos^2\alpha}=1$

32. $\dfrac{\csc^4\alpha-1}{\csc^2\alpha-1}=2+\cot^2\alpha$

33. $\dfrac{\sin^3\beta+\sin\beta\cos^2\beta}{\cos\beta}=\tan\beta$

34. $\tan\beta\sec^2\beta-\tan^3\beta=\tan\beta$

35. $\cos^4 x-\sin^4 x=\cos^2 x-\sin^2 x$

36. $\sec^4 x-\tan^4 x=\sec^2 x+\tan^2 x$

37. $\dfrac{\tan^3\theta-1}{\tan\theta-1}=\sec^2\theta+\tan\theta$

38. $\dfrac{1+\cot^3\theta}{1+\cot\theta}=\csc^2\theta-\cot\theta$

39. $\sin^6\theta+\cos^6\theta=1-3\sin^2\theta\cos^2\theta$

40. $\sec^6\theta-\tan^6\theta=1+3\sec^2\theta\tan^2\theta$

C. exercises

41. Prove the determinant identity:

$$\begin{vmatrix}\tan\theta & \sec\theta\\ \sec\theta & \tan\theta\end{vmatrix}=\begin{vmatrix}\cot\theta & \csc\theta\\ \csc\theta & \cot\theta\end{vmatrix}$$

42. Prove the matrix identity:

$$\begin{bmatrix}\cos\theta & \sin\theta\\ \sec\theta & \csc\theta\end{bmatrix}\begin{bmatrix}\cos\theta\\ \sin\theta\end{bmatrix}=\begin{bmatrix}1\\ 2\end{bmatrix}$$

43. Prove the logarithmic identity:

$$\log(\tan \alpha) + \log(\sec \alpha) = \log(\sin \alpha) - 2 \log(\cos \alpha)$$

44. **brain teaser** Show that the set of points (x, y) that satisfy the following system of equations is a unit circle with center at the origin.

$$\begin{cases} x = \sin \theta \\ y = \cos \theta \end{cases}$$

4.2 ADDITION FORMULAS

Certain fundamental identities were examined in Chapter 2 and reviewed in the previous section. Those expressions dealt exclusively with one angle. It is natural to wonder at this point if there are identities that relate the functions of more than one angle. The interdependence of these functions has already been seen in the fundamental identities. We now encounter some extremely striking and particularly important expressions involving two angles.

Consider an angle γ, in standard position and a unit circle as shown in Figure 4.1. If d is the distance AB, then, using the distance formula:

$$\begin{aligned} d &= \sqrt{(\cos \gamma - 1)^2 + (\sin \gamma - 0)^2} \\ d^2 &= \cos^2 \gamma - 2 \cos \gamma + 1 + \sin^2 \gamma \\ &= \cos^2 \gamma + \sin^2 \gamma + 1 - 2 \cos \gamma \\ &= 1 + 1 - 2 \cos \gamma \\ &= 2 - 2 \cos \gamma \end{aligned}$$

Figure 4.1

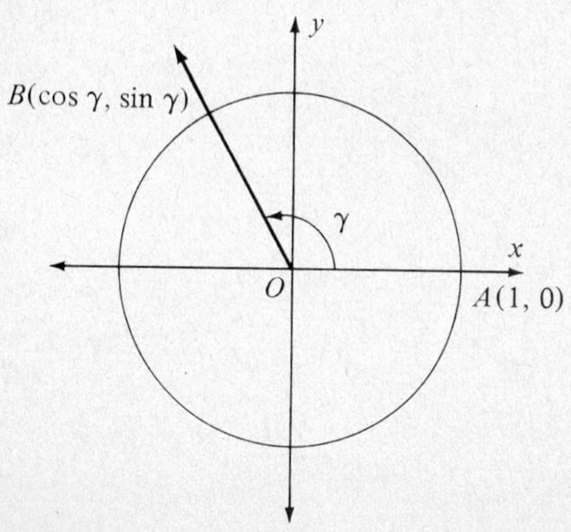

Figure 4.2

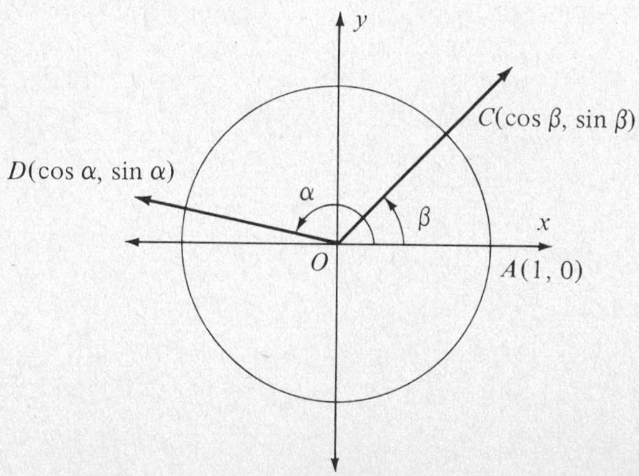

Now if γ is rotated to a new position, as shown in Figure 4.2, $AB = d = CD$, *where* $\gamma = \alpha - \beta = \angle COD = \angle AOB$. The distance d can then be found using α and β.

$$d = \sqrt{(\cos \alpha - \cos \beta)^2 + (\sin \alpha - \sin \beta)^2}$$

$$d^2 = \cos^2 \alpha - 2 \cos \alpha \cos \beta + \cos^2 \beta + \sin^2 \alpha - 2 \sin \alpha \sin \beta + \sin^2 \beta$$

$$= (\cos^2 \alpha + \sin^2 \alpha) + (\cos^2 \beta + \sin^2 \beta) - 2(\cos \alpha \cos \beta + \sin \alpha \sin \beta)$$

$$= 2 - 2(\cos \alpha \cos \beta + \sin \alpha \sin \beta)$$

Because the distance has been found in two different ways, the two expressions for d^2 may be equated.

$$2 - 2 \cos \gamma = 2 - 2(\cos \alpha \cos \beta + \sin \alpha \sin \beta)$$

$$-2 \cos \gamma = -2(\cos \alpha \cos \beta + \sin \alpha \sin \beta)$$

$$\cos \gamma = \cos \alpha \cos \beta + \sin \alpha \sin \beta$$

(1) $$\boldsymbol{\cos (\alpha - \beta) = \cos \alpha \cos \beta + \sin \alpha \sin \beta}$$

One immediate application of this identity is to find the exact value of the cosine of certain angles, as shown in the following example.

EXAMPLES

1. Find the *exact* value of cos 15°.

$$\cos 15° = \cos(60° - 45°)$$

$$= \cos 60° \cos 45° + \sin 60° \sin 45°$$

$$= \frac{1}{2} \cdot \frac{1}{\sqrt{2}} + \frac{\sqrt{3}}{2} \cdot \frac{1}{\sqrt{2}}$$

$$= \frac{1}{2\sqrt{2}} + \frac{\sqrt{3}}{2\sqrt{2}} = \frac{1 + \sqrt{3}}{2\sqrt{2}}$$

$$\boldsymbol{\cos 15° = \frac{\sqrt{2} + \sqrt{6}}{4}}$$

2. Derive an identity for $\cos(x - \pi)$.
   Use (1) and let $\alpha = x$ and $\beta = \pi$.

$$\cos(x - \pi) = \cos x \cos \pi + \sin x \sin \pi$$

$$= \cos x(\mathbf{-1}) + \sin x(\mathbf{0})$$

$$\boldsymbol{\cos(x - \pi) = -\cos x}$$

Example 2 illustrates how $\cos(\alpha - \beta)$ can be used to derive other identities. Such a derivation emphasizes the fact that these identities are true for any choice of angles. In the following, note how the careful choice of replacements for α and β lead to a series of interesting results.

First, using (1), let $\alpha = \pi/2$ and $\beta = \theta$.

$$\cos\left(\frac{\pi}{2} - \theta\right) = \cos\frac{\boldsymbol{\pi}}{\mathbf{2}} \cos\theta + \sin\frac{\boldsymbol{\pi}}{\mathbf{2}} \sin\theta$$

$$= \mathbf{0} \cdot \cos\theta + \mathbf{1} \cdot \sin\theta$$

$$\textbf{(2)} \quad \mathbf{cos}\left(\frac{\boldsymbol{\pi}}{\mathbf{2}} - \boldsymbol{\theta}\right) = \mathbf{sin}\ \boldsymbol{\theta}$$

Again using (1), let $\alpha = \pi/2$ and $\beta = \pi/2 - \theta$.

$$\cos\left[\frac{\pi}{2} - \left(\frac{\pi}{2} - \theta\right)\right] = \cos\frac{\boldsymbol{\pi}}{\mathbf{2}} \cos\left(\frac{\pi}{2} - \theta\right) + \sin\frac{\boldsymbol{\pi}}{\mathbf{2}} \sin\left(\frac{\pi}{2} - \theta\right)$$

$$\cos\left[\left(\frac{\pi}{2} - \frac{\pi}{2}\right) + \theta\right] = \mathbf{0} \cdot \cos\left(\frac{\pi}{2} - \theta\right) + \mathbf{1} \cdot \sin\left(\frac{\pi}{2} - \theta\right)$$

$$\textbf{(3)} \quad \mathbf{cos}\ \boldsymbol{\theta} = \mathbf{sin}\left(\frac{\boldsymbol{\pi}}{\mathbf{2}} - \boldsymbol{\theta}\right)$$

Now (2) and (3) can be used to find a similar form for tangent.

$$\tan\left(\frac{\pi}{2} - \theta\right) = \frac{\sin[(\pi/2) - \theta]}{\cos[(\pi/2) - \theta]}$$

$$= \frac{\cos\theta}{\sin\theta}$$

$$\textbf{(4)} \quad \mathbf{tan}\left(\frac{\boldsymbol{\pi}}{\mathbf{2}} - \boldsymbol{\theta}\right) = \mathbf{cot}\ \boldsymbol{\theta}$$

This group of results is referred to as **phase identities,** because of its graphic interpretation.

PHASE IDENTITIES

$$\sin\left(\frac{\pi}{2} - \theta\right) = \cos\theta$$

$$\cos\left(\frac{\pi}{2} - \theta\right) = \sin\theta$$

$$\tan\left(\frac{\pi}{2} - \theta\right) = \cot\theta$$

Returning to (1), let $\alpha = 0$ and $\beta = \theta$.

$$\cos(\mathbf{0} - \theta) = \cos \mathbf{0} \cos \theta + \sin \mathbf{0} \sin \theta$$

$$\cos(-\theta) = \mathbf{1} \cdot \cos \theta + \mathbf{0} \cdot \sin \theta$$

(5) $\boldsymbol{\cos(-\theta) = \cos \theta}$

Using (2) again, let $\theta = -\theta$.

$$\begin{aligned}\sin(-\theta) &= \cos\left[\frac{\pi}{2} - (-\theta)\right] \\ &= \cos\left(\frac{\pi}{2} + \theta\right) \\ &= \cos\left(\theta + \frac{\pi}{2}\right) \\ &= \cos\left[\theta - \left(-\frac{\pi}{2}\right)\right] \\ &= \cos\theta \cos\left(-\boldsymbol{\frac{\pi}{2}}\right) + \sin\theta \sin\left(-\boldsymbol{\frac{\pi}{2}}\right) \\ &= \cos\theta(\mathbf{0}) + \sin\theta(\mathbf{-1})\end{aligned}$$

(6) $\boldsymbol{\sin(-\theta) = -\sin \theta}$

Combining (5) and (6),

$$\begin{aligned}\tan(-\theta) &= \frac{\sin(-\theta)}{\cos(-\theta)} \\ &= \frac{-\sin\theta}{\cos\theta}\end{aligned}$$

(7) $\boldsymbol{\tan(-\theta) = -\tan \theta}$

$$\sin(-\theta) = -\sin\theta$$
$$\cos(-\theta) = \cos\theta$$
$$\tan(-\theta) = -\tan\theta$$

Although it was possible to develop two groups of useful identities from the cosine of the difference of two angles, it remains to find the sum and corresponding expressions for sine.

Using (1), let $\beta = -\beta$.

$$\cos[\alpha - (-\boldsymbol{\beta})] = \cos\alpha\cos(-\boldsymbol{\beta}) + \sin\alpha\sin(-\boldsymbol{\beta})$$

$$\cos(\alpha + \beta) = \cos\alpha\,\mathbf{cos}\,\boldsymbol{\beta} + \sin\alpha(-\mathbf{sin}\,\boldsymbol{\beta})$$

(8) $\boldsymbol{\cos(\alpha + \beta) = \cos\alpha\cos\beta - \sin\alpha\sin\beta}$

Using (1) again, let $\alpha = [(\pi/2) - \alpha]$.

$$\cos\left[\left(\frac{\pi}{2} - \alpha\right) - \beta\right] = \cos\left(\frac{\pi}{2} - \alpha\right)\cos\beta + \sin\left(\frac{\pi}{2} - \alpha\right)\sin\beta$$

$$\cos\left[\frac{\pi}{2} - (\boldsymbol{\alpha} + \boldsymbol{\beta})\right] = \mathbf{sin}\,\boldsymbol{\alpha}\cos\beta + \mathbf{cos}\,\boldsymbol{\alpha}\sin\beta$$

(9) $\boldsymbol{\sin(\alpha + \beta) = \sin\alpha\cos\beta + \cos\alpha\sin\beta}$

Finally, let $\beta = -\beta$ in (9).

$$\sin\alpha + (-\beta) = \sin\alpha\cos(-\beta) + \cos\alpha\sin(-\beta)$$

$$\sin(\alpha - \beta) = \sin\alpha\cos\beta + \cos\alpha(-\sin\beta)$$

(10) $\boldsymbol{\sin(\alpha + \beta) = \sin\alpha\cos\beta - \cos\alpha\sin\beta}$

ADDITION FORMULAS

$$\sin(\alpha + \beta) = \sin\alpha\cos\beta + \cos\alpha\sin\beta$$

$$\sin(\alpha - \beta) = \sin\alpha\cos\beta - \cos\alpha\sin\beta$$

$$\cos(\alpha + \beta) = \cos\alpha\cos\beta - \sin\alpha\sin\beta$$

$$\cos(\alpha - \beta) = \cos\alpha\cos\beta + \sin\alpha\sin\beta$$

EXAMPLES

3. Find $\sin(\alpha - \beta)$ if α and β are acute, $\tan\alpha = \frac{7}{24}$, and $\tan\beta = \frac{15}{8}$.

$$\begin{aligned} c &= \sqrt{24^2 + 7^2} & k &= \sqrt{8^2 + 15^2} \\ &= \sqrt{576 + 49} & &= \sqrt{64 + 225} \\ &= \sqrt{625} & &= \sqrt{289} \\ &= 25 & &= 17 \end{aligned}$$

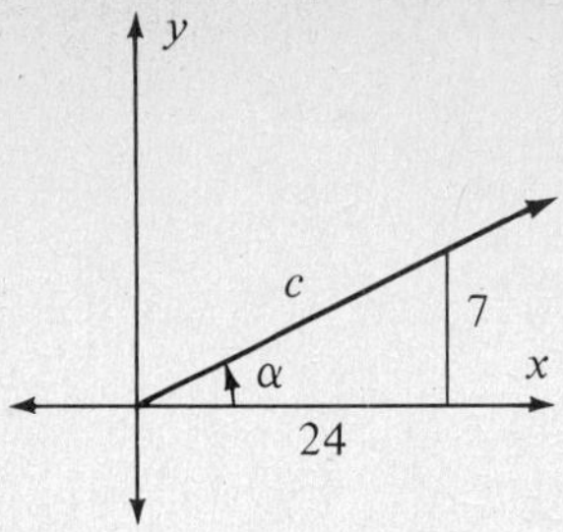

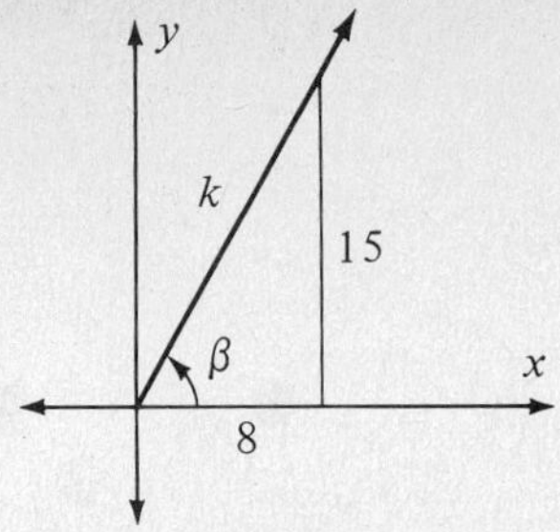

Thus

$$\sin\alpha = \tfrac{7}{25} \quad \text{and} \quad \sin\beta = \tfrac{15}{17}$$

$$\cos\alpha = \tfrac{24}{25} \qquad\qquad \cos\beta = \tfrac{8}{17}$$

$$\sin(\alpha - \beta) = \sin\beta\cos\beta - \sin\beta\cos\alpha$$

$$= (\tfrac{7}{25})(\tfrac{8}{17}) - (\tfrac{15}{17})(\tfrac{24}{25}) = \tfrac{56}{425} - \tfrac{360}{425}$$

$$\mathbf{\sin(\alpha - \beta) = -\tfrac{304}{425}}$$

4. $\sin A = \frac{1}{3}$, $\cos B = \frac{1}{4}$, A in QII and B in QI. Find $\cos(A + B)$.

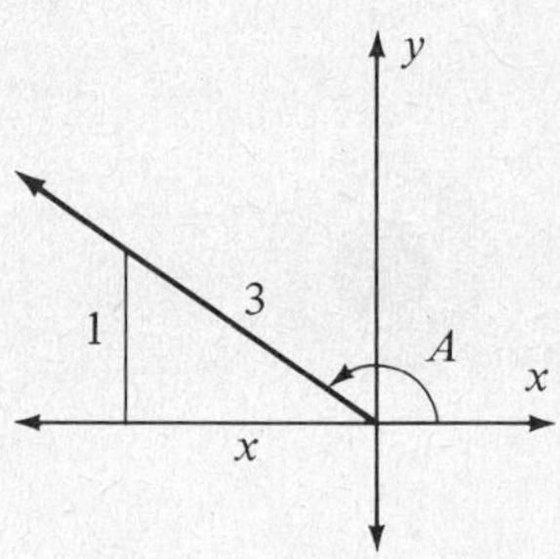

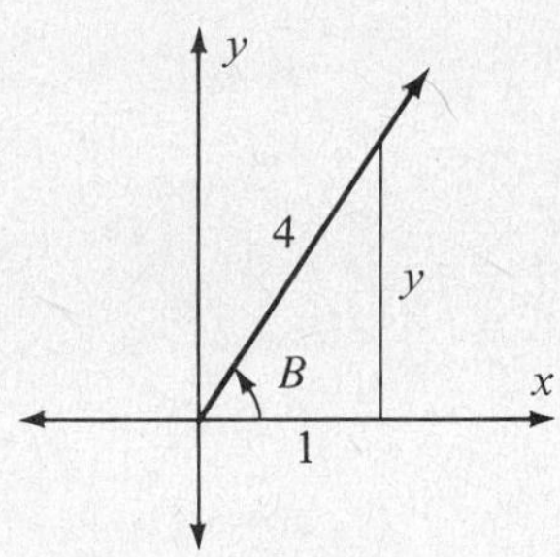

$$x = -\sqrt{3^2 - 1^2} \qquad\qquad y = \sqrt{4^2 - 1^2}$$

$$= -\sqrt{8} = -2\sqrt{2} \qquad\qquad = \sqrt{15}$$

$$\cos A = \frac{-2\sqrt{2}}{3} \qquad\qquad \sin B = \frac{\sqrt{15}}{4}$$

$$\cos(A + B) = \cos A\cos B - \sin A\sin B$$

$$= \left(\frac{-2\sqrt{2}}{3}\right)\left(\frac{1}{4}\right) - \left(\frac{1}{3}\right)\left(\frac{\sqrt{15}}{4}\right)$$

$$\mathbf{\cos(A + B) = \frac{-2\sqrt{2} - \sqrt{15}}{12}}$$

5. Prove that $\tan x + \tan y = \sin(x + y)/(\cos x \cos y)$.

$$\tan x + \tan y = \frac{\sin(x + y)}{\cos x \cos y}$$

$$= \frac{\sin x \cos y + \sin y \cos x}{\cos x \cos y}$$

$$= \frac{\sin x \cos y}{\cos x \cos y} + \frac{\sin y \cos x}{\cos x \cos y}$$

$$= \frac{\sin x}{\cos x} + \frac{\sin y}{\cos y}$$

$$\mathbf{\tan x + \tan y = \tan x + \tan y}$$

6. Prove that $\sin(A + B) \cos(A - B) = \sin A \cos A + \sin B \cos B$.

$\sin A \cos A + \sin B \cos B =$

$$= \sin(A + B) \cos(A - B)$$

$$= (\sin A \cos B + \cos A \sin B)(\cos A \cos B + \sin A \sin B)$$

$$= \sin A \cos A \cos^2 B + \cos^2 A \sin B \cos B + \sin^2 A \sin B \cos B + \sin A \sin^2 B \cos A$$

$$= \sin A \cos A(\cos^2 B + \sin^2 B) + \sin B \cos B(\sin^2 A + \cos^2 A)$$

$$\mathbf{= \sin A \cos A + \sin B \cos B}$$

Application to Graphing

Combining functions by addition of y coordinates, functions were encountered of the form:

$$a_1 \cos Bx + a_2 \sin Bx = A \cos B(x - h)$$

That is, under certain conditions, two waves of the same frequency form another wave with that frequency. Using the newly developed addition formulas,

$$A \cos B(x - h) = A \cos(Bx - Bh)$$

$$= A(\cos Bx \cos Bh + \sin Bx \sin Bh)$$

$$= A \cos Bx \cos Bh + A \sin Bx \sin Bh$$

which is equal to

$$= a_1 \cos Bx + a_2 \sin Bx$$

if and only if $a_1 = A \cos Bh$ and $a_2 = A \sin Bh$. Squaring these expressions for a_1 and a_2, we get:

$$\begin{aligned} a_1^2 + a_2^2 &= A^2 \cos^2 Bh + A^2 \sin^2 Bh \\ &= A^2(\cos^2 Bh + \sin^2 Bh) \\ &= A^2 \\ A &= \sqrt{a_1^2 + a_2^2} \end{aligned}$$

Additionally, the ratio of a_1 and a_2 produces

$$\frac{a_2}{a_1} = \frac{A \sin Bh}{A \cos Bh} = \frac{\sin Bh}{\cos Bh} = \tan Bh$$

$a_1 \cos Bx + a_2 \sin Bx = A \cos B(x - h)$ if and only if $A = \sqrt{a_1^2 + a_2^2}$ and $\tan Bh = a_2/a_1$.

The use of this identity may now avoid graphing by addition or ordinates if the sum is actually a simple cosine wave.

EXAMPLES

Express as a cosine if possible.

7. $y = \sin 2x + \cos 2x$

$$y = A \cos B(x - h)$$

where $A = \sqrt{1^2 + 1^2} = \sqrt{2}$, $B = 2$, and $\tan Bh = 1/1 = 1$

$$Bh = \frac{\pi}{4} \quad \text{so} \quad h = \frac{\pi}{8}$$

Many values of Bh will yield $\tan Bh = 1$; for convenience we take

$$-\frac{\pi}{2} < Bh < \frac{\pi}{2}$$

Thus

$$\mathbf{y = \sqrt{2} \cos 2\left(x - \frac{\pi}{8}\right)}$$

which can be sketched without resort to addition of ordinates.

8. $y = 3 \sin 4x - \sqrt{3} \cos 4x$

$$y = A \cos B(x - h)$$

where $A = \sqrt{2^2 + (\sqrt{3})^2} = \sqrt{4 + 3} = \sqrt{7}$, $B = 4$, and $\tan Bh = -\sqrt{3}/3$ or $-1/\sqrt{3}$. (Remember that the two functions must have the same frequency.)

$$4h = -\frac{\pi}{3}$$

$$h = -\frac{\pi}{12}$$

Therefore

$$\mathbf{y = \sqrt{7} \cos 4 \left(x + \frac{\pi}{12}\right)}$$

exercise set 4.2

A. exercises

Find A, B, or C in Exercises 1–10.

1. $\sin A = -\sin \theta$
2. $\cos B = \cos(-\theta)$
3. $\cos x = \sin C$
4. $\sin A = -\cos y$
5. $\sin B = \sin 2x \cos 3x - \sin 3x \cos 2x$

See the summary of addition formulas.

6. $\cos C = \cos 2x \cos 3x - \sin 2x \sin 3x$
7. $\cos 40° \cos 10° - \sin 40° \sin 10° = \cos A$
8. $\sin 25° \cos 65° + \sin 25° \sin 65° = \sin B$
9. $\cos A \cos 2x + \sin A \sin 2x = \cos x$
10. $\sin 3x \cos B - \sin B \cos 3x = \sin x$

Find the exact value of each function in Exercises 11–20.

11. $\cos 15°$
12. $\sin 75°$
13. $\cos 75°$
14. $\cos 165°$
15. $\sin 195°$
16. $\sin 105°$
17. $\sec 255°$
18. $\csc 285°$
19. $\tan 15°$
20. $\tan 75°$

See Example 1. Hint: $165 = 315 - 150$, or reference angle for 165° is??

Obtain formulas for the functions in Exercises 21–30.

21. $\cos(\theta - 180°)$

22. $\sin(x + \pi)$

See Example 2.

23. $\sin\left(x + \frac{\pi}{6}\right)$

24. $\cos(x - 60°)$

25. $\sin(\theta + 60°)$

26. $\cos\left(x - \frac{\pi}{6}\right)$

27. $\tan(x - \pi)$

28. $\tan(x + \pi)$

29. $\cos\left(\theta + \frac{\pi}{4}\right)$

30. $\sin\left(\theta - \frac{\pi}{4}\right)$

B. exercises

Use the given information in Exercises 31–34 to find the required function.

31. $\tan \alpha = \frac{3}{4}$ and $\tan \beta = \frac{7}{24}$. Find $\sin(\alpha + \beta)$ if α and β are acute.

See Examples 3 and 4.

32. $\tan \alpha = \frac{5}{12}$ and $\tan \beta = \frac{8}{15}$. Find $\cos(\alpha - \beta)$ if α and β are acute.

33. Find $\sin(A - B)$ if A and B are QII, $\sin A = \frac{24}{25}$, and $\cos B = -\frac{15}{17}$.

34. Find $\cos(A + B)$ if A and B are in QIII, $\sin A = -\frac{12}{13}$, and $\cos B = -\frac{9}{41}$.

Prove the identities in Exercises 35–44.

35. $\cos(\alpha + \beta) + \cos(\alpha - \beta) = 2 \cos \alpha \cos \beta$

See Examples 5 and 6.

36. $\sin(\alpha + \beta) + \sin(\alpha - \beta) = 2 \sin \alpha \cos \beta$

37. $\sin(\alpha + \beta) \cos \beta - \cos(\alpha + \beta) \sin \beta = \sin \alpha$

38. $\cos(\alpha - \beta) \cos \beta - \sin(\alpha - \beta) \sin \beta = \cos \alpha$

39. $\sin(A + B) \sin(A - B) = \sin^2 A - \sin^2 B$

40. $\cos(A + B) \cos(A - B) = \cos^2 A + \sin^2 B$

41. $\tan(\gamma - \delta) - \tan(\delta - \gamma) = \dfrac{2(\tan \gamma - \tan \delta)}{1 + \tan \gamma \tan \delta}$

42. $\dfrac{\tan(\gamma + \delta) - \tan \delta}{1 + \tan(\gamma + \delta) \tan \delta} = \tan \gamma$

43. $\dfrac{\cos(x + h) - \cos x}{h} = \cos x \left(\dfrac{\cos h - 1}{h}\right) - \sin x \left(\dfrac{\sin h}{h}\right)$

44. $\dfrac{\sin(x + h) - \sin x}{h} = \cos x \left(\dfrac{\sin h}{h}\right) - \sin x \left(\dfrac{1 - \cos h}{h}\right)$

Express each of the expressions in Exercises 45–50 as a single cosine wave and sketch.

45. $y = \sin x - \cos x$

46. $y = \sin 2x + \cos 2x$

47. $y = \sin 3x + \sqrt{3} \cos 3x$

48. $y = \sqrt{3} \sin 5x - \cos 5x$

49. $y = 3 \sin 4x + 3 \cos 4x$

50. $y = 5 \sin 3x - 5 \cos 3x$

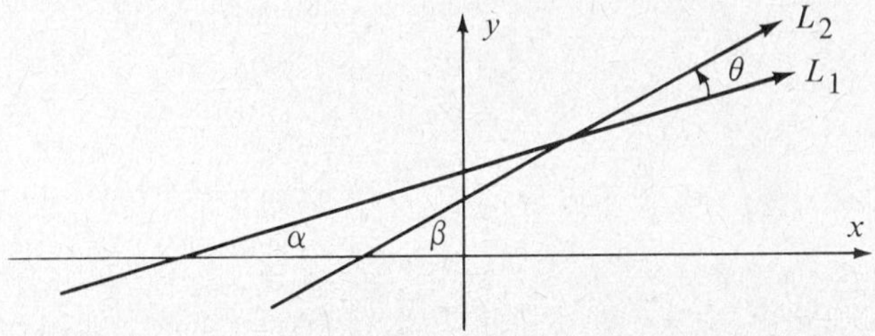

Figure 4.3

The slopes of lines L_1 and L_2 in Figure 4.3 are m_1 and m_2, respectively. Recall that the slope of a line is the tangent of the angle with the horizontal. Exercises 51–56 refer to this information.

51. Prove that if θ is the angle from L_1 to L_2, then

$$\tan \theta = \frac{m_2 - m_1}{1 + m_1 m_2}$$

52. Find the corresponding formula for the angle ϕ from L_2 to L_1.

Use the results of Exercises 51 and 52 in Exercises 53–56.

53. Find the acute angle between $x - 2y + 3 = 0$ and $3x - y - 2 = 0$.

54. Find the acute angle between $x - 3y + 2 = 0$ and $2x - y + 3 = 0$.

55. Find the obtuse angle between $y = \sqrt{3}x + \sqrt{2}$ and $\sqrt{3}y - x = 1$.

56. Find the obtuse angle between $y = \sqrt{3}x + 2$ and $\sqrt{3}x + y = 1$.

Exercises 57 and 58 refer to the statement from Chapter 2 that a trigonometric ratio of an acute angle is equal to the coratio of the complement of that angle.

57. Write this relationship for sine and prove it.

58. Express this relationship for tangent and prove it.

Prove the tangent formulas in Exercises 59 and 60.

59. $\tan(x + y) = \dfrac{\tan x + \tan y}{1 - \tan x \tan y}$

60. $\tan(x - y) = \dfrac{\tan x - \tan y}{1 + \tan x \tan y}$

C. exercises

61. Prove the matrix identity:

$$\begin{bmatrix} \cos\alpha & \sin\alpha \\ -\sin\alpha & \cos\alpha \end{bmatrix}\begin{bmatrix} \cos\beta & \sin\beta \\ -\sin\beta & \cos\beta \end{bmatrix} = \begin{bmatrix} \cos(\alpha+\beta) & \sin(\alpha+\beta) \\ -\sin(\alpha+\beta) & \cos(\alpha+\beta) \end{bmatrix}$$

62. Prove the matrix identity:

$$\begin{bmatrix} \sin A & \cos A \\ \cos A & \sin A \end{bmatrix}\begin{bmatrix} \sin B & \cos B \\ \cos B & -\sin B \end{bmatrix} = \begin{bmatrix} \cos(A-B) & \sin(A-B) \\ \sin(A+B) & \cos(A+B) \end{bmatrix}$$

4.3 MULTIPLE ANGLE FORMULAS

The addition formulas open a whole new direction. Here we take the addition formulas and let the two angles be the same to get the double-angle formulas.

$$\sin(A + B) = \sin A \cos B + \sin B \cos A$$

$$\sin(x + x) = \sin x \cos x + \sin x \cos x$$

$$\mathbf{\sin 2x = 2 \sin x \cos x}$$

This can now be repeated for cosine and tangent.

$$\cos(A + B) = \cos A \cos B - \sin A \sin B$$

$$\cos(x + x) = \cos x \cos x - \sin x \sin x$$

$$\mathbf{\cos 2x = \cos^2 x - \sin^2 x}$$

$$\tan(A + B) = \frac{\tan A + \tan B}{1 - \tan A \tan B}$$

$$\tan(x + x) = \frac{\tan x + \tan x}{1 - \tan x \tan x}$$

$$\mathbf{\tan 2x = \frac{2 \tan x}{1 - \tan^2 x}}$$

HISTORICAL ASIDE

Johann Müller (1436–1476) was also known by the Latin name Regiomontanus. He was the first European to write a treatise on trigonometry independent of astronomy. A twentieth-century commentator, F. Rudio, observes that: "The silent work of the great Regiomontanus in his chamber at Nuremberg . . . made possible the discovery of America by Columbus." Curiously, the feat that gained him admiration in his time was not his mathematics. He designed a mechanical eagle to honor Emperor Maximilian I. The eagle beat its wings and was so ingenuious that it was considered one of the wonders of the time.

EXAMPLES

Express each as a single function of one angle.

1. $2 \sin 2\alpha \cos 2\alpha$

$$2 \sin 2\alpha \cos 2\alpha = \sin 2(2\alpha)$$

$$\mathbf{2 \sin 2\alpha \cos 2\alpha = \sin 4\alpha}$$

2. $3 \sin^2 3\beta - 3 \cos^2 3\beta$

$$\begin{aligned} 3 \sin^2 3\beta - 3 \cos^2 3\beta &= -3(\cos^2 3\beta - \sin^2 3\beta) \\ &= -3[\cos 2(3\beta)] \end{aligned}$$

$$\mathbf{3 \sin^2 3\beta - 3 \cos^2 3\beta = -3 \cos 6\beta}$$

3. $\dfrac{\tan \theta/2}{1 - \tan^2 \theta/2}$

$$\begin{aligned} \frac{\tan \theta/2}{1 - \tan^2 \theta/2} &= \frac{1}{2}\left(\frac{2 \tan \theta/2}{1 - \tan^2 \theta/2}\right) \\ &= \frac{1}{2} \tan 2\left(\frac{\theta}{2}\right) \\ &= \mathbf{\frac{1}{2} \tan \theta} \end{aligned}$$

The **double-angle formulas** can be used to find functions of an angle if something is known about one-half of the angle.

EXAMPLES

4. Find sin 60° and cos 60° using the functions of 30°.

$$\begin{aligned} \sin 60^\circ &= \sin 2(30^\circ) \\ &= 2 \sin 30^\circ \cos 30^\circ \\ &= 2\left(\frac{1}{2}\right)\left(\frac{\sqrt{3}}{2}\right) \\ &= \mathbf{\frac{\sqrt{3}}{2}} \end{aligned}$$

$$\begin{aligned} \cos 60^\circ &= \cos 2(30^\circ) \\ &= \cos^2 30^\circ - \sin^2 30^\circ \\ &= \left(\frac{\sqrt{3}}{2}\right)^2 - \left(\frac{1}{2}\right)^2 \end{aligned}$$

$$= \frac{3}{4} - \frac{1}{4} = \frac{2}{4}$$

$$\mathbf{= \frac{1}{2}}$$

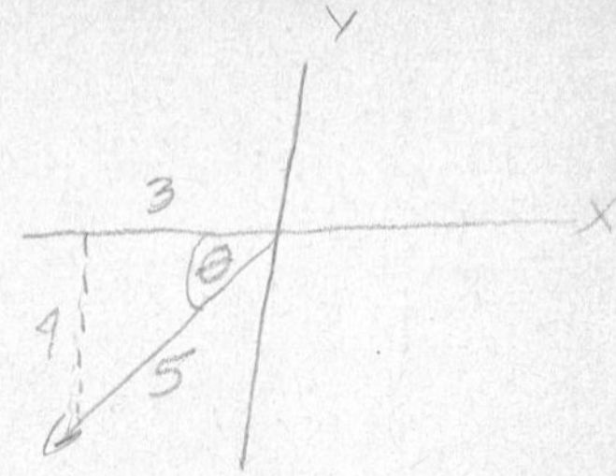

5. Find the functions of 2θ if $\sin \theta = -\frac{4}{5}$ and θ is in QIII.
   First, $\cos \theta = -\frac{3}{5}$ and $\tan \theta = \frac{4}{3}$ in QIII. (Make a sketch!)

$$\sin 2\theta = 2 \sin \theta \cos \theta \qquad \csc 2\theta = \frac{1}{\sin 2\theta}$$

$$= 2(-\tfrac{4}{5})(-\tfrac{3}{5}) \qquad \mathbf{= \tfrac{25}{24}}$$

$$\mathbf{= \tfrac{24}{25}}$$

(Notice that the fundamental identities are constantly useful.)

$$\cos 2\theta = \cos^2 \theta - \sin^2 \theta \qquad \sec 2\theta = \frac{1}{\cos 2\theta}$$

$$= (-\tfrac{3}{5})^2 - (-\tfrac{4}{5})^2 \qquad \mathbf{= -\tfrac{25}{7}}$$

$$= \tfrac{9}{25} - \tfrac{16}{25}$$

$$\mathbf{= -\tfrac{7}{25}}$$

$$\tan 2\theta = \frac{\sin 2\theta}{\cos 2\theta} \qquad \cot 2\theta = \frac{1}{\tan 2\theta}$$

$$= \frac{\frac{24}{25}}{-\frac{7}{25}} \qquad \mathbf{= -\tfrac{7}{24}}$$

$$\mathbf{= -\tfrac{24}{7}}$$

Now, $\cos 2x$ can be written in terms of $\cos x$ or $\sin x$ as needed. This gives you greater flexibility when using these expressions.

$$\cos 2x = \cos^2 x - \sin^2 x$$

$$= \cos^2 x - (1 - \cos^2 x)$$

$$= \cos^2 x - 1 + \cos^2 x$$

$$\mathbf{\cos 2x = 2 \cos^2 x - 1}$$

These three forms of $\cos 2x$ are used interchangeably, because at one time you'll need cosines and at other times sines.

or

$$= 2(1 - \sin^2 x) - 1$$

$$= 2 - 2 \sin^2 x - 1$$

$$\mathbf{\cos 2x = 1 - 2 \sin^2 x}$$

These double-angle formulas may be extended to produce formulas for any multiple of the angle, although the double angle is used more frequently.

EXAMPLES

6. Derive a formula for $\sin 3\alpha$ in terms of $\sin \alpha$.

$$\begin{aligned}\sin 3\alpha &= \sin(\alpha + 2\alpha)\\ &= \sin \alpha \cos 2\alpha + \sin 2\alpha \cos \alpha\\ &= \sin \alpha(1 - 2\sin^2 \alpha) + (2 \sin \alpha \cos \alpha)\cos \alpha\\ &= \sin \alpha - 2\sin^3 \alpha + 2 \sin \alpha \cos^2 \alpha\\ &= \sin \alpha - 2\sin^3 \alpha + 2\sin \alpha(1 - \sin^2 \alpha)\\ &= \sin \alpha - 2\sin^3 \alpha + 2 \sin \alpha - 2\sin^3 \alpha\end{aligned}$$

$$\boldsymbol{\sin 3\alpha = 3 \sin \alpha - 4 \sin^3 \alpha}$$

7. Derive a formula for $\sin 5\beta$ in terms of $\sin \beta$.

$$\begin{aligned}\sin 5\beta &= \sin(2\beta + 3\beta)\\ &= \sin 2\beta \cos 3\beta + \sin 3\beta \cos 2\beta\\ &= (2 \sin \beta \cos \beta)(4\cos^3 \beta - 3 \cos \beta)^* + (3 \sin \beta - 4 \sin^3 \beta)(1 - 2\sin^2 \beta)\\ &= 2 \sin \beta \cos^2 \beta(4 \cos^2 \beta - 3) + 3 \sin \beta - 10 \sin^3 \beta + 8 \sin^5 \beta\\ &= 2 \sin \beta(1 - \sin^2 \beta)(4 - 4\sin^2 \beta - 3) + 3 \sin \beta - 10 \sin^3 \beta + 8 \sin^5 \beta\\ &= 2 \sin \beta(1 - 5 \sin^2 \beta + 4 \sin^3 \beta) + 3 \sin \beta - 10 \sin^3 \beta + 8 \sin^5 \beta\\ &= 2 \sin \beta - 10 \sin^3 \beta + 8 \sin^5 \beta + 3 \sin \beta - 10 \sin^3 \beta + 8 \sin^5 \beta\end{aligned}$$

$$\boldsymbol{\sin 5\beta = 5 \sin \beta - 20 \sin^3 \beta + 16 \sin^5 \beta}$$

8. Prove that $2 \sin 4\delta \cos 3\delta = \sin 7\delta + \sin \delta$.

Choose the right side because 7δ is the greatest multiple. Keeping an eye on the other side, is there a way to change 7δ and δ into 3δ and 4δ?

$$\begin{aligned}2 \sin 4\delta \cos 3\delta &= \sin(3\delta + 4\delta) + \sin(4\delta - 3\delta)\\ &= \boldsymbol{\sin 3\delta \cos 4\delta} + \sin 4\delta \cos 3\delta\\ &\quad + \sin 4\delta \cos 3\delta - \boldsymbol{\sin 3\delta \cos 4\delta}\end{aligned}$$

$$\boldsymbol{2 \sin 4\delta \cos 3\delta = 2 \sin 4\delta \cos 3\delta}$$

* The identity for $\cos 3\beta$ will be proved in the exercises.

Half-angle Formulas

It is also convenient to be able to express the functions of half an angle in terms of functions of the full angle. To obtain these formulas we simply substitute $x/2$ for θ in $\cos 2\theta$.

$$\cos 2\theta = 1 - 2\sin^2\theta$$

$$\cos 2\left(\frac{x}{2}\right) = 1 - 2\sin^2\frac{x}{2}$$

$$\cos x = 1 - 2\sin^2\frac{x}{2}$$

$$2\sin^2\frac{x}{2} = 1 - \cos x$$

$$\sin^2\frac{x}{2} = \frac{1 - \cos x}{2}$$

$$\mathbf{\sin\frac{x}{2} = \pm\sqrt{\frac{1 - \cos x}{2}}}$$

The sign of the radical will depend on the quadrant of $x/2$. Similarily,

$$\mathbf{\cos\frac{x}{2} = \pm\sqrt{\frac{1 + \cos x}{2}}}$$

$$\mathbf{\tan\frac{x}{2} = \pm\sqrt{\frac{1 - \cos x}{1 + \cos x}}}$$

Tangent is obtained as the ratio of sine and cosine.

The tangent is obtained as the ratio of sine to cosine and by rationalizing the denominator or the numerator, and can also be written as

$$\tan\frac{x}{2} = \frac{1 - \cos x}{\sin x} \quad \text{or} \quad \tan\frac{x}{2} = \frac{\sin x}{1 + \cos x}$$

EXAMPLES

9. Use the functions of 30° to find cos 15°.

$$\cos 15^\circ = \cos\left(\frac{30^\circ}{2}\right)$$

$$= \sqrt{\frac{1 + \cos 30^\circ}{2}}$$

Radical is positive because 15° is in QI.

$$= \sqrt{\frac{1 + \frac{\sqrt{3}}{2}}{2}}$$

$$\mathbf{\cos 15^\circ = \sqrt{\frac{2 + \sqrt{3}}{4}}}$$

Compare with Section 1.3, Exercise 80, and Section 4.2, Exercise 11.

10. Prove that $8 \sin^4 \theta/2 = 3 - 4 \cos \theta + \cos 2\theta$.

$$\begin{aligned} 3 - 4 \cos \theta + \cos 2\theta &= 8 \sin^4 \frac{\theta}{2} \\ &= 8\left(\pm\sqrt{\frac{1 - \cos \theta}{2}}\right)^4 \\ &= 8\left(\frac{1 - \cos \theta}{2}\right)^2 \\ &= 2(1 - 2 \cos \theta + \cos^2 \theta) \\ &= 2 - 4 \cos \theta + 2 \cos^2 \theta \\ &= (2 + \mathbf{1}) - 4 \cos \theta + (2 \cos^2 \theta - \mathbf{1}) \end{aligned}$$

$$\mathbf{3 - 4 \cos \theta + \cos 2\theta = 3 - 4 \cos \theta + \cos 2\theta}$$

exercise set 4.3

A. exercises

Express each of the expressions in Exercises 1–10 as a single function of one angle.

See Examples 1–3.

1. $2 \sin 3A \cos 3A$
2. $6 \sin B \cos B$
3. $\cos^2 2x - \sin^2 2x$
4. $\sin^2 y - \cos^2 y$
5. $\dfrac{4 \tan \theta}{1 - \tan^2 \theta}$
6. $\dfrac{\tan 2\theta}{1 - \tan^2 2\theta}$
7. $\sin \dfrac{\alpha}{2} \cos \dfrac{\alpha}{2}$
8. $1 - 2 \sin^2 3\beta$
9. $2 \cos^2 \alpha - 1$
10. $10 \sin^2 \beta - 5$

Find the exact value of the given function in Exercises 11–16 using the specified angle.

See Examples 4 and 9.

11. $\tan 60°$ using $30°$
12. $\sin 120°$ using $60°$
13. $\cos \dfrac{\pi}{2}$ using $\dfrac{\pi}{4}$
14. $\cos \dfrac{\pi}{2}$ using π
15. $\sin 75°$ using $150°$
16. $\cos 105°$ using $210°$

Find sine, cosine, and tangent of 2θ with the information given in Exercises 17–22.

17. $\cos\theta = \frac{5}{13}$, θ is acute

18. $\tan\theta = \frac{9}{40}$, θ in QI

See Example 5.

19. $\cot\theta = -\frac{5}{12}$, $\sin\theta > 0$

20. $\sin\theta = -\frac{16}{65}$, θ in QIII

21. $\sec\theta = \frac{17}{15}$, $(3\pi/2) < \theta < 2\pi$

22. $\csc\theta = -\frac{25}{7}$, $\cos\theta > 0$

Find sine, cosine, and tangent of $\theta/2$ with the information given in Exercises 23–26.

23. $\sin\theta = -\frac{13}{85}$, $\tan\theta > 0$

24. $\cos\theta = -\frac{12}{37}$, $\tan < 0$

See Examples 5 and 9.

25. $\tan\theta = \frac{3}{4}$, $\sec\theta < 0$

26. $\tan\theta = -\frac{8}{15}$, $\csc\theta < 0$

B. exercises

Derive formulas for the functions given in Exercises 27–32 in terms of the same function of θ.

27. $\cos 3\theta$

28. $\sin 4\theta$

See Examples 6 and 7.

29. $\cos 4\theta$

30. $\cos 5\theta$

31. $\sin 6\theta$

32. $\cos 6\theta$

Prove the identities in Exercises 33–60.

33. $2\cos^2\alpha = 1 + \cos 2\alpha$

34. $2\sin^2\alpha = 1 - \cos 2\alpha$

35. $\cos 2\beta + 2\sin^2\beta = 1$

36. $\sin 2\beta \csc\beta = 2\cos\beta$

37. $\tan\frac{x}{2} = \csc x - \cot x$

38. $\frac{\sin x}{2} = \sin\frac{x}{2}\cos\frac{x}{2}$

39. $\sin\theta - \cos\theta\tan\frac{\theta}{2} = \tan\frac{\theta}{2}$

40. $\tan\frac{\theta}{2}\sin\theta = \frac{\tan\theta - \sin\theta}{\sin\theta\sec\theta}$

41. $\cot 2\lambda = \frac{\cot^2\lambda - 1}{2\cot\lambda}$

42. $\sec 2\lambda = \frac{\sec^2\lambda}{2 - \sec^2\lambda}$

43. $\cot A \sin 2A = 1 + \cos 2A$

44. $(\cos A - \sin A)^2 = 1 - \sin 2A$

45. $\cot B - \tan B = 2\cot 2B$

46. $\tan^2 B + 2\cot 2B\tan B = 1$

47. $\tan\gamma + \tan\frac{1}{2}\gamma = \csc\gamma - 2\cot 2\gamma$

48. $\tan\frac{1}{2}\gamma + 2\sin^2\gamma + \cos 2\gamma = 1 - \cot\gamma + \csc\gamma$

49. $2(\sin 2\delta\cos\delta - \cos 2\delta\sin\delta)^2 = 1 - \cos 2\delta$

50. $2(\cos 3\delta \cos \delta + \sin 3\delta \sin \delta)^2 = 1 + \cos 4\delta$

51. $\dfrac{1 - \cos 2x}{\sin 2x} = \tan x$

52. $\dfrac{2 \tan x}{1 + \tan^2 x} = \sin 2x$

53. $\dfrac{\sin 2\alpha}{\sin \alpha} - \dfrac{\cos 2\alpha}{\cos \alpha} = \sec \alpha$

54. $\dfrac{\sin 3\alpha}{\sin \alpha} - \dfrac{\cos 3\alpha}{\cos \alpha} = 2$

55. $\dfrac{\cos 2\beta - \sin 2\beta}{\sin \beta \cos \beta} = \cot \beta - \tan \beta - 2$

56. $\dfrac{1 + \tan \beta/2}{1 + \cot \beta/2} = \csc \beta + \cot \beta$

57. $\tan^2 \theta = \dfrac{\sec 2\theta - 1}{\sec 2\theta + 1}$

58. $\dfrac{1 - \cos 2\theta}{1 + \cos 2\theta} = \tan^2 \theta$

59. $\dfrac{\sin 5x}{\sin x} - \dfrac{\cos 5x}{\cos x} = 4 \cos 2x$

60. $\dfrac{\sin 3x}{\cos x} \dfrac{\cos 3x}{\sin x} = 2 \cot 2x$

C. exercises

61. Show that $\sin \theta > \frac{1}{2} \sin 2\theta$ when θ is acute.

62. If θ is acute and $\tan \theta = (n^2 - 1)/2n$, find $\sin 2\theta$ and $\cos 2\theta$.

63. Prove the matrix identity:

$$\begin{bmatrix} \cos A & \sin A \\ \sin A & \cos A \end{bmatrix} \begin{bmatrix} \sin A & \cos A \\ \cos A & -\sin A \end{bmatrix} = \begin{bmatrix} \sin 2A & \cos 2A \\ 1 & 0 \end{bmatrix}$$

4.4 TRIGONOMETRIC EQUATIONS

The statement, $2 \sin x = \sin 2x$, is *not* an identity. That is, it is true for *some* values of x but *not* for others.

If $x = 0$,	If $x = \dfrac{\pi}{4}$
$2 \sin 0 = \sin 2(0)$	$2 \sin \dfrac{\pi}{4} = \sin 2\left(\dfrac{\pi}{4}\right)$
$2 \sin 0 = \sin 0$	$2 \sin \dfrac{\pi}{4} = \sin \dfrac{\pi}{2}$
$2 \cdot 0 = 0$	$2 \cdot \dfrac{\sqrt{2}}{2} \neq 1$
True	False

Such an equality is called a **conditional equation.** The conditions under which it is true are the roots or solutions of the equation. If we sketch the graphs of $y = 2 \sin x$ and $y = \sin 2x$ as in Figure 4.4, they will intersect when $2 \sin x = \sin 2x$. Note that there are infinitely many points of intersection. Hence there are infinitely many solutions to the equation. However, over any closed interval there are a finite number, as those marked on $0 \leq x < 2\pi$ in Figure 4.4.

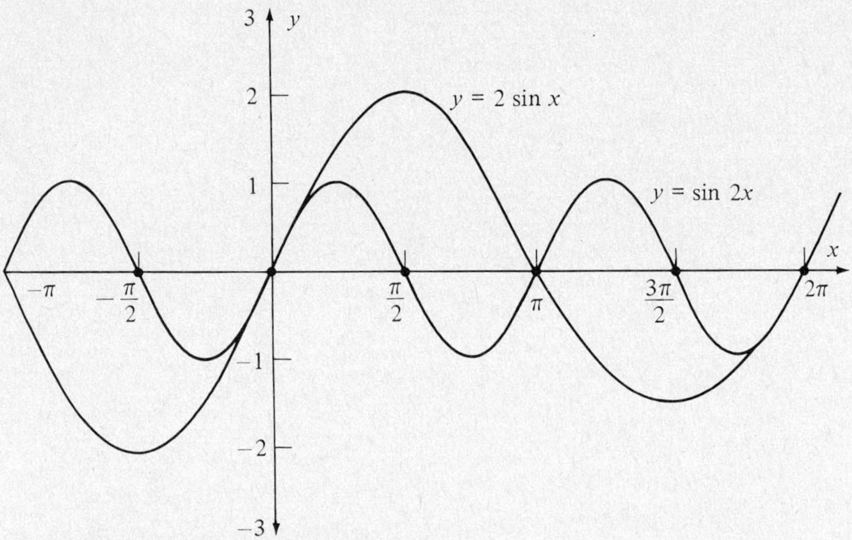

Figure 4.4 Graphs of y = 2 sin x and y = sin 2x.

Like proving identities, there is no general method for solving trigonometric equations that will work in all cases. There are only broad guidelines similar to those for proving identities. Consider the example of $2 \sin x = \sin 2x$.

EXAMPLE

1. Solve $2 \sin x = \sin 2x$.

$2 \sin x = \sin 2x$	Reduce to functions of a single angle x.
$2 \sin x = 2 \sin x \cos x$	
$2 \sin x - 2 \sin x \cos x = 0$	Isolate functions.
$2 \sin x(1 - \cos x) = 0$	Factor.
$2 \sin x = 0$ or $1 - \cos x = 0$	Set factors equal to 0.
$x = 0 + k\pi$ $\quad \cos x = 1$	Solve individually.
$= k\pi$ $\quad x = 0 + 2k\pi$	where k is an integer
$= 2k\pi$	
$\{k\pi\}$	

It is usually sufficient to find all solutions on some prespecified interval.

Notice that the first step was to get rid of the multiple angle. All angles were then the same, and at that point the equation was essentially of the form

$$\mathbf{2A = 2 \quad A \quad B}$$
$$2 \sin x = 2 \sin x \cos x$$

From this point until the values of A and B are known, the solution of $2A = 2AB$ is like any other algebraic equation. Only the first and the last steps actually involve trigonometry. Keep that in mind in the examples that follow. You use the techniques of algebraic equation solving with trigonometric functions as variables. The trigonometric identities often help to simplify the equation and allow the use of the algebraic techniques. These points are best illustrated by example.

EXAMPLES

Solve the following for all nonnegative values of the variable less than 2π.

2. $\sin \alpha = \cos \alpha$

$\dfrac{\sin \alpha}{\cos \alpha} = \dfrac{\cos \alpha}{\cos \alpha}$ Use the ratio identity to reduce to a single function.

$\tan \alpha = 1$

$\alpha = \dfrac{\pi}{4}, \dfrac{5\pi}{4}$

$\left\{\dfrac{\pi}{4}, \dfrac{5\pi}{4}\right\}$ or [45°, 225°]

3. $\sin 2\beta + 2 \sin \beta = \cos \beta + 1$

$$2 \sin \beta \cos \beta + 2 \sin \beta = \cos \beta + 1$$
$$2 \sin \beta(\cos \beta + 1) = \cos \beta + 1$$
$$2 \sin \beta(\cos \beta + 1) - (\cos \beta + 1) = 0$$
$$(2 \sin \beta - 1)(\cos \beta + 1) = 0$$

$\sin \beta = \dfrac{1}{2}$ $\qquad$ $\cos \beta = -1$

$\beta = \dfrac{\pi}{6}, \dfrac{5\pi}{6}$ $\qquad$ $\beta = \pi$

$\left\{\dfrac{\pi}{6}, \dfrac{5\pi}{6}, \pi\right\}$ or {30°, 150°, 180°}

You say that noticing the factor of $\cos \beta + 1$ in Example 3 was another "then a miracle occurs." Actually it might have been easier to notice if it looked more algebraic. Consider Example 3 again.

$$\sin 2\beta + 2 \sin \beta = \cos \beta + 1$$

$$2 \sin \beta \cos \beta + 2 \sin \beta = \cos \beta + 1$$

At this point let $x = \sin \beta$ and $y = \cos \beta$. Just seeing it in algebraic form may help!

$$\mathbf{2xy + 2x = y + 1}$$

In this form, notice that x has been isolated, so it is natural to factor out x.

$$\mathbf{2x(y + 1) = y + 1}$$

We could divide by $y + 1$, but recall that this might result in lost roots, so bring everything to the left.

$$\mathbf{2x(y + 1) - (y + 1) = 0}$$

$$\mathbf{(2x - 1)(y + 1) = 0}$$

Now x and y can be replaced by $\sin \beta$ and $\cos \beta$ and the solution continued. Let such a substitution be the very last resort. Identities and factoring also play a large part in many of the following examples.

HISTORICAL ASIDE

Leonard Euler (1707–1783) was a giant in his time. He was a prolific writer with great impact. He is responsible, as mentioned earlier, for the present use of π. Euler also introduced the notation f(x) for functions, e for the natural base, i for the imaginary unit, and Σ for summation. So prolific was he that although he died in 1793, the journal of the Petersburg Academy contained one or more of his articles down to 1818. His complete works would fill an estimated 16,000 pages. Only Lagrange can compete with Euler as the greatest mathematician of the eighteenth century

EXAMPLES

Solve the following for all nonegative values of the variable less than 360°.

4. $3 \cos \theta + \sin \theta = 1$

   $3 \cos \theta = 1 - \sin \theta$

   Now square both sides so that the Pythagorean identities can be used. However, recall that squaring may introduce extraneous roots, so the results must be checked.

$$(3 \cos \theta)^2 = (1 - \sin \theta)^2$$

$$9 \cos^2 \theta = 1 - 2 \sin \theta + \sin^2 \theta$$

$$9(1 - \sin^2 \theta) = 1 - 2 \sin \theta + \sin^2 \theta$$

$$-10 \sin^2 \theta + 2 \sin \theta + 8 = 0$$

$$5 \sin^2 \theta - \sin \theta - 4 = 0$$

$$(\sin\theta - 1)(5\sin\theta + 4) = 0$$

$$\sin\theta = 1 \qquad \sin\theta = -\frac{4}{5}$$

$$\theta = \frac{\pi}{2} \qquad \theta = \arcsin\left(-\frac{4}{5}\right)$$

or $\quad 90° \qquad \approx 233.13°$ or $306.87°$

However, 233.13° is extraneous, thus the solution set is

{90°, 306.87°}

The extraneous solution is found by checking in the original equation.

5. $2\cos 3x + 1 = 0$

$$2\cos 3x = -1$$

$$\cos 3x = -\frac{1}{2}$$

$$3x = \frac{2\pi}{3}, \frac{4\pi}{3}, \frac{8\pi}{3}, \frac{10\pi}{3}, \frac{14\pi}{3}, \frac{16\pi}{3}$$

$$x = \frac{2\pi}{9}, \frac{4\pi}{9}, \frac{8\pi}{9}, \frac{10\pi}{9}, \frac{14\pi}{9}, \frac{16\pi}{9}$$

Here we include all values of $3x$ such that when we divide by 3 we shall get *all* values of x less than 2π or 360°.

or **40°, 80°, 160°, 200°, 280°, 320°**

6. $2(\cos^2\theta + 1) = 5\cos\theta$

$$2\cos^2\theta + 2 = 5\cos\theta$$

$$2\cos^2\theta - 5\cos\theta + 2 = 0$$

$$(\cos\theta - 2)(2\cos\theta - 1) = 0$$

$$\cos\theta = 2 \qquad \cos\theta = \tfrac{1}{2}$$

No solution, because cosine never exceeds 1. $\qquad$ $\boldsymbol{\theta = 60°, 300°}$

In this example, notice that not every factor will produce solutions. The range of the trigonometric functions comes directly into play.

7. $\cos^2\alpha + 5\sin\alpha + 2 = 0$

$$(1 - \sin^2\alpha) + 5\sin\alpha + 2 = 0$$

$$\sin^2\alpha - 5\sin\alpha - 3 = 0$$

which is a nonfactorable quadratic in $\sin\alpha$; that is,

$$x^2 - 5x - 3 = 0$$

where $x = \sin \alpha$. Recall that if

$$ax^2 + bx + c = 0$$

then

$$x = \frac{-b \pm \sqrt{b^2 - 4ac}}{2a}$$

Thus the **quadratic formula** is used to solve for $\sin \alpha$.

$$\sin \alpha = \frac{5 \pm \sqrt{25 + 12}}{62}$$

$$\sin \alpha \approx 5.5414, \ -0.5414$$

5.5414 is rejected. Why?

$$\alpha \approx \mathbf{32.78°, 147.22°}$$

8. $\dfrac{2 \tan \beta}{1 - \tan^2 \beta} = 1$

If this looks familiar, you *might* think to use:

$$\tan 2\beta = 1$$

$$2\beta = 45°, 225°, 405°, 585°$$

$$\boldsymbol{\beta} = \mathbf{22°30', 112°30', 202°30', 292°30'}$$

9. *Alternative solution for Exercise 8.* If you did not notice the double-angle identity, you might begin by clearing the fraction.

$$\frac{2 \tan \beta}{1 - \tan^2 \beta} = 1$$

$$2 \tan \beta = 1 - \tan^2 \beta$$

$$\tan^2 \beta + 2 \tan \beta - 1 = 0$$

which does not factor with integers, so the quadratic formula must be used.

$$\tan \beta = \frac{-2 \pm \sqrt{4 + 4}}{2} = \frac{-2 \pm 2\sqrt{2}}{2}$$

$$\beta = -1 \pm \sqrt{2}$$

$$\beta = \arctan(-1 \pm \sqrt{2})$$

$$\boldsymbol{\beta} = \mathbf{22°30', 112°30', 202°30', 292°30'}$$

$-1 \pm \sqrt{2}$ is not a value for one of the special angles, so we use degree measure and give the result to the nearest minute. Notice that these values are identical with those of Example 8.

SUGGESTIONS FOR SOLVING TRIGONOMETRIC EQUATIONS

To simplify the trigonometric equation:

1. Be alert to the use of the identities.
2. Reduce to functions of the same angle when multiple angles are used.
3. Reduce all terms to the same function when several functions are present.
4. Consider squaring both sides to introduce Pythagorean relationships when all terms are linear.

To solve the simplified equation:

5. Bring all nonzero terms to one side and attempt to factor.
6. As a last resort, replace each distinct function by a different variable and consider the algebraic alternatives.

EXAMPLE

Solve for all nonnegative values of the variable less than 2π.

10. $\sin 5A \tan 2A = 0$

Here you might be tempted to reduce to functions of a common angle, *but* there is already a *product*.

$\sin 5A = 0$ or $\tan 2A = 0$

$$5A = 0 + \pi k \qquad\qquad 2A = 0 + \pi k$$

$$A = 0 + \frac{\pi}{5}k \qquad\qquad A = 0 + \frac{\pi}{2}k$$

$$A = 0, \frac{\pi}{5}, \frac{2\pi}{5}, \frac{3\pi}{5}, \ldots \qquad\qquad A = 0, \frac{\pi}{2}, \pi, \frac{3\pi}{2}$$

$$\left\{0, \frac{\pi}{5}, \frac{2\pi}{5}, \frac{\pi}{2}, \frac{3\pi}{5}, \frac{4\pi}{5}, \pi, \frac{6\pi}{5}, \frac{7\pi}{5}, \frac{3\pi}{2}, \frac{8\pi}{5}, \frac{9\pi}{5}\right\}$$

exercise set 4.4

A. exercises

Solve for all values of the variable in Exercises 1–10.

1. $\sin \alpha + 1 = 0$
2. $2 \sin \alpha - 1 = 0$ See Example 1.
3. $\dfrac{1 + \tan \beta}{1 - \tan \beta} = 1$
4. $\dfrac{\cos \beta - 1}{3} = \dfrac{\cos \beta - 2}{5}$

5. $4\cos^2\gamma - 3 = 0$ 6. $3\tan^2\gamma - 1 = 0$

7. $2\cos^2\delta - 1 = 0$ 8. $2\sin^2\delta - \sin\delta = 0$

9. $\sin 2\theta \sin\theta - \cos\theta = 0$ 10. $\sin^2\theta + \cos 2\theta = 0$

Solve for all nonnegative variables less than 360° in Exercises 11–20.

See Examples 4–9.

11. $\tan^2\alpha - \tan\alpha = 0$ 12. $2\cos^2\alpha + \cos\alpha = 0$

13. $\cos^2\beta - \cos\beta - 2 = 0$ 14. $\sec^2\beta - \sec\beta - 2 = 0$

15. $\tan^2\gamma + 2\tan\gamma = 3$ 16. $2\sin^2\gamma = \sin + 3$

17. $2\sec^2\delta = \tan^2\delta + 4$ 18. $\cot^2\delta = 3 - 2\csc^2\delta$

19. $\cos 2\theta - 2\sin^2\theta - 1 = \sqrt{3}$ 20. $2\cos^2\theta + \cos 2\theta - 2 = 0$

In Exercises 21–40, solve for all nonnegative values of the variable less than 2π.

See Examples 2, 3, and 10.

21. $\cos 4A = 0$ 22. $\sin 7B = 0$

23. $\tan 5C = 0$ 24. $\sec 7D = 0$

See Exercise 56 ("Trig Talk").

25. $\sin^2 2E + 2\sin 2E + 1 = 0$ 26. $\cos^2 5G + 1 = 2\cos 5G$

27. $2\cos H + \sqrt{3} = 0$ 28. $2\sin I + 1 = 0$

29. $2\cos 4L - 1 = 0$ 30. $2\sin 3M - \sqrt{3} = 0$

B. exercises

Ø is the letter "oh."

31. $2\sin^2 2N - \cos 2N = 2$ 32. $2\cos^2\frac{Ø}{3} = \sqrt{3}\cos\frac{Ø}{3}$

33. $2\cot^2 5R - \csc^2 5R = 0$ 34. $2\sin^2 3S = 3\cos 3S + 3$

35. $4\sin^2 T + 2(\sqrt{2} - \sqrt{3})\sin T = \sqrt{6}$

36. $3\sec^2 2U + (3\sqrt{2} + 2\sqrt{3})\sec 2U + 2\sqrt{6} = 0$

37. $3\sin V - \sin 3V = \frac{1}{2}$ 38. $6\cos W + 2\cos 3W + 1 = 0$

39. $(\cos 7X - 1)(\cos^2 7X - 4) = 0$

40. $4\sin^3 3Y - 4\sin^2 3Y + \sin 3Y = 0$

In Exercises 41–50, solve for the nonnegative values of the variables less than 360°. Estimate approximate values to the nearest thousandth of a degree.

See Examples 4, 7, and 9.

41. $4 \cos^2 x - \cos x = 3$

42. $\sec^2 x + 8 = 6 \sec x$

43. $3 \tan^2 \theta - 8 \tan \theta - 3 = 0$

44. $12 \sin^2 \theta + 5 \sin \theta - 2 = 0$

45. $\tan \lambda - 4 \cot \lambda = 0$

46. $2 \sec \lambda + 5 \csc \lambda = 0$

47. $\tan \gamma + 3 \cot \gamma = 4$

48. $2 \cos \gamma + \sin \gamma = 1$

49. $5 \sin^2 \delta = 2 \sin \delta + 1$

50. $\cos^2 2\delta - 2 \cos 2\delta - 1 = 0$

Exercises 51–55 deal with the following formula. If a right circular cylinder is truncated by an oblique plane, the truncation has volume V.

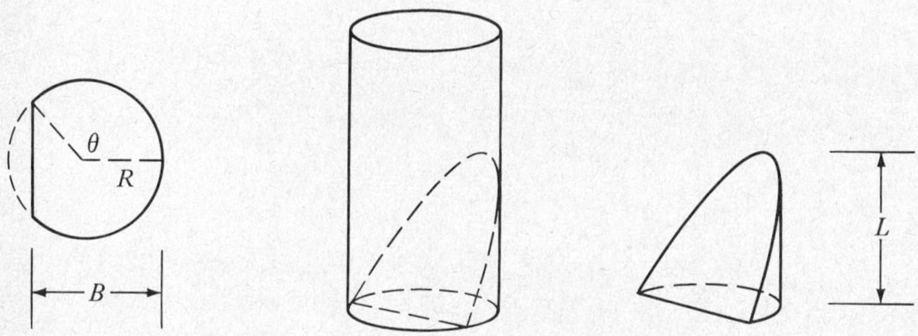

$$V = \frac{R^3 L}{B}\left[\sin \theta - \frac{\sin^3 \theta}{3} - \theta \cos \theta\right]$$

Note that θ must be in radian measure. Why?

51. **calculator problem** Find the volume of a 3-ft-tall, 100° truncation of a 1-ft, 3-in.-radius cylinder if B is approximately 2 ft $2\frac{3}{8}$ in.

52. **calculator problem** Find the volume of a 114° truncation of a cylinder with radius of 63 cm, where $L \approx 197$ cm and $B \approx 105$ cm.

C. exercises

53. **brain teaser** If the face of the cut is at 45° with the base of the cylinder, then $B = L$. Simplify the formula for this special case.

54. **brain teaser** B seems to be dependent on R and θ. Write an equivalent formula without B.

Watch out for more than one case.

55. Suppose that ϕ is the angle the face of the cut makes with the base of the cylinder. Simplify the formula, reducing the number of variables by introducing ϕ into the formula.

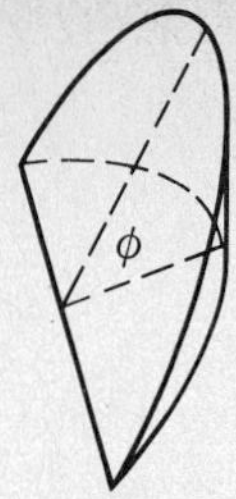

56. trig talk Use Exercises 21–40 to fill the blanks with the capital letter that corresponds to the value below the blank. Solve the equations for the *smallest positive* value of the variable. When all the blanks are filled, the quotation may be read.

“. . . $\underset{\frac{\pi}{9}}{\_\_}\ \underset{\frac{\pi}{8}}{\_\_}\ \underset{\frac{\pi}{3}}{\_\_}\ \underset{\frac{5\pi}{6}}{\_\_}\ \underset{\frac{3\pi}{4}}{\_\_}\ \underset{\frac{\pi}{9}}{\_\_}\ \underset{\frac{\pi}{8}}{\_\_}\ \underset{\frac{\pi}{3}}{\_\_}\ \underset{\frac{7\pi}{6}}{\_\_}\ \underset{\frac{\pi}{5}}{\_\_}\ \underset{\frac{\pi}{8}}{\_\_}\ \underset{\frac{\pi}{12}}{\_\_}$

$\underset{\frac{7\pi}{6}}{\_\_}\ \underset{\frac{\pi}{4}}{\_\_}\ \underset{\frac{\pi}{6}}{\_\_}\ \underset{\frac{3\pi}{4}}{\_\_}\ \underset{\frac{\pi}{4}}{\_\_}\ \underset{\frac{\pi}{3}}{\_\_}\ \underset{\frac{7\pi}{6}}{\_\_}\ \underset{\frac{\pi}{2}}{\_\_}\ \underset{\frac{\pi}{4}}{\_\_}\qquad \underset{\frac{7\pi}{6}}{\_\_}\ \underset{\frac{2\pi}{9}}{\_\_}$

$\underset{\frac{\pi}{4}}{\_\_}\ \underset{\frac{\pi}{2}}{\_\_}\ \underset{\frac{\pi}{3}}{\_\_}\qquad \underset{\frac{\pi}{20}}{\_\_}\ \underset{\frac{3\pi}{4}}{\_\_}\ \underset{\frac{\pi}{8}}{\_\_}\ \underset{\frac{2\pi}{9}}{\_\_}\ \underset{\frac{\pi}{2}}{\_\_}\ \underset{\frac{\pi}{4}}{\_\_}\ \underset{\frac{7\pi}{6}}{\_\_}\ \underset{\frac{\pi}{4}}{\_\_}\ \underset{\frac{2\pi}{5}}{\_\_}$

$\underset{\frac{\pi}{7}}{\_\_}\ \underset{\frac{3\pi}{8}}{\_\_}\ \underset{\frac{\pi}{3}}{\_\_}\qquad \underset{\frac{7\pi}{6}}{\_\_}\ \underset{\frac{\pi}{9}}{\_\_}\ \underset{\frac{\pi}{8}}{\_\_}\ \underset{\frac{2\pi}{5}}{\_\_}\ \underset{\frac{7\pi}{6}}{\_\_}\ \underset{\frac{\pi}{4}}{\_\_}\ \underset{\frac{\pi}{8}}{\_\_}\ \underset{\frac{\pi}{3}}{\_\_}\ \underset{\frac{7\pi}{6}}{\_\_}\ \underset{\frac{\pi}{2}}{\_\_}\ \underset{\frac{\pi}{4}}{\_\_}$.” (A. De Morgan)

4.5 CHAPTER REVIEW

Fill in the words or symbols that make the statements in Exercises 1–10 complete and correct.

1. (4.1) $\dfrac{\sin\theta}{\cos\theta} =$ ________ or ________.
2. (4.1) $\sin^2\theta + \cos^2\theta =$ ________.
3. (4.1) A suggested strategy for proving identities is “When in doubt, change everything to ________________ and ________________.”
4. (4.2) ________ $= \sin x \cos y + \sin y \cos x$.
5. (4.2) ________ $= \sin x \sin y + \cos x \cos y$.

6. (4.2) $a_1 \cos Bx + a_2 \sin Bx =$ ______________________________
   if and only if $A = \sqrt{a_1^2 + a_2^2}$ and $\tan Bh = a_2/a_1$.
7. (4.3) $\cos^2 \theta - \sin^2 \theta =$ __________.
8. (4.3) $\sin \dfrac{x}{2} =$ __________.
9. (4.3) __________ $= \dfrac{2 \tan A}{1 - \tan^2 A}$.
10. (4.4) One might square both sides of a trigonometric equation in order to introduce ______________________________.

Rewrite each of the expressions in Exercises 11 and 12 as a single function of one angle.

11. (4.2) $\sin 3x \cos 2x - \sin 2x \cos 3x$
12. (4.3) $2 \cos^2 3y - 1$
13. (4.2) If α is in QII, β is in QIII, $\sin \alpha = \frac{3}{5}$, and $\tan \beta = \frac{8}{15}$, find $\cos(\alpha + \beta)$.
14. (4.3) If α is acute and $\cos \alpha = \frac{12}{13}$, find $\sin \alpha/2$ and $\sin 2\alpha$.

Prove the identities in Exercises 15–19.

15. (4.1) $\dfrac{1 - \sin^2 \theta}{1 + \cot^2 \theta} = \sin^2 \theta \cos^2 \theta$
16. (4.2) $\dfrac{\sin(\alpha + \beta)}{\sin(\alpha - \beta)} = \dfrac{\tan \alpha + \tan \beta}{\tan \alpha - \tan \beta}$
17. (4.2) $\sqrt{2} \cos\left(\theta - \dfrac{3\pi}{4}\right) = \sin \theta - \cos \theta$
18. (4.3) $\sin 3A = 3 \sin A - 4 \sin^3 A$
19. (4.3) $\sin 2B(1 + \cot^2 B) = 2 \cot B$
20. (4.4) Solve for all the values of the variable.
    (a) $\cos 2x + \cos x = -1$
    (b) $\sin^2 y = 2 \cos y + \frac{1}{4}$

oblique triangles five

5.1 THE LAW OF SINES

The trigonometric functions can be thought of as ratios of the sides of a right triangle, and the Pythagorean theorem applies only to right triangles. These facts allowed the solution of right triangles for missing sides and angles in Chapter 1. However, what of triangles that do not contain a right angle? Such triangles are called **oblique,** and this chapter will develop methods to deal with **oblique triangles.** An oblique triangle with three acute angles is an **acute triangle,** and one with two acute angles and an obtuse angle is called an **obtuse triangle.**

Any plane region bounded by line segments can be subdivided into a finite number of triangles. The surveyor, for one, takes advantage of this division to find distances and areas with a minimum of information. First, then, consider the area of a triangle.

Area of a Triangle

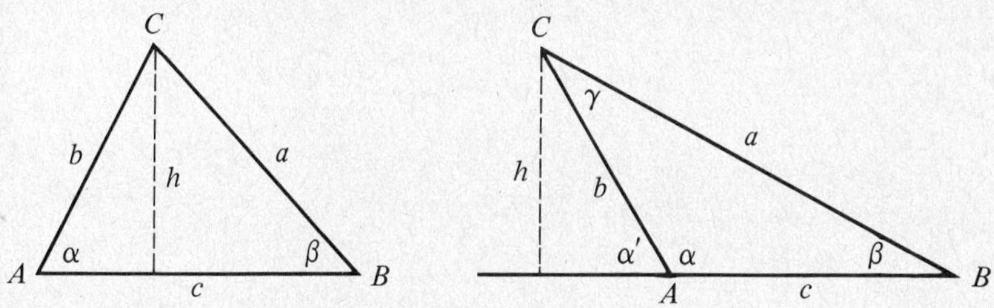

Figure 5.1 Oblique triangles.

The area A of a triangle is $\frac{1}{2}$(base)(height). Considering $\triangle ABC$ in Figure 5.1, the area of the oblique triangle is given by

$$A = \tfrac{1}{2}ch$$

However h can be found in terms of a or b.

$$\frac{h}{b} = \sin \alpha \qquad \text{or} \qquad \frac{h}{a} = \sin \beta$$

$$h = b \sin \alpha \qquad\qquad h = a \sin \beta$$

These values of h may be substituted into the expression for area to obtain

$$A = \tfrac{1}{2}c(b \sin \alpha) = \tfrac{1}{2}c(a \sin \beta)$$

An altitude could be erected on side a or b, instead of on side c as in Figure 5.1. Then the area could be found in terms of γ, giving three equivalent forms:

$$A = \tfrac{1}{2}bc \sin \alpha = \tfrac{1}{2}ac \sin \beta = \tfrac{1}{2}ab \sin \gamma$$

THE AREA OF A TRIANGLE

The area of any triangle is given by one-half the product of the lengths of two sides and the sine of the included angle.

EXAMPLES

1. If $\gamma = 63°$, $a = 197$, and $b = 246$, what is the area of $\triangle ABC$?

$$A = \tfrac{1}{2}ab \sin \gamma$$
$$= \tfrac{1}{2}(197)(246) \sin 63°$$

[63] [sin] [×] [246] [×] [197] [÷] [2] [=]

$$\approx 21{,}589.979$$
$$\approx 21{,}590$$

The area is approximately 21,590 square units.

2. A surveyor measures two sides of a triangular plot of land. The sides are 326 ft and 284 ft. If the angle between them is 39.45°, what is the area of the parcel?

$$A = \tfrac{1}{2}bc \sin \alpha$$
$$= \tfrac{1}{2}(326)(284) \sin 39.45°$$
$$\approx 29{,}414.150$$

This is in square feet. There are 43,560 ft^2 in an acre, the more common unit of area for land measure. So,

$$A \approx \frac{29{,}414.150}{43{,}560} \approx 0.6752559 \approx 0.68$$

The parcel is approximately 0.68 acres.

Law of Sines

The area of a triangle can be used to develop a very important result. First recall the three forms of the area.

$$\tfrac{1}{2}cb \sin \alpha = \tfrac{1}{2}ca \sin \beta = \tfrac{1}{2}ab \sin \gamma \qquad \text{Areas}$$

$$cb \sin \alpha = ca \sin \beta = ab \sin \gamma \qquad \text{Multiply by 2}$$

$$\frac{cb \sin \alpha}{abc} = \frac{ca \sin \beta}{abc} = \frac{ab \sin \gamma}{abc} \qquad \text{Divide by } abc$$

$$\frac{\sin \alpha}{a} = \frac{\sin \beta}{b} = \frac{\sin \gamma}{c} \qquad \text{Simplify}$$

or

$$\frac{a}{\sin \alpha} = \frac{b}{\sin \beta} = \frac{c}{\sin \gamma}$$

This says that the sides of a triangle are proportional to the sines of the angles opposite. You may even have suspected that the sides and angles were proportional in some way, because the largest angle is always opposite the longest side, and so on. Now we see that although the sides and *angles* are *not* proportional, the sides and the *sines of the angles* are proportional. The proportion we have developed is called the **law of sines** and is extremely useful in solving a triangle with *certain* information known.

THE LAW OF SINES

In any triangle, the sides are proportional to the sines of the angles opposite.

$$\frac{a}{\sin \alpha} = \frac{b}{\sin \beta} = \frac{c}{\sin \gamma}$$

$$\frac{a}{\sin \alpha} = \frac{b}{\sin \beta}$$

$$\frac{a}{\sin \alpha} = \frac{c}{\sin \gamma}$$

$$\frac{b}{\sin \beta} = \frac{c}{\sin \gamma}$$

The law of sines can be used in several situations. First, consider the case in which two angles and one side are known. A second case will be considered in the next section.

Two Angles and a Side

If any two angles are known, the third may be found, so any two may be given and any one of the three sides. With this information, a proportion may be formed by the law of sines, as the following examples illustrate.

You may remember the AAS and ASA cases from plane geometry.

EXAMPLES

3. Find side c in $\triangle ABC$ if $\angle A = 25°$, $\angle B = 74°$ and $a = 56$.

Any two angles are sufficient, because the third is easily found.

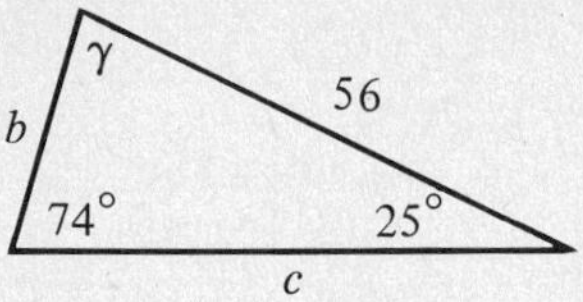

$$\angle C = 180° - \angle A - \angle B$$
$$= 180° - 25° - 74°$$
$$= 81°$$

$$\frac{c}{\sin C} = \frac{a}{\sin A}$$

$$\frac{c}{\sin 81°} = \frac{56}{\sin 25°}$$

$$c = \frac{56 \sin 81°}{\sin 25°}$$

$$\boxed{81}\ \boxed{\sin}\ \boxed{\times}\ \boxed{56}\ \boxed{\div}\ \boxed{25}\ \boxed{\sin}\ \boxed{=}$$

$$\approx 130.87590$$

$$\boldsymbol{c \approx 131}$$

4. Solve $\triangle ABC$ if $\alpha = 42°$, $\beta = 67°$, and $a = 43$.

The proportion may be inverted when we wish to find a side.

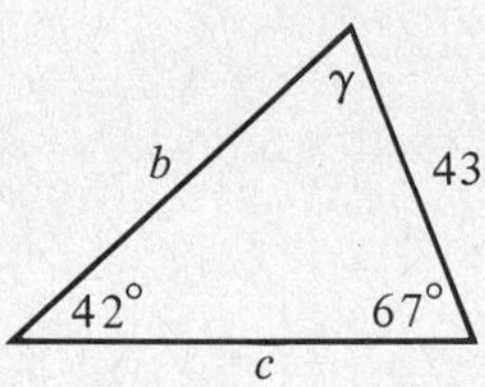

$$\frac{\sin \alpha}{a} = \frac{\sin \beta}{b}$$

$$\frac{a}{\sin \alpha} = \frac{b}{\sin \beta}$$

$$\frac{43}{\sin 42°} = \frac{b}{\sin 67°}$$

$$b = \frac{43 \sin 67°}{\sin 42°}$$

$$\approx 59.153935$$

$$\boldsymbol{b \approx 59}$$

Knowing two angles, we may find the third.

$$\gamma = 180° - \alpha - \beta$$
$$= 180° - 42° - 67°$$
$$\boldsymbol{\gamma = 71°}$$

With the third angle, a proportion may be formed to find the remaining side.

$$\frac{a}{\sin \alpha} = \frac{c}{\sin \gamma}$$

$$\frac{43}{\sin 42^\circ} = \frac{c}{\sin 71^\circ}$$

$$c = \frac{43 \sin 71^\circ}{\sin 42^\circ}$$

$$\approx 60.761379$$

$$\boldsymbol{c \approx 61}$$

The continuing use of the eight-place figures allows you to check the use of your calculator along with the example.

5. Find a if $\beta = 28.41^\circ$, $\gamma = 114.23^\circ$, and $b = 125.3$.

$$\alpha = 180^\circ - \beta - \gamma$$

Hence $\alpha = 37.36^\circ$.

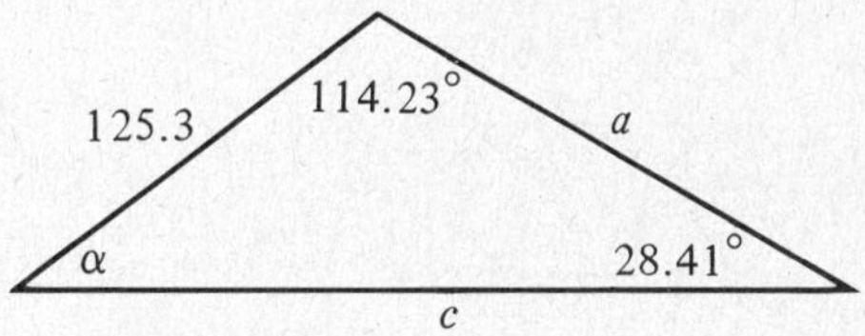

$$\frac{a}{\sin \alpha} = \frac{b}{\sin \beta}$$

$$\frac{a}{\sin 37.36^\circ} = \frac{125.3}{\sin 28.41^\circ}$$

$$a = \frac{125.3 \sin 37.36^\circ}{\sin 28.41^\circ}$$

$$\approx 159.81135$$

$$\boldsymbol{a \approx 159.8}$$

Check this calculation yourself so that you are certain you can do a similar problem.

6. Find the area of $\triangle ABC$ in Example 5 if the dimensions are in inches.

Because sides a and b are known, $\frac{1}{2}ab \sin \gamma$ is used.

$$A = \tfrac{1}{2}ab \sin \gamma$$

$$= \tfrac{1}{2}(159.8)(125.3) \sin 114.23^\circ$$

$$\approx 9129.5131$$

$$\approx 9130$$

The area is approximately 9130 in.^{2}.

7. A tree stands on an embankment that is inclined 12° with the horizontal. Forty-five feet downhill from the base of the tree, the angle

of elevation to the top of the tree is 39°. What is the height of the tree?

Make a sketch!

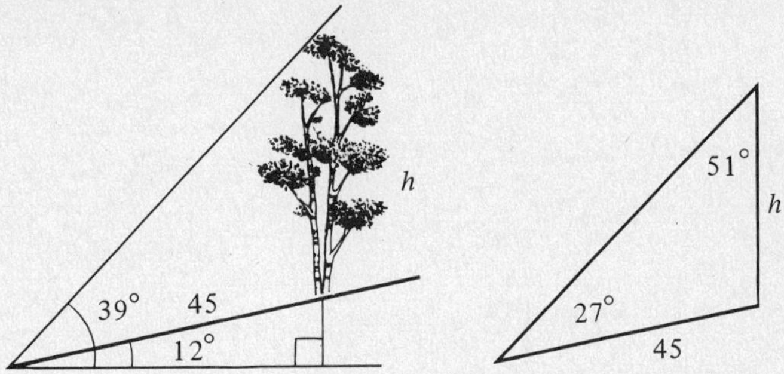

$$\frac{h}{\sin 27°} = \frac{45}{\sin 51°}$$

$$h = \frac{45 \sin 27°}{\sin 51°}$$

$$\approx 26.287947$$

$$\approx 26.3$$

The tree is approximately 26.3 ft high.

exercise set 5.1

A. exercises

Find the area of the triangles specified in Exercises 1–4.

1. $a = 35$ cm, $b = 76$ cm, $\gamma = 47°$ See Example 1.

2. $a = 12$ mi, $c = 7$ mi, $\beta = 23°$

3. $b = 112$ km, $c = 94$ km, $\alpha = 103°$

4. $c = 137$ ft, $a = 123$ ft, $\beta = 121°$

Find the required side in the triangles specified in Exercises 6–10.

5. $a = 50$, $\alpha = 65°$, $\beta = 40°$, $b = ?$ See Examples 3–5.

6. $b = 7.07$, $\alpha = 30°$, $\beta = 45°$, $a = ?$

7. $c = 9.56$, $\beta = 60°$, $\alpha = 60°$, $b = ?$

8. $\alpha = 50°$, $\beta = 75°$, $c = 60$, $a = ?$

9. $a = 550$, $\alpha = 10°$, $\beta = 46°$, $c = ?$

10. $b = 375$, $\beta = 122°$, $\gamma = 34°$, $a = ?$

Use the information given in Exercises 11–20 to find the required side as illustrated in Figure 5.2.

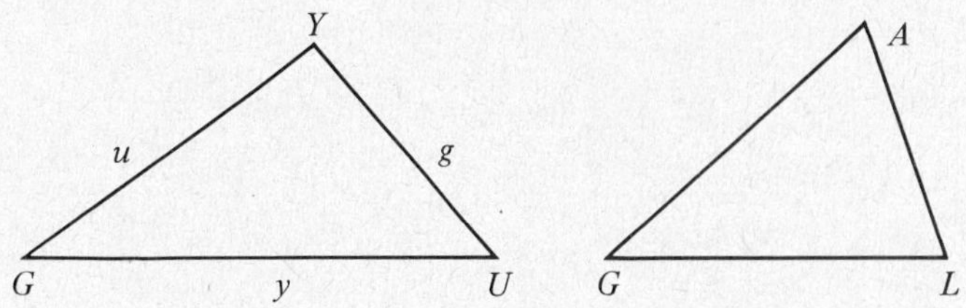

Figure 5.2

Be careful, both $\triangle GUY$ and $\triangle GAL$ contain an $\angle G$ and side g.

11. $\angle G = 32°$, $\angle U = 41°$, $g = 12.4$, $y = ?$ See Examples 3–5.

12. $\angle Y = 114°$, $\angle U = 39°$, $u = 24.1$, $g = ?$

13. $\angle Y = 109.8°$, $\angle G = 28.7°$, $y = 1.74$, $g = ?$

14. $\angle U = 40.3°$, $\angle Y = 117.2°$, $y = 7.53$, $u = ?$

15. $\angle G = 41°10'$, $\angle A = 47°40'$, $a = 124.9$, $l = ?$

16. $\angle L = 83°20'$, $\angle G = 38°50'$, $g = 93.7$, $a = ?$

17. $a = 1.065$, $\angle L = 79.43°$, $\angle A = 47.19°$, $g = ?$

18. $g = 2.109$, $\angle A = 50.23°$, $\angle L = 71.31°$, $a = ?$

19. $y = 1362$, $\angle G = 29°48'$, $\angle U = 31°37'$, $g = ?$

20. $u = 2587$, $\angle L = 81°29'$, $\angle G = 40°57'$, $l = ?$

In Exercises 21–30, find the area of the triangle specified by the previous exercise named. You will need the results of that exercise to compute the area.

21. Exercise 11   22. Exercise 12   See Example 6.

23. 13   24. 14   25. 15   26. 16

27. 17   28. 18   29. 19   30. 20

B. exercises

31. A surveyor measures the frontage of a triangular lot as 235 ft. The angles at the frontage corners prove to be 84.32° and 45.53°. What does the surveyor compute for the area of the lot?

See Example 2.

Make a sketch!

32. The two angles on the street side of a large triangular plot are 71.52° and 32.29°. The street frontage is 315 ft. What is the area of the plot?

33. Two observers are 1760 yd apart. They sight an object that as they face each other is at 37.45° and 49.63° with the line between them. How far is the object from the closer observer?

See Example 7.

34. One person stands at each end of a 4200-ft bridge across a strait. An island is visible to each person. The angle between the line of sight of the island and the bridge is 42.07° for one observer and 63.52° for the other. How far is the island from the farther observer?

35. The longer diagonal of a parallelogram is 35 cm, and it makes angles of 25° and 32° with the sides. Find the sides of the figure.

36. The longer diagonal of a trapezoid is 142 mm and makes an angle of 28° with one of the bases. Find the length of the shorter base if it also makes an angle of 33° with one of the sides.

37. A ship is sighted from two points 10 km apart along a straight shoreline. Facing each other, the line of sight of the first observer is 34°40′ with the shore and the second 29°10′. Find the distance from the first observer to the ship.

38. A motorist sights a mountain from two points along a straight stretch of highway. First is is 39°50′ with the highway ahead. Fifty miles down the road, it is 77°20′ with the road behind. How far away is the mountain at the second sighting?

39. A surveyor climbs 109 ft up a 21.24° embankment. If the angle of depression to the opposite bank of the river is 5.12°, what is the width of the river?

40. A bridge crosses a river at a 72.34° angle with the downriver shoreline. A surveyor walks 255 ft downriver and sights the end of the bridge on the opposite bank. If the angle with the upriver shore is 84.52°, what is the overwater span of the bridge?

HISTORICAL ASIDE

Nicolas Copernicus (1473–1543) is best known for his heliocentric theory of the universe. His work, however, compelled further development of trigonometry. Copernicus himself wrote a treatise on trigonometry. His student, Georg Rhaeticus (1514–1576), spent a dozen years compiling extensive tables of trigonometric functions. One was a 15-place table of sines for every 10″. Rhaeticus was primarily responsible for the publication of Copernicus' great work, and the first copy was presented to him just before his death.

C exercises

41. **brain teaser** Use the law of sines to show that the area A of any triangle is given by

$$A = \frac{c^2 \sin \alpha \sin \beta}{2 \sin \gamma}$$

where $\alpha + \beta + \gamma = 180°$.

42. **brain teaser** Prove that the area A of a *segment* of a circle (shaded region in Figure 5.3) is given by

$$A = \frac{r^2}{2}(\theta - \sin\theta)$$

where θ is in radian measure.

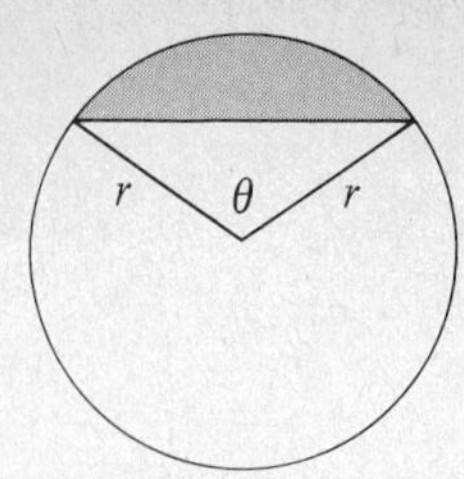

Figure 5.3
Segment of a circle.

43. Given that the diagonals of a quadrilateral meet at angle θ and have lengths m and n, prove that the area A is given by

$$A = \tfrac{1}{2}mn\sin\theta$$

5.2 THE AMBIGUOUS CASE

There is a second set of data that can be used with the law of sines. In the first case, any two angles and a side produces a unique result: a second side. Given two sides and the nonincluded angle, however, the result is ambiguous. That is, there are four distinct possibilities, as illustrated by Figures 5.4 and 5.5

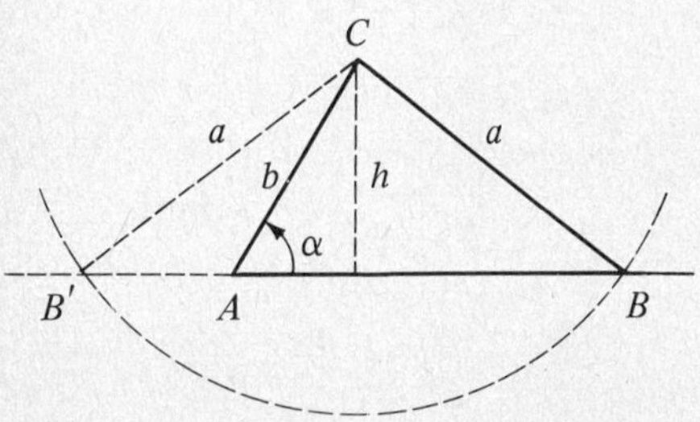

Figure 5.4 Case I, $a > b$; the nonadjacent side is longer.

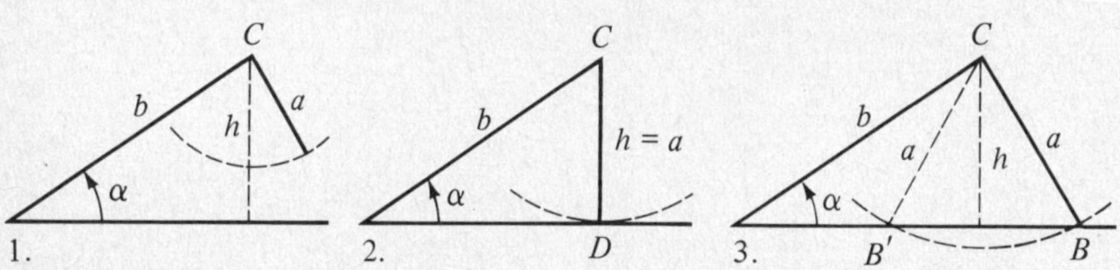

Figure 5.5 Case II, $a < b$; the nonadjacent side is shorter.

If the given angle is obtuse, there is a single solution.

I. **If $a > b$, a unique triangle is determined,** because a intersects the third side in one and only one point, which determines a triangle. See Figure 5.4.

II. **If $a < b$ and**

   1. **If $a < h$, no triangle is possible,** because a cannot intersect the third side. The height is $b \sin \alpha$, so

a, b, and α represent any two sides and a nonincluded angle. a, b, and β, or b, c, and γ, could be used just as well.

$$\sin \beta = \frac{b \sin \alpha}{a} > 1$$

The calculator will signal *error* when you press arcsin or $\sin^{-1}$.

$\sin \beta$ cannot exceed 1, so there is no solution.

2. **If $a = h$, a unique right triangle is determined,** because a coincides with h and is perpendicular to the base. If $a = b \sin \alpha$, then

$$\sin \beta = \frac{b \sin \alpha}{a} = 1$$

$\sin \beta = 1$ means that $\beta = 90°$ and a right triangle is determined.

3. **If $a > h$, two distinct triangles are determined,** because a can intersect the base in two different points, one on either side of h. If $a > b \sin \alpha$, then

$$\sin \beta = \frac{b \sin \alpha}{a} < 1$$

However, β could be acute or obtuse, because $\sin \beta$ is positive in QI *and* QII. Both β and $180° - \beta$ are solutions.

It seems that to apply the law of sines in the ambiguous case, we proceed as usual ***except*** when the nonadjacent side is shorter. In that case, if we obtain a nonright angle solution, its supplement is also a solution. A sketch is often helpful in these situations.

THE AMBIGUOUS CASE

Given a, b, and an acute angle α,

I. **If** a > b **then** one triangle

II. **If** a < b **then**

- **1.** a < b sin α — no triangle
- **2.** a = b sin α — one right triangle
- **3.** a > b sin α — two triangles

What happens when $a = b$? Must the law of sines apply?

EXAMPLES

Find the required part, if possible.

1. If $a = 8$, $b = 10$, and $\alpha = 41°$, then $\beta = ?$

First note that $a > b$, so there is a unique solution.

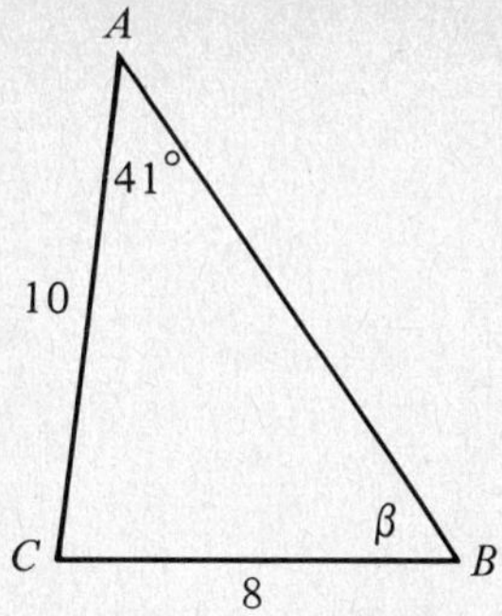

$$\frac{\sin\beta}{b} = \frac{\sin\alpha}{a}$$

$$\frac{\sin\beta}{10} = \frac{\sin 41°}{8}$$

$$\sin\beta = \frac{10\sin 41°}{8}$$

$$\beta = \arcsin\left(\frac{10\sin 41°}{8}\right)$$

$$\approx 55.092180$$

$$\boldsymbol{\beta \approx 55.1°}$$

2. If $b = 12.7$, $c = 69.3$, and $\gamma = 74.14°$, then $\delta = ?$

First, $c > b$, so there is a unique solution. Now b is the nonadjacent side for γ. Make a sketch!

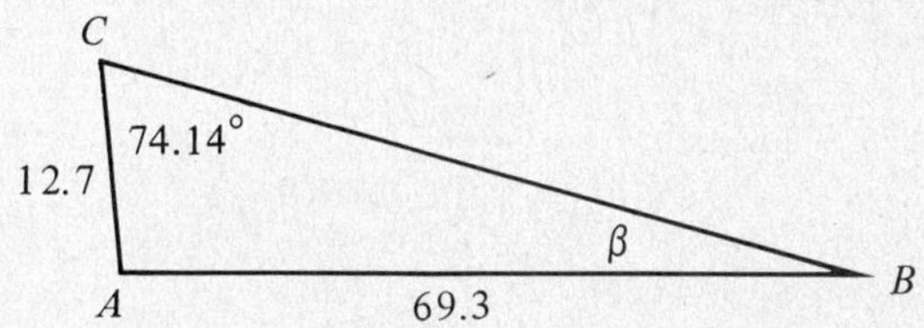

$$\frac{\sin\beta}{b} = \frac{\sin\gamma}{c}$$

$$\frac{\sin\beta}{12.7} = \frac{\sin 74.14°}{69.3}$$

$$\sin\beta = \frac{12.7\sin 74.14°}{69.3}$$

$$\beta = \arcsin\left(\frac{12.7\sin 74.14°}{69.3}\right)$$

$$\approx 10.153437°$$

$$\boldsymbol{\beta \approx 10.15°}$$

3. If $a = 13.1$, $c = 40.2$, and $\alpha = 68.2°$, find γ.

First, c is the nonadjacent side to α, and the solution is ambiguous.

$$a < c \qquad \frac{\sin\gamma}{c} = \frac{\sin\alpha}{a}$$

$$\frac{\sin\gamma}{40.2} = \frac{\sin 68.2°}{13.1}$$

$$\sin \gamma = \frac{40.2 \sin 68.2°}{13.1}$$

$$\sin \gamma \approx 2.849247$$

The sine may not exceed 1. Therefore,

No solution

4. Given $a = 5.9$, $b = 12.9$, and $\alpha = 27°13'$, find β.

$$a < b \qquad \frac{\sin \beta}{b} = \frac{\sin \alpha}{a}$$

$$\frac{\sin \beta}{12.9} = \frac{\sin 27°13'}{5.9}$$

$$\sin \beta \approx 0.9999831$$

$$\approx 89.66725°$$

$$\boldsymbol{\beta \approx 89°40'}$$

or

$$\mathbf{180° - \boldsymbol{\beta} \approx 90°20'}$$

Notice this is very nearly a right angle. A four-place table would be useless here, because since 0.9999831 is 1.0000 to the nearest ten-thousandth.

5. Given $\beta = 30°$, $b = 19.52$, and $c = 39.04$, find γ.

$$b < c \qquad \frac{\sin \gamma}{39.04} = \frac{\sin 30°}{19.52} \qquad \text{The ambiguous case}$$

$$\sin \gamma = \frac{39.04(\frac{1}{2})}{19.52} = 1$$

$$\boldsymbol{\gamma = 90°} \qquad \text{One solution: a right triangle}$$

6. If $\alpha = 38°43'$, $a = 9.72$, $b = 11.84$, then $\beta = ?$

$$b < c \qquad \frac{\sin \beta}{11.84} = \frac{\sin 38°43'}{9.72}$$

$$\sin \beta = \frac{11.84 \sin 38°43'}{9.72}$$

$$\sin \beta \approx 0.76188895$$

$$\beta \approx 49.631008$$

$$\boldsymbol{\beta \approx 49°38'}$$

or

$$\mathbf{180° - \boldsymbol{\beta} \approx 130°22'}$$

The third angle could now be found; each of the two possibilities would produce a different value for the third side.

7. Reno is 348 mi N41°25′W of Las Vegas. Fresno, California, is due south of Reno and 264 mi from Las Vegas. Find the distance from Reno to Fresno and the direction of Las Vegas from Fresno.

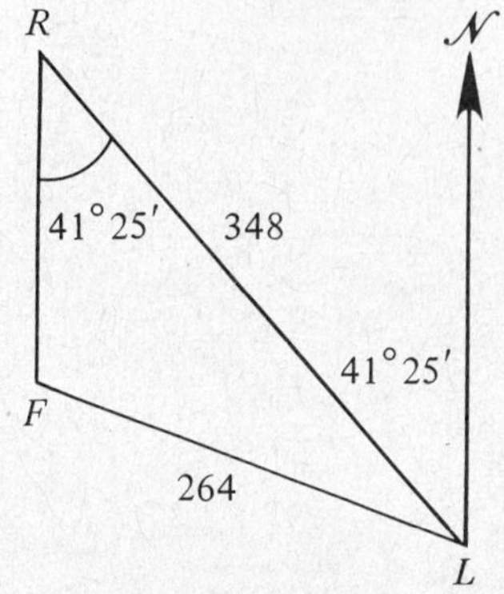

$$\frac{\sin F}{RL} = \frac{\sin R}{FL}$$

$$\frac{\sin F}{348} = \frac{\sin 41°25'}{264}$$

$$\sin F = \frac{348 \sin 41°25'}{264}$$

$$\approx 0.87201686$$

$$F \approx 60.693864$$

$$\approx 60°42' \quad \text{or} \quad 119°28' \qquad \text{We take the obtuse value.}$$

$$L \approx 180° - F - R$$

$$\approx 180° - 119°28' - 41°25' \qquad \angle FLN = 41°25' + \angle L$$

$$\approx 19°7' \qquad\qquad = 60°32'$$

$$\frac{RF}{\sin L} = \frac{FL}{\sin R}$$

$$RF = \frac{264 \sin 19°7'}{\sin 41°25'}$$

$$\approx 130.69412$$

$$\approx 131$$

Fresno is approximately 131 mi south of Reno and is N60°32′W of Las Vegas.

exercise set 5.2

A. exercises

Determine the number of solutions in each triangle in Exercises 1–6. Do not solve.

1. $a = 80,\ b = 100,\ \alpha = 30°$

2. $a = 50,\ c = 100,\ \alpha = 30°$

3. $b = 40,\ c = 100,\ \beta = 30°$

4. $a = 11,\ c = 13,\ \gamma = 69°$

5. $c = 75,\ b = 70,\ \gamma = 60°$

6. $b = 84,\ a = 134,\ \beta = 52°$

Solve for the required part in Exercises 7–10.

See Examples 1–6.

7. Find β in Exercise 1.

8. Find γ in Exercise 2.

9. Find α in Exercise 4.

10. Find β in Exercise 5.

The required part in Exercises 11–30 refers to the figures shown. Find all results to the nearest unit or degree.

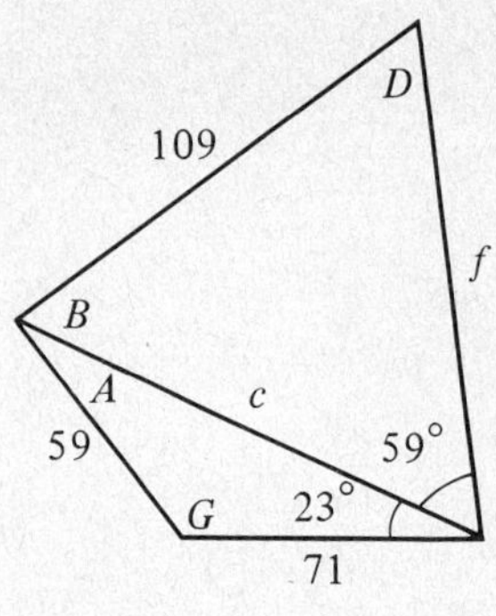

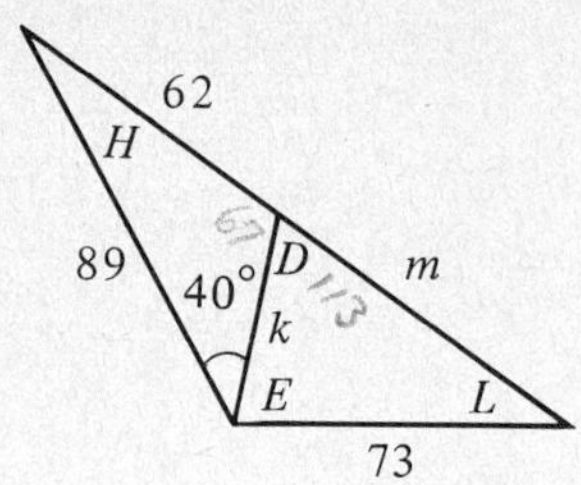

D should have the same value in each figure.

11. A	12. B	13. c	14. D
15. E	16. f	17. G	18. H
19. k	20. L	21. m	22. I

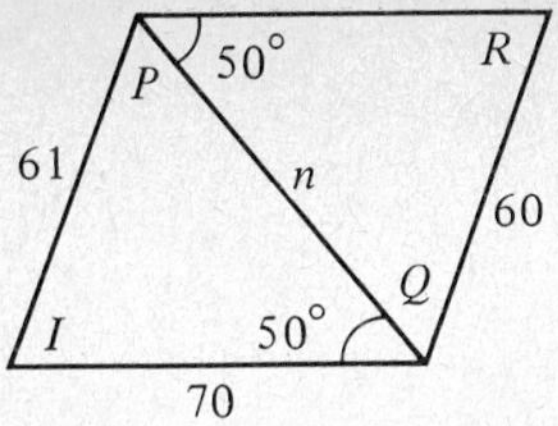

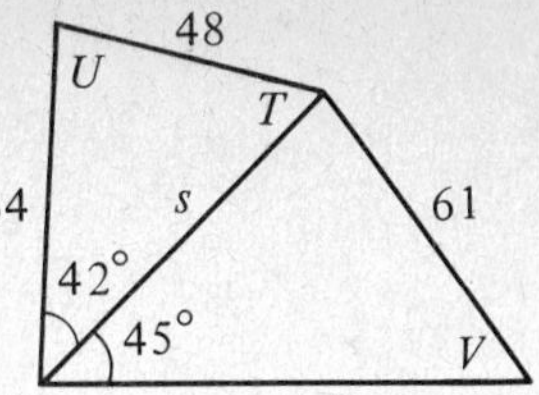

23. n 24. P 25. Q 26. R

27. s 28. T 29. U 30. V

B. exercises

31. The shorter diagonal of a parallelogram makes angles of 45° and 80° with the sides. What is the longer side if the shorter one is 71.8 cm?

32. The longer diagonal of a parallelogram makes angles 13° and 5° with the sides. What is the shorter side if the longer one is 100 cm?

33. A shipping route is a straight 54-mi run between two ports. However, 54 mi out of port the captain discovers that the ship has been traveling 3° off course. How far is the ship from its destination?

34. An artillery shell is fired 2 mi toward a target. An observer reports that the distance was correct, but the aim was off 2°. By how many *feet* was the target missed?

35. Fort Worth, Texas, is 926 mi N7°31′E of Mexico City, Mexico. San Diego, California, is due west of Fort Worth and 1547 mi from Mexico City. What is the direction from San Diego to Mexico City?

36. Pittsburgh, Pennsylvania, is 549 mi N32°44′E of Atlanta, Georgia. Miami, Florida, is due south of Pittsburgh and 625 mi from Atlanta. What is the direction from Miami to Atlanta?

37. The faces of Kha-ef-Re's pyramid in the Nile valley are inclined at 53°4′ with the horizontal. Five hundred feet directly back from the base, the angle of elevation to the top of the pyramid is 28°53′. What is the length of the face from base to pinnacle?

38. If the base of Kha-ef-Re's pyramid in the Nile valley is 708 ft, determine the slant height of the face in a different way to check the results of Exercise 37.

HISTORICAL ASIDE

One must assume in problems about distances between cities that the earth is flat over that region. Of course, we live on a sphere. For such problems, there is a spherical geometry and a spherical trigonometry. Working on the surface of a sphere rather than a plane is much more difficult. In fact, the 9th edition of the Encyclopedia Britannica admits that "perhaps to the student there is no part of elementary mathematics that is so repulsive as is spherical trigonometry."

C. exercises

39. **trig talk** Use Exercises 11–30 to fill the blanks with the letters that correspond to the values below the blanks. The letter "oh" has

already been inserted. When all the blanks are filled, the quotation may be read.

"___ ___ ___, ___ ___ ___ ___ ___ ___ ___ ___ O ___
68 63 68 69 63 27 79 78 28 74 74 63

___ ___ ___ ___ ___ ___ ___ O ___ ___ ___ ___ ___
63 27 79 78 79 63 27 67 63 27 28 63

___ O ___ ___ ___ ___ ___ ___ ___ ___ ___ O ___ ___ ___ ___."
69 34 53 79 69 63 27 79 62 71 54 34 79 78 (H. Maschke)

40. **brain teaser** Prove that if a circle is circumscribed about a triangle, then the radius r of the circle is given by

$$r = \frac{a}{2 \sin \alpha} = \frac{b}{2 \sin \beta} = \frac{c}{2 \sin \gamma}$$

5.3 THE LAW OF COSINES

The law of sines provides the means to deal with several cases of solving a triangle. However, several cases remain unresolved. Let us go back to the oblique triangle as shown in Figure 5.1 and place it in a coordinate system.

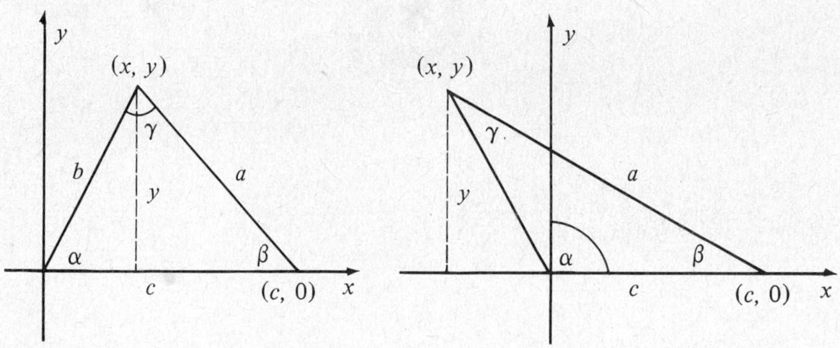

Figure 5.6 Oblique triangles.

In Figure 5.6 the distance between $(c, 0)$ and (x, y) is a. Now, x and y can be found in terms of the sides and angles of the triangle. So let us use the distance formula to find that distance.

$$a = \sqrt{(x - c)^2 + (y - 0)^2}$$

$$a^2 = (x - c)^2 + y^2$$

Simplify and substitute, then rearrange.

$$\frac{x}{b} = \cos \alpha \quad \text{and} \quad \frac{y}{b} = \sin \alpha$$

$$x = b \cos \alpha \quad \text{and} \quad y = b \sin \alpha$$

$$\begin{aligned} a^2 &= x^2 - 2cx + c^2 + y^2 \\ &= b^2 \cos^2 \alpha - 2bc \cos \alpha + c^2 + b^2 \sin^2 \alpha \\ &= b^2 \cos^2 \alpha + b^2 \sin^2 \alpha + c^2 - 2bc \cos \alpha \\ &= b^2(\cos^2 \alpha + \sin^2 \alpha) + c^2 - 2bc \cos \alpha \end{aligned}$$

$$\boldsymbol{a^2 = b^2 + c^2 - 2bc \cos \alpha}$$

Finally, factor and use an identity.

Reorienting the triangle with β and then γ at the origin produces two other forms:

$$\boldsymbol{b^2 = a^2 + c^2 - 2ac \cos \beta}$$

$$\boldsymbol{c^2 = a^2 + b^2 - 2ab \cos \gamma}$$

Together, these three equations form what we call the **law of cosines.**

THE LAW OF COSINES

If a, b, and c are the sides of a triangle that are opposite angles α, β, and γ, then

$$a^2 = b^2 + c^2 - 2bc \cos \alpha$$

$$b^2 = a^2 + c^2 - 2ac \cos \beta$$

$$c^2 = a^2 + b^2 - 2ab \cos \gamma$$

Two Sides and the Included Angle

Now, given b, c, and α, a can be found. The law of sines dealt only with the case of the nonincluded angle. Let us illustrate this case with examples.

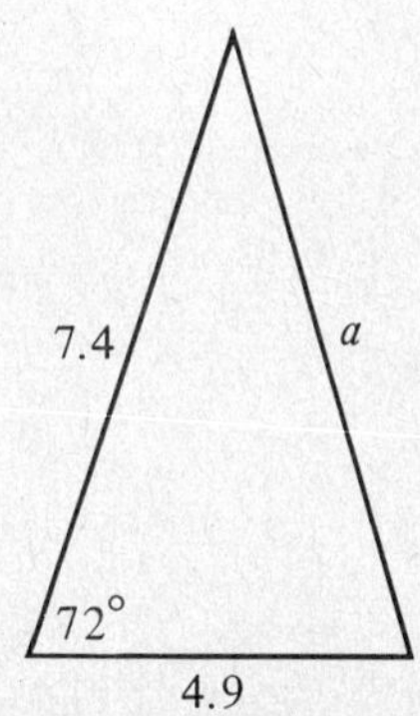

EXAMPLES

1. Given $b = 7.4$, $c = 4.9$, and $\alpha = 72°$, find a.

$$a^2 = b^2 + c^2 - 2bc \cos \alpha$$

$$\alpha^2 = (7.4)^2 + (4.9)^2 - 2(7.4)(4.9) \cos 72°$$

$$a = \sqrt{7.4^2 + 4.9^2 - 14.8(4.9)\cos 72^\circ}$$

$$\approx 7.5073355$$

$$\boldsymbol{a \approx 7.5}$$

See also the calculator key scheme below.

Using a calculator, there are many steps in this last example. Let us chart the key sequence in blocks of calculations indicated by the flow chart.

Once a is known in Example 1, the law of sines can be used to find the other parts of the triangle.

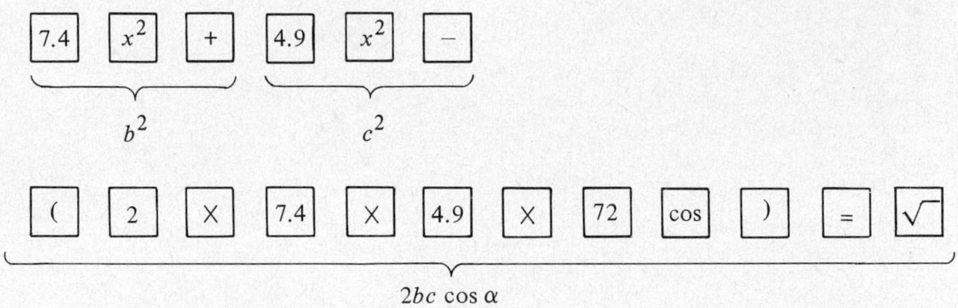

Note that in using the calculator, there are groups of operations that can be blocked together. Be certain that you check your answer against the given data to see that it is reasonable. A quick sketch is a good way to "keep your results honest."

EXAMPLE

2. From a spot on the mainland, two islands are 3.25 mi and 4.65 mi. If the angle of sight between the islands is 109°42′, what is the distance between them?

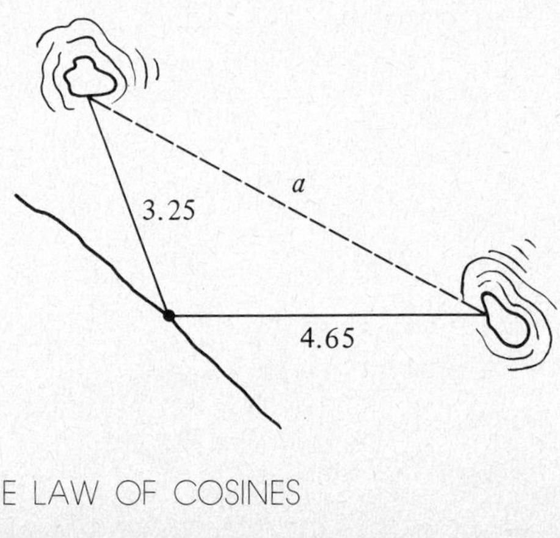

Notice that cosine will be negative. The calculator will, of course, remember.

$$a^2 = (3.25)^2 + (4.65)^2 - 2(3.25)(4.65)\cos 109°42'$$

$$a = \sqrt{3.25^2 + 4.65^2 - 6.5(4.65)\cos 109°42'}$$

$$\approx \sqrt{42.373704}$$

$$\approx 6.5095088 \approx 6.51$$

The islands are approximately 6.51 mi apart.

(Don't forget to state an answer to the problem.)

Three Sides of a Triangle

If the three sides of an oblique triangle are known, any of the three angles can be determined with the law of cosines.

EXAMPLE

3. What are the angles in a triangle with sides 5, 8, and 11?

$$a^2 = b^2 + c^2 - 2bc\cos\alpha$$

First, solve for $\cos\alpha$.

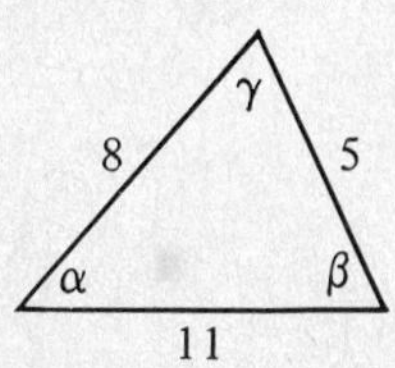

$$2bc\cos\alpha = b^2 + c^2 - a^2$$

$$\cos\alpha = \frac{b^2 + c^2 - a^2}{2bc}$$

$$= \frac{8^2 + 11^2 - 5^2}{2(8)(11)} = \frac{64 + 121 - 25}{176} = \frac{160}{176}$$

$$= 0.909\overline{090}$$

$$\alpha \approx 24.619977°$$

$$\boldsymbol{\alpha \approx 24.62°}$$

Now that α is known, the law of sines can be used.

$$\frac{\sin\beta}{b} = \frac{\sin\alpha}{a}$$

$$\frac{\sin\beta}{8} = \frac{\sin 24.619977°}{5}$$

(Use your *best* value for α in computing β.)

$$\sin\beta \approx 0.66655646$$

$$\beta \approx 41.801844°$$

$$\boldsymbol{\beta \approx 41.80°}$$

The third angle is easily found.

$$\gamma = 180° - \alpha - \beta$$

$$\gamma \approx 180° - 24.62° - 41.80°$$

$$\gamma \approx \mathbf{113.58°}$$

Two Sides and the Nonincluded Angle

You will recognize this as the ambiguous case from the law of sines. Here we rework a couple of examples from the previous section so that the results can be checked against those obtained with the law of sines. Using the law of sines, a second solution in the ambiguous case is sometimes ignored. Using the law of cosines, both solutions are explicit, but the law of sines is much easier to use.

EXAMPLES

4. If $\alpha = 38°43'$, $a = 9.72$, and $b = 11.84$, find c.
   (See Example 6 from Section 5.2.)

$$a^2 = b^2 + c^2 - 2bc \cos \alpha$$

Solving for c produces a quadratic in c.

$$c^2 - (2b \cos \alpha)c + (b^2 - a^2) = 0$$

Unless exact data is known and the quadratic factors, the quadratic formula is necessary. Recall that if

$$Ax^2 + Bx + C = 0,$$

then

$$x = \frac{-B \pm \sqrt{B^2 - 4AC}}{2A}$$

Therefore,

$$c = \frac{2b \cos \alpha \pm \sqrt{(2b \cos \alpha)^2 - 4(b^2 - a^2)}}{2}$$

$$= b \cos \alpha \pm \sqrt{(b \cos \alpha)^2 - (b^2 - a^2)}$$

This is not an easy calculation. Be sure to compute the discriminant (the radical in the formula) first, so that you can compute both values of c without recomputing everything.

$$c = 11.84 \cos 38°43' \pm \sqrt{(11.84 \cos 38°43' - (11.84^2 - 9.72^2)}$$
$$= 9.2381422 \pm \sqrt{39.636071}$$
$$= 9.2381422 \pm 6.2957185$$
$$= 2.9424237 \quad \text{or} \quad 15.533861$$

c is either 2.94 or 15.53, to the nearest hundredth.

5. If $a = 9.72$, $b = 11.84$, and $c = 2.94$, find β. (Using the results of Example 6 in Section 5.2, this should be 130°22′.)

$$b^2 = a^2 + c^2 - 2ac \cos \beta$$
$$\cos \beta = \frac{a^2 + c^2 - b^2}{2ac}$$
$$= \frac{9.72^2 + 2.94^2 - 11.84^2}{2(9.72)(2.94)}$$
$$= -0.64849108$$
$$\beta \approx 130.42793$$
$$\boldsymbol{\beta \approx 130°26'}$$

Notice the 4′ difference computing β in this way. Is this a serious error? Why does it occur?

exercise set 5.3

A. exercises

Find the required part of the triangles in Exercises 1–10.

See Examples 1–5.

1. $a = 30$, $b = 54$, $\gamma = 46°$, $c = ?$
2. $a = 33$, $c = 30$, $\beta = 35°$, $b = ?$
3. $a = 7$, $b = 8$, $c = 5$, $\alpha = ?$
4. $a = 4$, $b = 7$, $c = 10$, $\gamma = ?$
5. $b = 77.99$, $c = 83.39$, $\alpha = 72°15'$, $a = ?$
6. $a = 9.56$, $c = 8.57$, $\beta = 45°0'$, $b = ?$
7. $a = 21$, $b = 24$, $c = 27$, $\beta = ?$
8. $a = 56$, $b = 70$, $c = 60$, $\alpha = ?$
9. $a = 18$, $b = 20$, $\alpha = 55.4°$, $c = ?$
10. $b = 19$, $c = 18$, $\gamma = 15.5°$, $a = ?$

B. exercises

The figures shown will require both the law of sines and the law of cosines to determine all the required parts. Find all results to the nearest tenth of a unit or degree in Exercises 11–30. (See Exercise 41.)

Note that to find any one quantity may require finding several others.

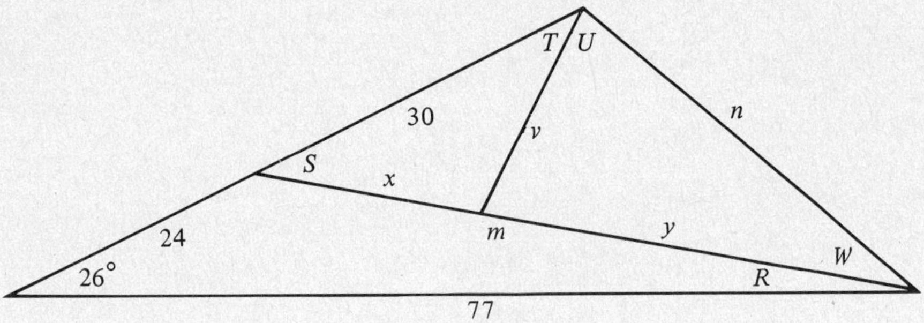

11. a 12. B 13. C 14. D

15. E 16. F 17. g 18. h

19. I 20. L 21. m 22. n

23. R 24. S 25. T 26. U

27. W 28. v 29. x 30. y

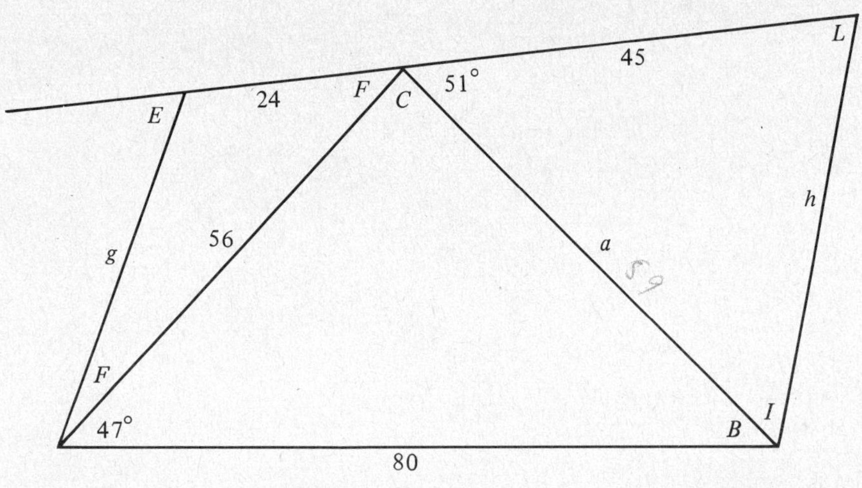

31. Two planes leave at the same time and fly in directions that make an angle of 123°. If the planes maintain speeds of 185 mi/h and 170 mi/h, how far apart are they after 2 h?

32. Two ships leave port at the same time and sail on courses 165° apart. How far apart are they after 12 h if they maintain an average of 10 knots and 12 knots on straight courses?

A "knot" is one nautical mile per hour.

33. In Cincinnati's Riverfront Stadium, the distance from home plate to straightaway center field is 404 ft. What is the distance from straightaway center to third base?

The bases are 90 ft apart.

34. The distance from home to straightaway centerfield in Boston's Fenway Park is 390 ft. How long is the throw from straightaway center to first base?

35. Madrid is 785 mi S19°W from London. Paris is 214 mi S29°E from London. Find the distance between Paris and Madrid. (Estimate to the nearest 10 mi.)

36. Saigon is 745 mi S7°E from Hanoi. Bangkok is 590 mi S32°W from Hanoi. Estimate the distance between Saigon and Bangkok to the nearest 10 mi.

37. Two straight roads intersect at an angle of 81°. A car is 1000 yd from the intersection. Another car is 1500 yd away on the other road. What is the distance between the cars?

38. A train is 2 mi from the junction of two lines. Another train is 3.25 mi from the junction on the other track. What is the distance between the two trains if the tracks are straight and meet at 68°?

39. The distance between two points that are on opposite sides of a lake must be determined. A surveyor finds a third point back from the lake such that the distance to both of the other points can be measured. The distances of 2015 and 1972 yd make an angle of 26° at the third point. Find the required distance.

40. A ladder 52 ft long leans against an inclined wall. The ladder is set 20 ft from the base of the wall and reaches 46 ft up its face. Find the angle of the wall with the vertical.

HISTORICAL ASIDE

Perhaps the first person to develop methods for solving plane and spherical triangles using all six functions was a French lawyer, Francois Vieté (1540–1603). Vieté worked on mathematics as a hobby and published most of his works at his own expense. In spite of the tremendous stature of his work, he gained considerable notoriety in a controversy over the Gregorian reform of the calendar in 1594. His position was most unscientific on the issue. Even the most learned of men show that they are human in matters of controversy.

C. exercises

41. **trig talk** Use Exercises 11–30 to fill the blanks with the letter that corresponds to the value below the blank. The letter "oh" has been inserted for you already. When all the blanks are filled, the quotation may be read.

"

____ ____ ____ ____ ____ ____ ____   ____ ____   ____ ____ ____

58.5 79.8 40.8 62.8 44.4 10.8 58.5   49.2 36.7   44.4 86.7 27.2

____ ____ ____ ____ ____ ____ ____   ____ ____   O ____ ____ ____ ____ ____

29.3 10.8 49.2 27.3 27.3 62.8 37.0   40.8 62.8   56.4 62.8 27.2 10.8 41.1

___ ___ ___   ___ ___ O ___ ___ ___ ___ ___   ___ ___

58.5 37.0 40.4   40.8 62.8   56.4 62.8 27.3 10.8 41.1   49.2 36.7

___ ___ ___   ___ ___ ___ ___ ___ ___ ___

44.4 86.7 27.3   22.4 49.2 40.8 86.7 10.8 62.8 40.4

___ ___ ___ ___ ___ ___ ___."

58.5 79.8 40.8 62.8 44.4 10.8 58.5   (Sophie Germain)

42. **brain teaser** Given a, b, and α, then

$$c = b \cos \alpha \pm \sqrt{b^2 \cos^2 \alpha - b^2 + a^2}$$

Show that this equation yields corresponding results for:

1. $a < b \sin \alpha$
2. $a = b \sin \alpha$
3. $a > b \sin \alpha$

of the ambiguous case.

43. **brain teaser** Use the law of cosines to prove that

$$a^2 + b^2 + c^2 = 2(ab \cos \gamma + ac \cos \beta + bc \cos \alpha)$$

in any triangle.

44. **brain teaser** Use Exercise 43 to prove that

$$\frac{\cos \alpha}{a} + \frac{\cos \beta}{b} + \frac{\cos \gamma}{c} = \frac{a^2 + b^2 + c^2}{2abc}$$

in any triangle.

45. **brain teaser** Prove that a, b, and c are *not* the sides of a triangle if $a + b < c$.

Figure 5.7
Vector representation.

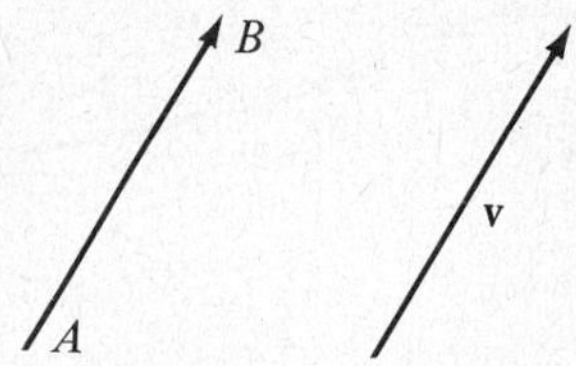

5.4 APPLICATIONS WITH VECTORS

Most physical quantities have only one dimension or magnitude. Height, weight, distance, area, volume, time, and others are represented by a single number, called a **scalar.** Other quantities have *direction* as well as *magnitude.* A force, for example, has magnitude and a direction of application. Such a quantity is called a **vector** quantity. Other vector quantities include velocity (speed and direction), acceleration, and displacement. A scalar is represented as a point on a number line or scale. A vector is represented as a directed line segment, where the segment has scalar length. See Figure 5.7.

A directed line segment when used to represent a vector quantity is called a vector

We denote a vector by a single letter, such as **v**, or with its starting and ending points, as **AB**.

Often vectors are indicated by arrows, such as $\vec{v}$ or $\overrightarrow{AB}$. We use boldface letters, such as **v** or **AB**, to denote vector quantities.

Addition of Vectors

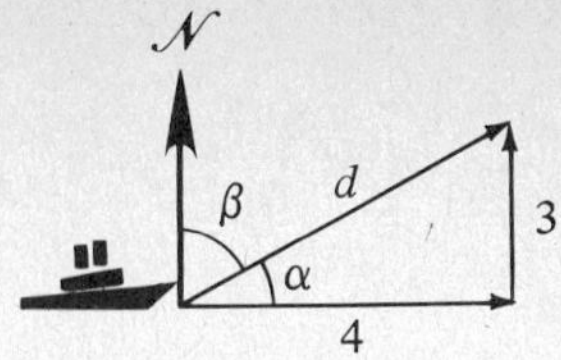

Figure 5.8

If two forces are applied simultaneously to the same body, there is a single force that would be the equivalent. That equivalent force is the sum of the two forces. Consider displacements. A ship travels east for 4 mi, turns, and travels north for 3 mi. The ship was displaced, first east, then north. Isn't there a single displacement in a northeastern direction that would be equivalent? Certainly there is! The diagram in Figure 5.8 shows the situation, and the distance d and the direction could be computed. In fact, $d = 5$ and $\beta \approx 36°19'$, so that **v** is a displacement of 5 mi N36°19′E.

ADDITION OF VECTORS

The sum of two vectors **v** and **w** is the vector **v** + **w** that extends from the foot of **v** to the head of **w** (where **w** is drawn from the head of **v**).
The sum is often called the **resultant** of the two vectors.

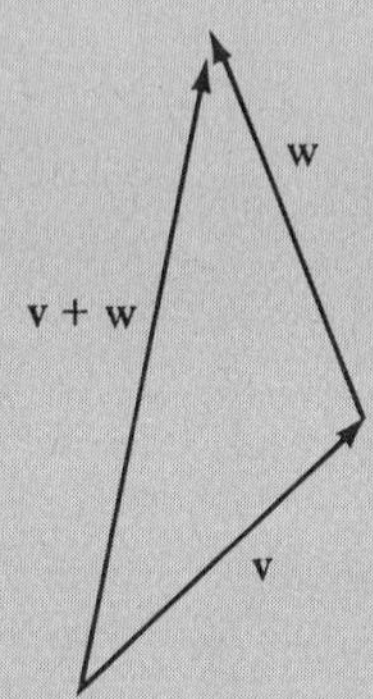

Geometrically, the sum of two vectors is the third side of the triangle formed by the first two. In physics, a so-called **parallelogram law** is used that is equivalent to this procedure. We keep our discussion here very informal and intuitive, as you will doubtless encounter a complete treatment of vectors if you continue your studies in mathematics or physics.

EXAMPLES

Figure 5.9 shows four very common applications of vectors and their peculiar method of indicating direction. Describe each vector quantity.

Solutions

1. A force **f** of 30 lb at 70°.
   The angle is measured counterclockwise from the positive x axis to the vector.

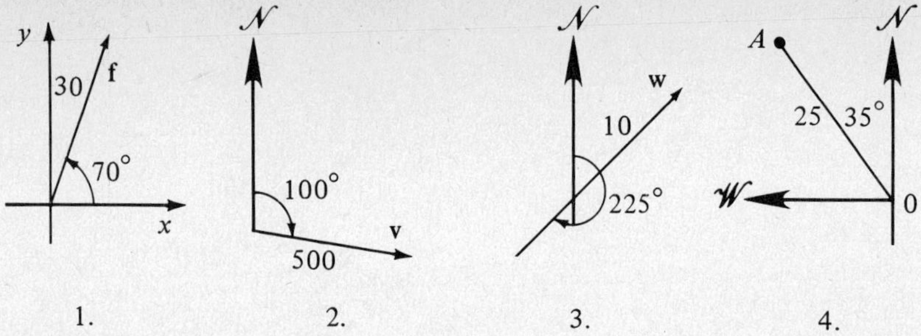

Figure 5.9 Designating vector quantities.

2. A velocity **v** of 500 mi/h at 100°.
   The angle, called heading, bearing, or course, is measured clockwise from north to the vector.
3. Wind velocity **w** of 10 knots from 225°.
   The angle is measured clockwise from north to the direction from which the wind is blowing.
4. A displacement **OA** from *O* to *A* of 25 km N35°W.
   The angle is measured from north (or south) toward west (or east).

Displacement

Notice the variety of ways that angles are measured in these applications. A displacement S20°E means turned 20° from south toward east. Consider an example and observe the importance of a sketch or diagram.

EXAMPLE

5. In a naval maneuver, one ship proceeds 20 mi N50°E and then 15 mi N10°W. A second ship begins at the same point but charts a straight course to rendezvous with the first ship. How far and in what direction did the second ship travel?

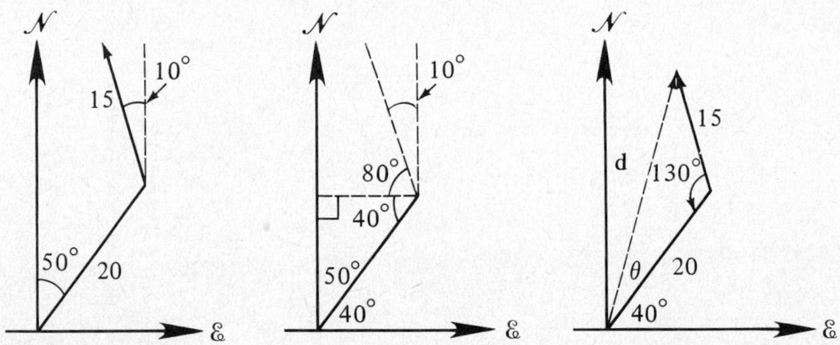

This is strictly a matter of displacement, because speeds are not considered.

First use the law of cosines to find d.

$$d^2 = 20^2 + 15^2 - 2(20)(15)\cos 130°$$
$$d \approx \sqrt{1010.6726}$$
$$\approx 31.791077$$

Then use the law of sines to determine direction.

$$\frac{\sin \theta}{15} = \frac{\sin 130°}{31.791077}$$
$$\sin \theta = \frac{15 \sin 130°}{31.791077}$$
$$\approx 0.36144314$$
$$\theta \approx 21.188851°$$
$$\approx 21°11'$$
$$40° + \theta = 60°11'$$

The second ship proceeds 31.8 mi N60°11′E.

Velocity

The motion of a boat in water or a plane in air is interesting because the water or air is itself moving. That is, current, tide, wind, or air stream throws one off a **heading** and on to a **course** or bearing that is the actual direction traveled. You point in one direction (heading) and are blown in a different direction (course). See Figure 5.10.

HISTORICAL ASIDE

Francois Vieté is considered the greatest French mathematician of the sixteenth century. He solved oblique triangles by dividing them into right triangles and is credited with the development of the law of tangents, which we will not study. The law of tangents was favored over the law of cosines while logarithms were the only practical means of computation. The law of cosines contains addition and subtraction that cannot be performed using logarithms. Vieté also used underlying trigonometric relationships to solve algebraic equations of higher degree.

Figure 5.10 Heading and course.

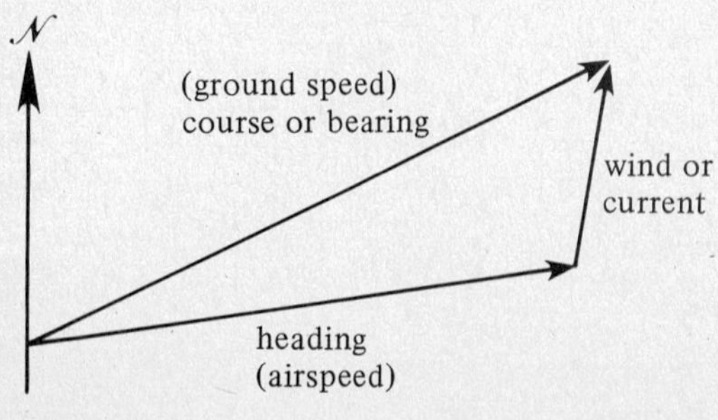

EXAMPLE

6. A pilot wishes to set a course of 170 mi/h at 130°. What is the heading and airspeed if there is a 24-mi/h wind from 200°?

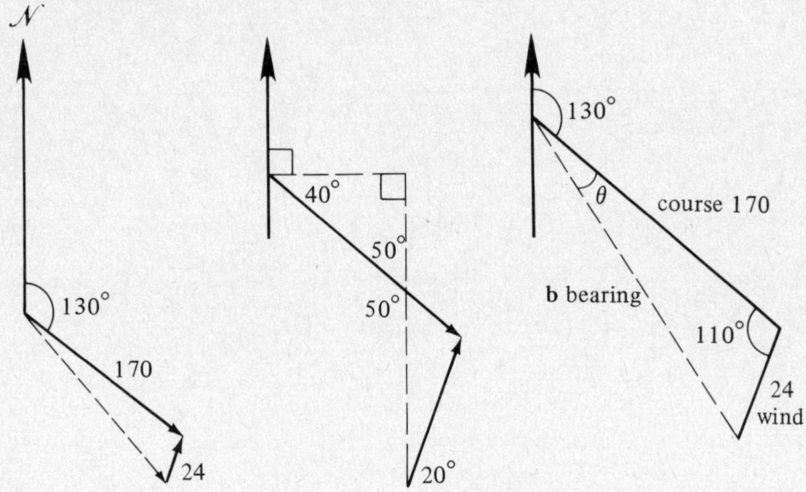

$$b^2 = 170^2 + 24^2 - 2(170)(24)\cos 110°$$
$$b \approx \sqrt{32266.884}$$
$$\approx 179.62985$$
$$\approx 180$$
$$\frac{\sin\theta}{24} = \frac{\sin 110°}{179.62985}$$
$$\sin\theta = \frac{24 \sin 110°}{179.62985}$$
$$\approx 0.12555052$$
$$\theta \approx 7.2125491°$$
$$\approx 7.21°$$
$$130° + \theta \approx 137.21°$$

The pilot needs an airspeed of 180 mi/h at 137.21° to stay on course.

Because of the approximate nature of the computation and the changeability of the wind, the bearing is frequently corrected.

7. A boat heads across a south-flowing river in the direction 32°. If the boat travels 9 mi/h in still water and the current is 4 mi/h, what is the actual speed and bearing of the boat?

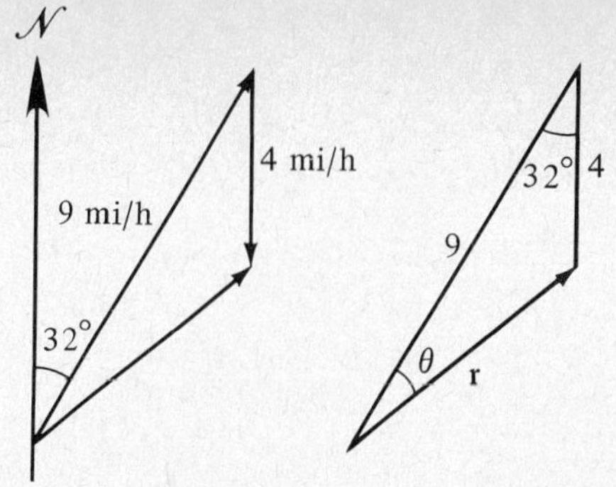

First use the law of cosines to find r.

$$r^2 = 4^2 + 9^2 - 2(4)(9)\cos 32^\circ$$
$$r \approx \sqrt{35.940537}$$
$$\approx 5.9950427$$

Knowing r, the law of sines may be used to find θ.

$$\frac{\sin\theta}{4} = \frac{\sin 32^\circ}{5.995}$$
$$\sin\theta \approx 0.35357415$$
$$\theta \approx 20.706083^\circ$$
$$\approx 20.71^\circ$$
$$32^\circ + \theta \approx 52.71^\circ$$

The boat actually travels approximately 5 mi/h on a course of 52.71°.

Force

Two forces acting in unison on a body may be resolved into a single force. If the body does not move, then it is resisting the forces with the same magnitude as the sum or resultant but in the opposite direction. Similarly, a single force may be broken into its component forces. A force system is said to be in equilibrium if the sum of the forces is zero.

EXAMPLES

8. A 550-lb weight is held in equilibrium by two lines at 30° and 45° with the vertical. Find the magnitude of the force exerted on each line.

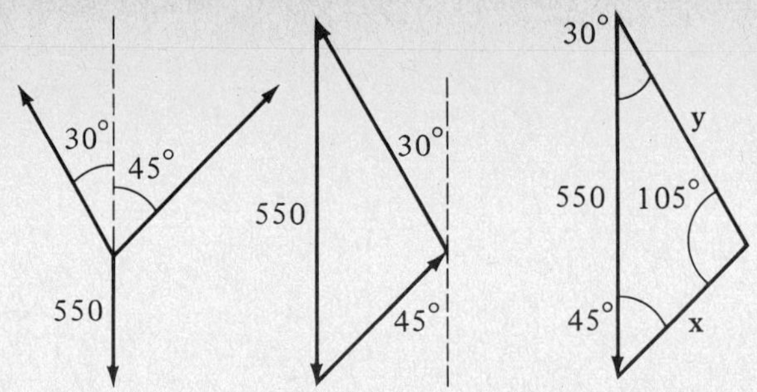

Why is it reasonable to assume that the lines are fixed on opposite sides of the vertical?

$$\frac{y}{\sin 45°} = \frac{550}{\sin 105°} \qquad \frac{x}{\sin 30°} = \frac{550}{\sin 105°}$$

$$y = \frac{550 \sin 45°}{\sin 105°} \qquad x = \frac{550 \sin 30°}{\sin 105°}$$

$$\approx 402.62794 \qquad \approx 284.70095$$

$$\approx 403 \qquad \approx 285$$

The 550-lb weight is held by a 403-lb force at 45° and 285 lb at 20°.

9. An 1800-lb automobile sits on a 22° incline. What force parallel to the incline (with the horizontal, of course) is necessary to keep the car at rest?

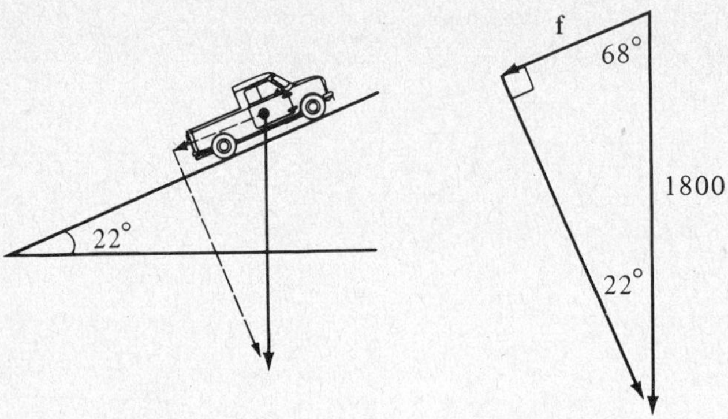

Break the force into component forces, one of which is the known weight of the car directed vertically down.

$$\frac{f}{1800} = \sin 22°$$

$$f = 1800 \sin 22°$$

$$\approx 674.29186$$

$$\approx 674$$

A force of approximately 674 lb keeps the 1800-lb car at rest.

What happens as the angle of the incline decreases?

exercise set 5.4

A. exercises

Sketch each of the vector quantities in Exercises 1–10 with special attention to direction.

1. A force **f** of 100 kg at 85° See Examples 1–4.

2. A 3-ton force **g** at 67°

3. A 600-lb force **p** at 130°

4. A 1000-kg force **q** at 220°

5. A velocity **v** of 140 mi/h at 60°

6. A velocity **w** of 225 km/h at 300°

7. A 30-knot wind from 40°

8. A 140° wind of 30 mi/h.

9. A 290-mi S30°E displacement.

10. A displacement of 500 km N60°W

B. exercises

Exercises 11–20 require a right triangle. Sketch the situation and solve completely.

11. A ship steams north 250 km, then turns west and travels another 100 km. How far away and what is the direction of the point of origin? See Example 5.

12. After a hiker walks 8 mi south, he turns east for another 5 mi. How far away and what is the direction of the starting point?

13. An airplane heads due west with an airspeed of 190 mi/h. A north wind blows at 30 mi/h. Find the bearing and ground speed of the plane. See Example 6.

14. A boat heads east across a south-flowing river. The current is 5 mi/h, and the speed of the boat in still water is 26 mi/h. Find the bearing and ground speed of the boat. See Example 7.

15. Forces of 150 and 175 lb act at right angles. Find the resultant force with the angle it makes with the larger force. See Example 8.

16. A 200-kg horizontal force and a 225-kg vertical force act on a body. Find the resultant with the angle it makes with the horizontal.

17. A force of 500 kg acts at 27° with the vertical. Find the magnitudes of the horizontal and vertical components of the force.

18. Find the magnitude of the horizontal and vertical components of a 200-lb force that acts at 19° with the horizontal.

19. A 350-lb weight rests on a plane inclined at 29°. What force parallel to the plane is necessary to keep the weight from moving? See Example 9.

20. A 165-lb force parallel to an inclined plane keeps a 2100-lb body from moving. What angle does the incline make with the horizontal?

Exercises 21–30 generally require an oblique triangle. Sketch the situation and solve completely.

21. Two ships leave port at the same time. One travels 24 knots and the other 19 knots. Four hours later they are 120 nautical miles apart. What is the angle between their courses? See Example 5.

One knot is one nautical mile per hour.

22. Two airliners leave San Francisco International at the same time, headed for Chicago and Houston. Two hours after take-off the planes are 470 mi apart. If they travel 425 mi/h and 450 mi/h, what is the angle between their courses?

23. A plane with airspeed of 210 mi/h contends with a strong west wind. It heads in the direction 159° but moves on the bearing 151°. What is the wind speed and the ground speed of the plane? See Example 6.

24. A boat whose rate in still water is 12 mi/h crosses a south-flowing river. It heads N42°W but moves N71°W. What is the rate of the current and the actual speed of the boat? See Example 7.

25. A river flows in the direction S28°W at the rate of 7 mi/h. Attempting to proceed due east, one must head N47°E. What is the speed of the boat in still water and the actual speed?

26. A pilot has a heading of 127°. A 29-mi/h wind from 40° blows the plane on to bearing 135°. Find the air and ground speeds of the plane.

27. A cruise tourist walks west across the deck of a ship moving north. The ship is moving at 22 mi/h and the person walks at 4 mi/h. Find the tourist's speed and direction relative to the ground.

28. A passenger walks north across the deck of a liner headed east. The person walks at 3 mi/h and the liner makes 25 mi/h. Find the passenger's direction and speed relative to the earth.

29. A ship proceeds 215 km S52°13′W from port before changing course to S7°55′E. After traveling 340 km on the new course, how far is the ship from port, and what is the direction of the new position from port?

30. A hiker travels 14 mi N14°25′W to where the night is spent in camp. The next day the hike is 20 mi N20°50′E. How far and in what direction is the hiker from the original starting position?

31. A child is pulling her little brother on a sled. The rope makes an angle of 36° with the level. If the girl exerts a pull of 35 lb on the rope, what force is being exerted on the sled to pull it forward?

See Example 8.

This, of course, ignores friction.

32. A 35-lb wagon holds a 63-lb child. The wagon tongue is held at 34° with the horizontal surface. How great a force on the tongue is necessary to get the wagon in motion?

33. How much force is wasted when a person must push with a force of 95 lb on the handle of a lawnmower, if it is held at 40° with the level ground?

34. A high window is opened with the aid of a window pole. If a person exerts 50 lb at an angle of 20° with the vertical, what vertical force will be necessary to open the window?

35. A large crane will withstand 30 tons tension. If the crane is extended at 75° with the horizontal, what is the greatest weight that can be lifted with a 3-ton safety margin?

36. A ship's boom is rated at 50 tons. If the boom must extend at 29° with the vertical to offload cargo, what is the greatest weight that can be lifted with a 5-ton safety margin?

37. A boy can just keep a barrel from rolling down a ramp by pushing with a force of 80 lb. If the barrel weighs 315 lb, what is the angle of elevation of the ramp?

See Example 9.

38. The sides of the pyramid of Khufu are 51°51′ with the horizontal. What force would have been necessary to hold a 2-ton block of stone from sliding backward during construction?

39. A theater marquee is hinged to the building and the weight supported by two cables attached to the building. If the cables make an angle of 35° with the side of the building and the marquee weighs 3400 lb, what is the tension on each cable?

40. A drawbridge is raised and lowered by two cables. When the bridge is down, the cables make an angle of 32° with a vertical tower. When the bridge is being raised, what is the maximum tension on each cable if the hinged bridge section weighs 16 tons?

C. exercises

41. **brain teaser** Prove that the sum of the squares of the lengths of the diagonals of a parallelogram is equal to the sum of the squares of the sides.

42. **peanuts problem** Suppose that Snoopy follows his own directions in the cartoon above. First, he moves 184.32 ft in the direction N81°2′W, then turns and moves again, and moves in a third direction. What is the distance and direction from his starting point after moving the third time?

5.5 CHAPTER REVIEW

Fill in the word or words that make the statements in Exercises 1–10 complete and correct.

1. (5.1) An ________ triangle contains no right angle.
2. (5.1) The area of any triangle is given by ________ the ________ of the lengths of two sides and the sine of the included angle.
3. (5.1) In any triangle, the sides are ________ to the sines of the opposite angles.

Exercises 4–7 refer to Figure 5.11.

Figure 5.11

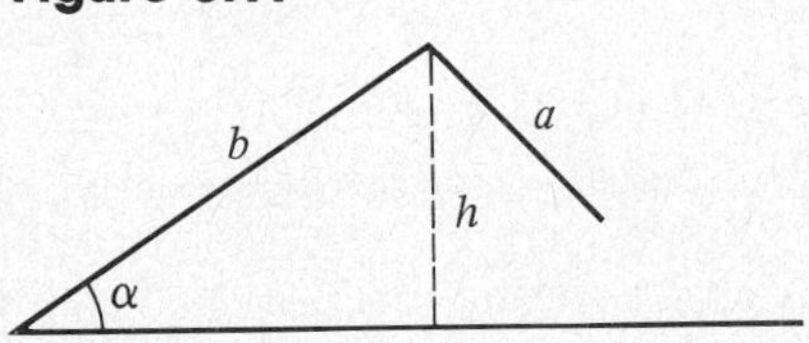

4. (5.2) Given a, b, and α, this is referred to as the ________ case.

5. (5.2) If $a < h$, ________ distinct triangle(s) is (are) determined.

6. (5.2) If $a = h$, a unique ________ triangle is determined.

7. (5.2) If $a > h$, ________ distinct triangle(s) is (are) determined.

8. (5.3) If three sides of a triangle are known, the law of ________ may be used to find any one of the angles.

9. (5.4) Heading, bearing, or course is measured________ from ________.

10. (5.4) The sum of two vectors is called their ________.

11. (5.1) Find the area of $\triangle ABC$ if:
    (a) $a = 154$, $c = 97$, $\beta = 51°$
    (b) $c = 46.9$, $b = 58.7$, $\beta = 132°$

12. Find the required part of $\triangle ABC$, if possible,
    (a) (5.1) $a = 63$, $\alpha = 109°$, $\beta = 43°$, $b = ?$
    (b) (5.1) $b = 13.15$, $\beta = 22°44'$, $\alpha = 58°32'$, $c = ?$
    (c) (5.2) $b = 2$, $c = 5$, $\beta = 30°$, $\gamma = ?$
    (d) (5.2) $a = 2.2$, $c = 2.6$, $\gamma = 69°$, $\beta = ?$
    (e) (5.2) $c = 15$, $b = 14$, $\gamma = 60°$, $\alpha = ?$
    (f) (5.3) $a = 20.1$, $b = 15.7$, $c = 9.5$, $\beta = ?$
    (g) (5.3) $a = 154$, $b = 97$, $\beta = 51°$, $c = ?$

Exercises 13–15 refer to the figure shown. Find the required part to the nearest tenth of a unit or hundredth of a degree.

13. (5.3) α

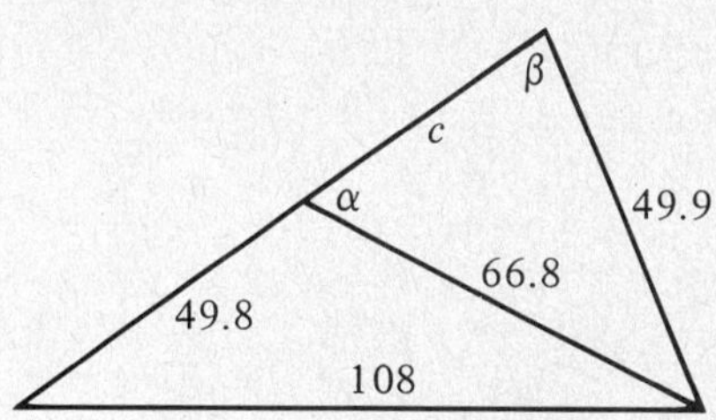

14. (5.2) β

15. (5.1) c

16. (5.1) A farmer is considering planting a new field that is triangular in shape. Two sides of the field are 574 ft and 713 ft and meet at 123°. How many acres is the new field?

17. (5.1) The shorter diagonal of a parallelogram is 72 in. and makes angles of 74° and 33° with the sides. Find the lengths of the sides.

18. (5.2) Philadelphia is 120 mi N42°E from Washington, D.C. Harrisburg is due north of Washington and 96 mi north and west of Philadelphia. How far is our nation's capital from the Pennsylvania capital?

19. (5.3) Two cities on a straight stretch of river are both visible from a mountain observation platform. The cities are 25 and 36 mi distant, and their lines of sight make a 21° angle. What is the distance between the cities?

20. (5.4) A forest ranger at an observation point sights a fire in the direction N74°E. A fire watcher spots the same fire at N57°E from his position 4 mi south of the ranger. How far is the fire from the closer of the two observers?

complex numbers and polar coordinates

six

6.1 OPERATIONS WITH COMPLEX NUMBERS

All quantities considered to this point have been real numbers. Trigonometry, however, can also be applied to complex numbers and does, in fact, give new meaning to certain concepts and explain others for the first time. First, recall the set of complex numbers.

THE SET OF COMPLEX NUMBERS

The set of complex numbers C is the set of numbers of the form $a + bi$, a and b real, and $i = \sqrt{-1}$, where a is called the **real part,** b is called the **imaginary part,** and i is the **imaginary unit.**

$a + bi$ is called the **standard form** of a complex number.

A direct result of the definition is the cyclic nature of powers of i.

$$i = \sqrt{-1}$$
$$i^2 = -1$$
$$i^3 = i^2 i = (-1)(i) = -i$$
$$i^4 = i^2 i^2 = (-1)(-1) = 1$$
$$i^5 = i^4 i = (1)(i) = i$$

$$\mathbf{i = \sqrt{-1}}$$
$$\mathbf{i^2 = -1}$$
$$\mathbf{i^3 = -i}$$
$$\mathbf{i^4 = 1}$$

Any higher integral power of i can be factored into factors of i^4 and a single factor of i, i^2, or i^3. Because $i^4 = 1$, every integral power of i is equal to one of these fundamental values.

EXAMPLES

Simplify the following.

1. $i^7 = i^4 i^3 = 1 \cdot i^3$

   $= i^3$

   $\mathbf{i^7 = -i}$

2. $i^{17} = i^{16} i = (i^4)^4 i$

   $= (1)^4 i = 1 \cdot i$

   $\mathbf{i^{17} = i}$

3. $i^{222} = i^{220}i^2$

$= (i^4)^{55}i^2$

$= 1^{55}(-1) = 1(-1)$

$\mathbf{i^{222} = -1}$

4. $i^{-7} = \frac{1}{i^7}\left(\frac{i}{i}\right)$

$= \frac{i}{i^8} = \frac{i}{1}$

$\mathbf{i^{-7} = i}$

Having defined i, the square root (or for that matter any even root) of a negative number can be found with the aid of the following.

$\sqrt{a}\sqrt{b} = \sqrt{ab}$ for real a and b not both negative.

EXAMPLES

5. $\sqrt{-16} = \sqrt{16}\sqrt{-1}$

$= \mathbf{4i}$

6. $\sqrt{-18} = \sqrt{9}\sqrt{2}\sqrt{-1}$

$= \mathbf{3\sqrt{2}i}$

Two complex numbers are equal if and only if the real parts are equal and the imaginary parts are equal. That is:

$a + bi = c + di$

if and only if $a = c$ and $b = d$.

This definition of equality allows us to perform operations with complex numbers. To add or subtract, the real parts are added or subtracted as well as the imaginary parts.

EXAMPLES

Simplify and write in standard form.

7. $(2 + 5i) + (3 - 4i) = 2 + 3 + 5i - 4i$

$= (2 + 3) + (5 - 4)i$

$= \mathbf{5 + i}$

HISTORICAL ASIDE

The Hindus were concerned with the concept of the square root of a negative as early as 1150 B.C., but the first to give it any serious attention was Girolamo Cardano (1501–1576). Cardano called them "sophistic," which means unnecessary or false, clever, and minute arguement. It was Decartes who, in 1637, first distinguished between real and imaginary numbers. He used this designation to settle the matter of their existence. Gauss wrote in 1863 that imaginary quantities have been "surrounded by a mysterious obscurity, (that) is to be attributed largely to an ill-adapted notation. If for instance, +1, −1, $\sqrt{-1}$ had been called direct, inverse, and lateral units, instead of positive, negative, and imaginary (or even impossible) such an obscurity would have been out of the question." It is still unfortunate that we refer to them as imaginary. They might be better referred to as "image-less," that is, without an image on the (real) number line.

8. $(1 - 6i) - (7 - 8i) = (1 - 7) + (-6 + 8)i$

$$= \mathbf{-6 + 2i}$$

9. Solve for x and y.

$$2x + 3yi = (5 + i) + (-6 + 2i)$$

$$2x + 3yi = -1 + 3i$$ Equate the real and imaginary parts.

$$2x = -1 \qquad 3y = 3$$

$$\mathbf{x = -\tfrac{1}{2}} \qquad \mathbf{y = 1}$$

To multiply or divide complex numbers, they are treated as binomials much the same as any radical expression except for the special cyclic property of i.

EXAMPLES

Simplify and write in standard form.

10. $(2 - i)i = 2i - i^2$

$$= 2i - (-1)$$

$$= \mathbf{1 + 2i}$$

11. $(2 + i)(1 + 3i) = 2 + 7i + 3i^2$

$$= 2 + 7i + 3(-1)$$

$$= 2 + 7i - 3$$

$$= \mathbf{-1 + 7i}$$

Notice that the operation is thc same as $(2 + x)(1 + 3x)$ except that $i^2 = -1$.

12. $(3 - 2i)(2 + i) = 6 - i - 2i^2$

$$= 6 - i - 2(-1)$$

$$= 6 - i + 2$$

$$= \mathbf{8 - i}$$

13. $\dfrac{5 + i}{i} = \dfrac{(5 + i)i}{i \cdot i} = \dfrac{5i + i^2}{i^2} = \dfrac{5i + (-1)}{-1}$

$$= \frac{5i - i}{-1}$$

$$= \mathbf{1 - 5i}$$

Here multiplying by i clears the denominator of imaginary units.

14. $$\frac{2}{3-i} = \frac{2(3+i)}{(3-i)(3+i)} = \frac{2(3+i)}{10}$$

$$= \frac{3+i}{5} = \mathbf{\frac{3}{5} + \frac{1}{5}}\boldsymbol{i}$$

The conjugate of $a + b$ is $a - b$, and their product is $a^2 - b^2$. Recall how the conjugate was used to "rationalize the denominator" when simplifying radical expressions.

15. $$\frac{3-2i}{2+i} = \frac{(3-2i)(2-i)}{(2+i)(2-i)}$$

$$= \frac{4-7i}{4+1}$$

$$\mathbf{= \frac{4}{5} - \frac{7}{5}}\boldsymbol{i}$$

16. Solve for x and y.

$$3x + 2yi = (3 + 2i)(1 - 2i)$$

$$3x + 2yi = 7 - 4i$$

and from the definition of equality, we set the real parts equal and the imaginary parts equal.

$$3x = 7 \quad \text{and} \quad 2y = -4$$

$$\boldsymbol{x = \tfrac{7}{3}} \qquad \boldsymbol{y = -2}$$

The extensive use of complex conjugates is evident from the previous examples. This useful technique relies on the differences of squares and the fact that the square of i is real.

$$(a + bi)(a - bi) = a^2 - b^2i^2$$

$$= a^2 - b^2(-1)$$

$$= a^2 + b^2$$

Hence multiplying by the conjugate will always produce a real number, because both a and b are real. Exercise Set 6.1 should provide sufficient practice to review and freshen skills with complex numbers and their operations.

exercise set 6.1

A. exercises

Simplify the powers of i in Exercises 1–10.

See Examples 1–4.

1. i^6
2. i^9
3. i^{77}
4. i^{57}
5. i^{276}
6. i^{142}
7. i^{-9}
8. i^{-15}
9. $i^5 + i^7 + i^9$
10. $i^4 + i^6 + i^8$

Write each quantity in Exercises 11–20 as a multiple of i.

See Examples 5 and 6.

11. $\sqrt{-4}$
12. $\sqrt{-9}$
13. $\sqrt{-121}$
14. $\sqrt{-144}$
15. $\sqrt{-27}$
16. $\sqrt{-8}$
17. $\sqrt{-48}$
18. $\sqrt{-72}$
19. $\sqrt{-120}$
20. $\sqrt{-140}$

Perform the indicated operations in Exercises 21–50, simplify, and write in standard form.

See Examples 7 and 8.

21. $(2 + 3i) + (3 - 2i)$
22. $(3 - 4i) + (4 + 5i)$
23. $(2 + 3i) - (3 - 2i)$
24. $(3 - 4i) - (4 + 5i)$
25. $(5 + 2i) + (5 - 2i)$
26. $(5 + 2i) - (5 - 2i)$

See Examples 10–12.

27. $(2 + i)(3 - i)$
28. $(1 - i)(-4 + i)$
29. $(1 + 2i)(-2 - 3i)$
30. $(3 + 4i)(1 - 3i)$
31. $(2 - 5i)(3 + 2i)$
32. $(3 + 7i)(2 - 3i)$
33. $i(3 - 4i)$
34. $i(5 + 3i)$
35. $(3 + i)^2$
36. $(-1 + 2i)^2$
37. $i(3 + 7i)(1 - 5i)$
38. $(2 - 9i)i(3 + i)$
39. $(2 + i)(1 - i)^2$
40. $(2 + i)^2(1 - i)$

41. $\dfrac{1}{1+i}$ 42. $\dfrac{1}{2-i}$ See Examples 13-15.

43. $\dfrac{i}{1-i}$ 44. $\dfrac{i}{2+i}$

45. $\dfrac{1-i}{1+i}$ 46. $\dfrac{2+i}{2-i}$

47. $\dfrac{2+i}{1-i}$ 48. $\dfrac{1+i}{2-i}$

49. $\dfrac{3-2i}{1+3i}$ 50. $\dfrac{-2+3i}{3-i}$

B. exercises

Solve each of the equations in Exercises 51–60 for x and y.

51. $x + yi = 4 + 2i - (2 + 5i)$ 52. $x + yi = 1 - 2i + (2 + 4i)$ See Examples 9 and 16.

53. $2x + yi = 6 + 5i$ 54. $x - 3yi = 5 - 6i$

55. $x + 2 = 5i - yi$ 56. $2i - yi = x - 5$

57. $x + yi = (2 - 3i)(3 + 5i)$ 58. $x + yi = (3 + 7i)(1 - 2i)$

59. $3x + 2yi = (1 + 7i)(5 - i)$ 60. $4x + 7yi = (3 + i)(1 - 5i)$

Simplify each expression in Exercises 61–76.

61. $G = (1 + i)^3(1 - i)i$ 62. $H = (1 - i)^2(1 + i)^2$ See Exercise 77.

63. $K = \dfrac{i(6 + 7i)(6 - 7i)}{(1 - 4i)(1 + 4i)}$ 64. $L = \dfrac{(2 + 11i)(2 - 11i)}{i(4 - 3i)(4 + 3i)}$

65. $M = 8i(1 - i) - 2(4 + 3i)$ 66. $N = i(8 - i) - 2(4i + 3)$

67. $P = (3 + 2i)(7i - 5) - (7i - 29)$

68. $R = (3 + 2i)(5i - 2) - (5i - 16)$

69. $Y = i(2 + i)(3 - 4i) - i(7 - 5i)$

70. $F = (1 + 2i)(2 - 3i) - (2 - 5i)(1 + 3i)$

71. $S = \dfrac{(1 - 3i)(1 + 2i)(1 + i)}{4i - 3}$ 72. $T = \dfrac{(1 + i)(2 + i)(1 - 2i)}{1 - 7i}$

73. $A = \dfrac{6}{7 - i} - \dfrac{7i}{3i + 4}$

74. $C = \dfrac{16}{9 - i} + \dfrac{2}{5 + 4i}$

75. $E = \dfrac{i}{1 - i} + \dfrac{1 - i}{2i}$

76. $I = \dfrac{i}{2 + i} - \dfrac{2i}{1 - 2i}$

C. exercises

77. **trig talk** Use Exercises 61–76 to fill the blanks with the letters that correspond to the values below the blanks. The letter "oh" has been inserted already. When the blanks are filled, the quotation may be read.

___ ___ ___   ___ ___ ___ ___ ___ ___ ___ ___ ___
$i \quad 4 \quad -1 \qquad 1 \quad 2i \quad -i \quad -4 \quad 1 \quad -5 \quad -i \quad 6i \quad 3i$

___ ___   O ___ ___   O ___   ___ ___ ___
$1 \quad -2i \qquad \quad -5 \quad -1 \qquad \quad -9 \qquad i \quad 4 \quad -1$

___ ___ ___ ___ ___ ___ ___ ___ ___ ___   O ___
$2i \quad -i \quad -2i \quad i \quad -1 \quad 6i \quad 5i \quad -1 \quad 3i \quad -2i \qquad \quad -9$

___ ___ ___ ___ ___ ___ ___ ___   ___ ___ ___ ___ ___ ___ ___.
$4i \quad 4 \quad 3i \quad -2i \quad 1 \quad 2 \quad -i \quad -5i \qquad -2i \quad 2 \quad 1 \quad -1 \quad -5 \quad 2 \quad -1$ (Thomas Hill)

78. **brain teaser** Prove that the set of complex numbers is closed for addition and subtraction.

Hint for Exercise 78: You must show that when you add two complex numbers, you get another number of the form $a + bi$.

79. **brain teaser** Prove that the set of complex numbers is closed for multiplication.

80. **brain teaser** Prove that the set of nonzero complex numbers is closed for division.

6.2 POLAR FORM OF COMPLEX NUMBERS

Because every complex number can be written in the form $a + bi$, each number can be associated with a unique ordered pair (a, b). The horizontal or a axis is called the **real axis** and the vertical or b axis the **imaginary axis.** In this way each complex number can be associated with a unique point in a rectangular coordinate system. Conversely, each point will be associated with a unique complex number. For this application the Cartesian plane is called the **complex plane.** See Figure 6.1.

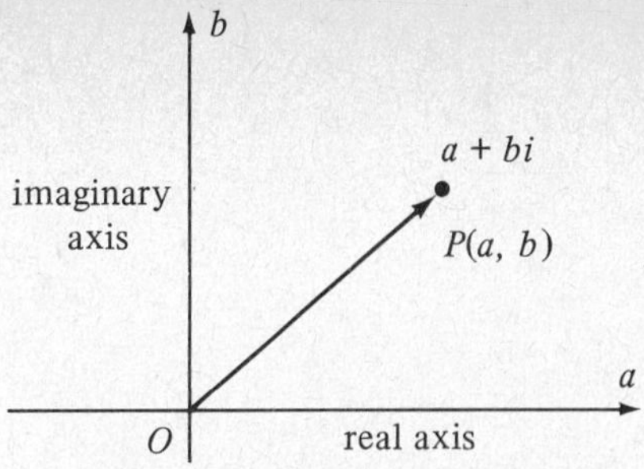

Figure 6.1 The complex plane.

EXAMPLES

Plot the numbers in the complex plane.

1. $5 + 7i$
2. $3 - 2i$
3. $5 + 0i$
4. $0 - 5i$

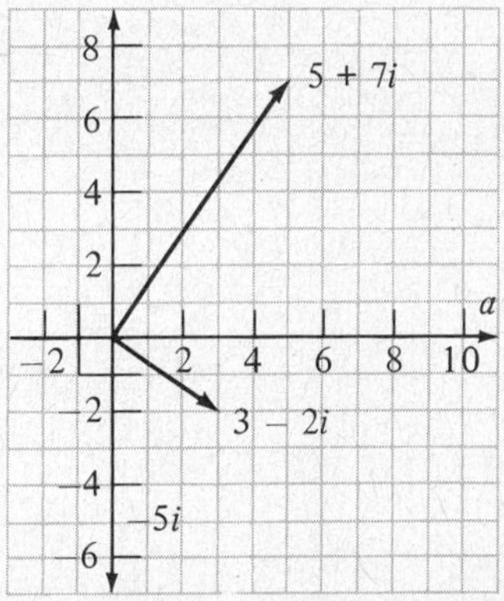

Each point determines a unique directed line segment, which starts from the origin and ends at the point. Recall that such a directed line segment is called a vector, **OP** in Figure 6.1. The sum of two complex numbers can be shown graphically similarly to the resultant of two vectors. The magnitude and direction of the vector quantity $x + yi$ can be given as r and θ. If r is the length of the segment from $(0, 0)$ to (x, y), then $r = \sqrt{x^2 + y^2}$. If the angle θ is the smallest positive angle in standard position, then the point (x, y) can be located uniquely with the ordered pair (r, θ). We say that the point (x, y) has **polar coordinates** (r, θ).

Observe from Figure 6.2 the following relationships between the two sets of coordinates.

$$r^2 = x^2 + y^2 \qquad \frac{x}{r} = \cos\theta \qquad \frac{y}{r} = \sin\theta$$

$$\mathbf{r = \sqrt{x^2 + y^2} \qquad x = r\cos\theta \qquad y = r\sin\theta}$$

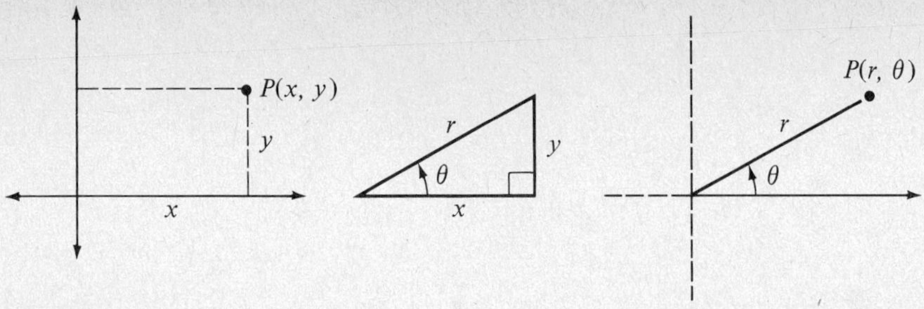

Figure 6.2 Rectangular and polar coordinates.

The complex number $x + yi$ can now be written in terms of r and θ.

$$x + yi = (r \cos \theta) + (r \sin \theta)i$$

$$\boldsymbol{x + yi = r(\cos \theta + i \sin \theta)}$$

This expression is often abbreviated as

$$\boldsymbol{r \operatorname{cis} \theta = r(\cos \theta + i \sin \theta)}$$

and is called the **polar form** (or **trigonometric form**) of the complex number $x + yi$. Here r is called the modulus and θ the argument of the number.

r is also referred to as the absolute value of the complex number $r \operatorname{cis} \theta$.

Once you know x and y, r and θ may be determined as shown above. Thus the number may be converted from rectangular form to polar form and conversely.

RECTANGULAR AND POLAR FORMS OF COMPLEX NUMBERS

$$x + yi = r \operatorname{cis}(\theta + 2k\pi)$$

or

$$r \operatorname{cis}(\theta + 360°k) \quad \text{for any integer } k$$

where

$$r = \sqrt{x^2 + y^2} \qquad x = r \cos \theta$$

$$\theta = \arctan \frac{y}{x} \qquad y = r \sin \theta$$

EXAMPLES

5. Write $1 + i$ in polar form.

$x = 1$ and $y = 1$ $\qquad \theta$ in QI

$$r = \sqrt{x^2 + y^2} \qquad \tan \theta = \frac{y}{x}$$

$$= \sqrt{1^2 + 1^2} \qquad = \frac{1}{1} = 1$$

$$= \sqrt{2} \qquad \theta = 45°$$

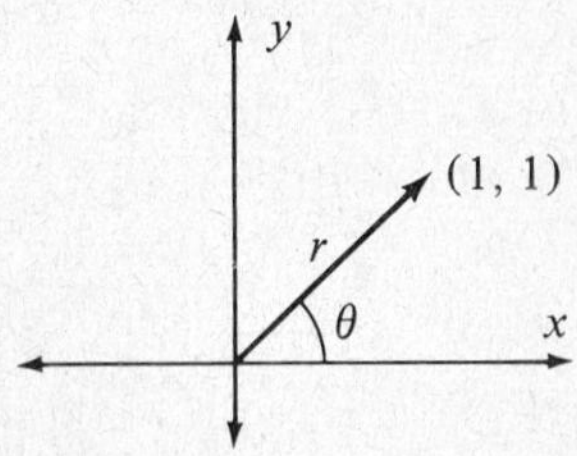

$$\mathbf{1 + i = \sqrt{2}(\cos 45° + i \sin 45°) = \sqrt{2} \text{cis } 45°}$$

or

$$\sqrt{2} \text{ cis } \frac{\pi}{4} \quad \text{or} \quad \sqrt{2} \text{ cis } \left(\frac{\pi}{4} + 2k\pi\right) \quad \text{or} \ldots$$

6. Write $\sqrt{3} - i$ in polar form.

$x = \sqrt{3}$ and $y = -1$ $\qquad \theta$ in QIV

$$r = \sqrt{(\sqrt{3})^2 + (-1)^2} \qquad \tan \theta = -\frac{1}{\sqrt{3}}$$

$$= \sqrt{3 + 1} = \sqrt{4} \qquad \theta = 330°$$

$$= 2$$

$$\mathbf{\sqrt{3} - i = 2 \text{ cis } 330°}$$

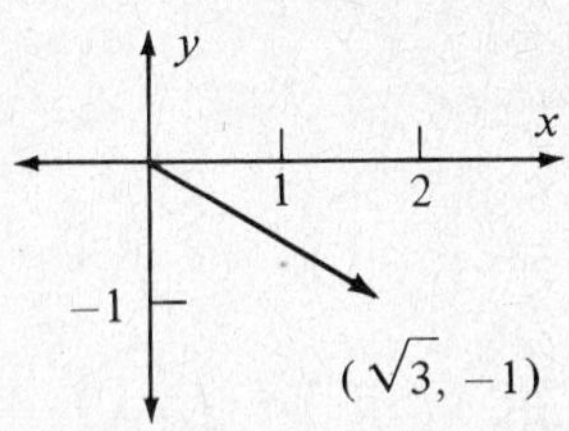

7. Write $-2i$ in polar form.

$$r = \sqrt{0^2 + (-2)^2}$$

$$= 2 \qquad \theta = 270°$$

$$\mathbf{-2i = 2 \text{ cis } 270°}$$

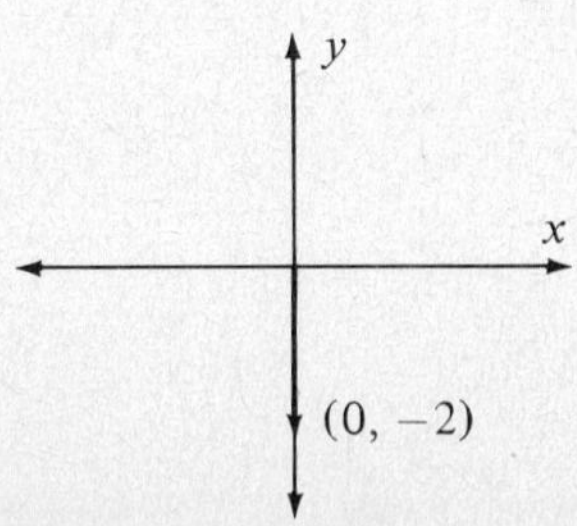

8. Write $5 \text{ cis } \frac{3\pi}{2}$ in rectangular form.

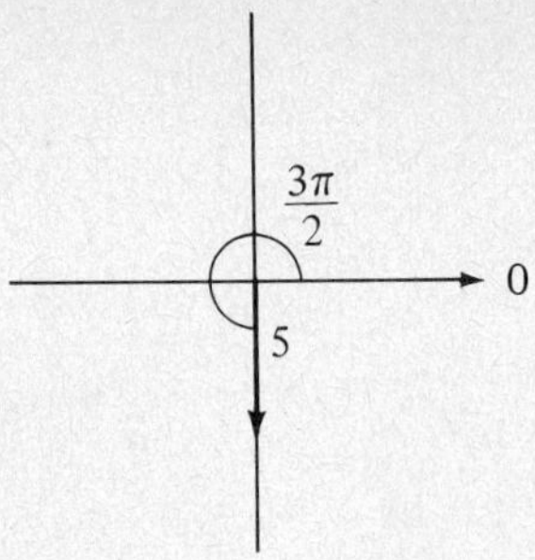

$$r = 5 \quad \text{and} \quad \theta = \frac{3\pi}{2}$$

$$\mathbf{x = r \cos \theta} \qquad \mathbf{y = r \sin \theta}$$

$$x = 5 \cos \frac{3\pi}{2} \qquad y = 5 \sin \frac{3\pi}{2}$$

$$= 5(0) \qquad = 5(-1)$$

$$\mathbf{5 \text{ cis } \frac{3\pi}{2} = 0 - 5i}$$

9. Write 3 cis 225° in rectangular form.

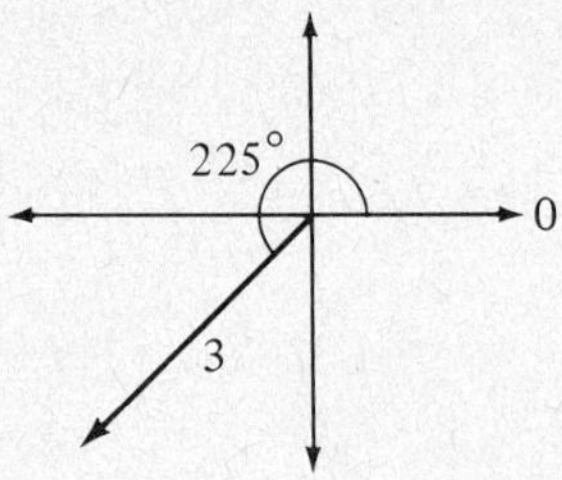

$$r = 3 \quad \text{and} \quad \theta = 225°$$

$$x = 3 \cos 225° \qquad y = 3 \sin 225°$$

$$= 3\left(-\frac{\sqrt{2}}{2}\right) \qquad = 3\left(-\frac{\sqrt{2}}{2}\right)$$

$$= \frac{-3\sqrt{2}}{2} \qquad = \frac{-3\sqrt{2}}{2}$$

$$\mathbf{3 \text{ cis } 225° = -\frac{3\sqrt{2}}{2} - \frac{3\sqrt{2}}{2}}\,i$$

10. Write $\sqrt{6} \text{ cis } \frac{2\pi}{3}$ in rectangular form.

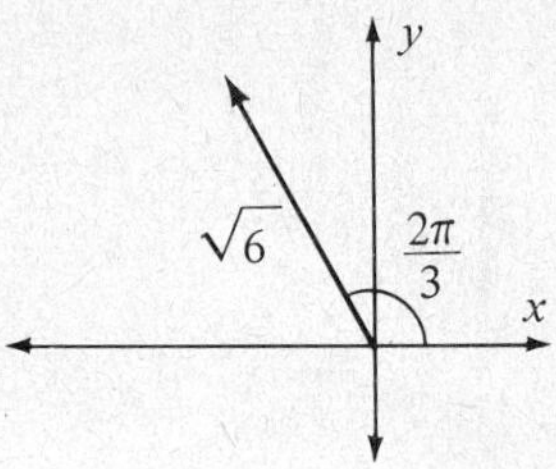

$$r = \sqrt{6} \quad \text{and} \quad \theta = \frac{2\pi}{3}$$

$$x = \sqrt{6} \cos \frac{2\pi}{3} \qquad y = \sqrt{6} \sin \frac{2\pi}{3}$$

$$= \sqrt{6}\left(-\frac{1}{2}\right) \qquad = \sqrt{6}\left(\frac{\sqrt{3}}{2}\right)$$

$$= -\frac{\sqrt{6}}{2} \qquad = \frac{3\sqrt{2}}{2}$$

$$\mathbf{\sqrt{6} \text{ cis } \frac{2\pi}{3} = -\frac{\sqrt{6}}{2} + \frac{3\sqrt{2}}{2}}\,i$$

Tables or a calculator yield approximations, so certain complex numbers in polar form can only be approximated in rectangular form and conversely.

EXAMPLES

11. Write $2 - 3i$ in polar form.

Make a sketch, with θ in QIV.

$$x = 2 \quad \text{and} \quad y = -3$$

$$r = \sqrt{2^2 + (-3)^2} \qquad \theta = \arctan -\tfrac{3}{2}$$

$$= \sqrt{13} \qquad \approx -56.309932^\circ$$

Note that the calculator gives a negative angle, and we wish the smallest positive angle.

$$\theta \approx 303.69006^\circ$$

$$\theta \approx 303^\circ 41'$$

$$\mathbf{2 - 3i \approx \sqrt{13}\ \text{cis}\ 303^\circ 41'} \quad \text{or} \quad 3.6056 \text{ cis } 303.68^\circ$$

12. Write $4 \text{ cis } 172^\circ$ in rectangular form.

$$r = 4 \quad \text{and} \quad \theta = 172^\circ$$

$$x = 4 \cos 172^\circ \qquad y = 4 \sin 172^\circ$$

$$\approx -3.9610722 \qquad \approx 0.55669240$$

$$\approx -3.96 \qquad \approx 0.56$$

$$\mathbf{4 \text{ cis } 172^\circ \approx -3.96 + 0.56i}$$

Certain calculator models have a rectangular-to-polar and polar-to-rectangular conversion feature. Their operation varies more widely brand to brand than other features. An owner's manual should be consulted for instructions. Use this feature to check your work, but be certain you understand the conversion as shown in the examples.

exercise set 6.2

A. exercises

Plot each of the numbers in Exercises 1–10 in the complex plane.

See Examples 1–4.

1. $3 + 4i$
2. $2 - 5i$
3. $-7 + i$
4. $-6 - 9i$
5. $0 + 3i$
6. $5 + 0i$
7. $5 - 3i$
8. $5 + 3i$
9. $-5 + 3i$
10. $-5 - 3i$

Write the numbers in Exercises 11–20 in polar form.

See Examples 5–7.

11. $1 + i$
12. $1 - i$
13. $2 + 0i$
14. $0 - 3i$
15. $\sqrt{3} - i$
16. $1 + i\sqrt{3}$
17. $2\sqrt{3} - 2i$
18. $\sqrt{2} - i\sqrt{2}$
19. $-3 - 3i\sqrt{3}$
20. $-2 + 2i\sqrt{3}$

Write the numbers in Exercises 21–30 in rectangular form.

See Examples 8–10.

21. $3(\cos 60° + i \sin 60°)$
22. $3\left(\cos \frac{\pi}{6} + i \sin \frac{\pi}{6}\right)$
23. $2 \text{ cis } 45°$
24. $5 \text{ cis } 90°$
25. $\sqrt{2} \text{ cis } \frac{3\pi}{4}$
26. $\sqrt{3} \text{ cis } \frac{7\pi}{4}$
27. $4 \text{ cis } 120°$
28. $3 \text{ cis } 210°$
29. $\sqrt{5} \text{ cis } 330°$
30. $\sqrt{3} \text{ cis } 150°$

B. exercises

calculator problems *Change the numbers in Exercises 31–40 to polar or rectangular form. State answers to the nearest minute or hundredth.*

See Examples 11 and 12.

31. $3 + 4i$
32. $5 - 12i$
33. $2 \text{ cis } 72°$

24. $5 \text{ cis } 28°$

35. $-8 + 15i$
36. $-24 - 7i$
37. $\sqrt{7} \text{ cis } 251°$
38. $\text{cis } 153°$
39. $\sqrt{2} - \sqrt{3}i$
40. $-20 + 21i$

6.3 DEMOIVRE'S THEOREM

The polar form of a complex number offers distinct advantages over the rectangular form for multiplication and division. This form also makes the taking of roots of complex numbers practical. First, consider the product of two complex numbers in polar form.

$$\begin{aligned}(r_1 \text{ cis } \theta_1)(r_2 \text{ cis } \theta_2) &= r_1r_2 \text{ cis } \theta_1 \text{ cis } \theta_2\\ &= r_1r_2(\cos \theta_1 + i \sin \theta_1)(\cos \theta_2 + i \sin \theta_2)\\ &= r_1r_2(\cos \theta_1 \cos \theta_2 + i^2 \sin \theta_1 \sin \theta_2\\ &\quad + i \cos \theta_1 \sin \theta_2 + i \sin \theta_1 \cos \theta_2)\\ &= r_1r_2[(\cos \theta_1 \cos \theta_2 - \sin \theta_1 \sin \theta_2)\\ &\quad + i(\cos \theta_1 \sin \theta_2 + \sin \theta_1 \cos \theta_2)]\\ &= r_1r_2[\cos(\theta_1 + \theta_2) + i \sin(\theta_1 + \theta_2)]\\ (r_1 \text{ cis } \theta_1)(r_2 \text{ cis } \theta_2) &= r_1r_2 \text{ cis}(\theta_1 + \theta_2)\end{aligned}$$

Simply put, to multiply $r_1 \text{ cis } \theta_1$ and $r_2 \text{ cis } \theta_2$, multiply r_1 and r_2, and add θ_1 and θ_2. That is, multiply the moduli and add the arguments.

The quotient is similar:

$$\frac{r_1 \text{ cis } \theta_1}{r_2 \text{ cis } \theta_2} = \frac{r_1}{r_2} \text{ cis}(\theta_1 - \theta_2)$$

COMPLEX PRODUCT AND QUOTIENT IN POLAR FORM

$$(r_1 \text{ cis } \theta_1)(r_2 \text{ cis } \theta_2) = r_1 r_2 \text{ cis}(\theta_1 + \theta_2)$$

To multiply two complex numbers in polar form, multiply the moduli and add the arguments.

$$\frac{r_1 \text{ cis } \theta_1}{r_2 \text{ cis } \theta_2} = \frac{r_1}{r_2} \text{ cis}(\theta_1 - \theta_2)$$

To divide two complex numbers in polar form, divide the moduli and subtract the arguments.

EXAMPLES

1. $(2 \text{ cis } 45°)(3 \text{ cis } 30°) = (2 \cdot 3) \text{ cis}(45° + 30°)$

   $= \mathbf{6 \text{ cis } 75°}$

2. $(\sqrt{2} \text{ cis } 137°)(2\sqrt{6} \text{ cis } 284°) = \sqrt{2} \cdot 2\sqrt{6} \text{ cis}(137° + 284°)$

   $= 4\sqrt{3} \text{ cis } 421°$ 421° is coterminal with 421° − 360° = 61°.

   $= \mathbf{4\sqrt{3} \text{ cis } 61°}$

3. $\dfrac{15 \text{ cis}(\pi/2)}{5 \text{ cis}(\pi/3)} = \dfrac{15}{5} \text{ cis}\left(\dfrac{\pi}{2} - \dfrac{\pi}{3}\right)$

   $= \mathbf{3 \text{ cis } \dfrac{\pi}{6}}$

4. $\dfrac{12 \text{ cis } 63°}{18 \text{ cis } 87°} = \dfrac{12}{18} \text{ cis}(63° - 87°)$

   $= \dfrac{2}{3} \text{ cis}(-24°)$ −24° is coterminal with −24° + 360° = 336°.

   $= \mathbf{\dfrac{2}{3} \text{ cis } 336°}$

Find $(r \text{ cis } \theta)^2$ and $(r \text{ cis } \theta)^3$.

$$\begin{aligned}(r \text{ cis } \theta)^2 &= (r \text{ cis } \theta)(r \text{ cis } \theta)\\ &= r \cdot r \text{ cis}(\theta + \theta)\\ &= r^2 \text{ cis } 2\theta\\ (r \text{ cis } \theta)^3 &= (r \text{ cis } \theta)(r \text{ cis } \theta)^2\\ &= (r \text{ cis } \theta)(r^2 \text{ cis } 2\theta)\\ &= r \cdot r^2 \text{ cis}(\theta + 2\theta)\\ &= r^3 \text{ cis } 3\theta\end{aligned}$$

Intuitively, it seems that

$$(r \text{ cis } \theta)^4 = r^4 \text{ cis } 4\theta$$

which is true, and the general result is called **DeMoivre's theorem.**

DEMOIVRE'S THEOREM

$$(\text{r cis } \theta)^n = \text{r}^n \text{ cis } n\theta$$

where n is a natural number

The theorem is proved by a technique called "mathematical induction." This type of proof is usually considered in another course.

EXAMPLES

5. Find $(1 + i\sqrt{3})^8$ in polar form.
   First convert to polar form.

$$(1 + i\sqrt{3})^8 = \left(2 \text{ cis } \frac{\pi}{3}\right)^8$$

Then apply DeMoivre's theorem.

$$\begin{aligned}&= 2^8 \text{ cis } 8\left(\frac{\pi}{3}\right)\\ &= 256 \text{ cis } \frac{8\pi}{3}\end{aligned}$$

$$\mathbf{(1 + i\sqrt{3})^8 = 256 \text{ cis } \frac{2\pi}{3}}$$

or

$$-128 + 128\sqrt{3}i$$

HISTORICAL ASIDE

Abraham DeMoivre (1667–1754) was a Frenchman who lived most of his life in England and was a close friend of Isaac Newton. His work was important in the development of actuarial mathematics and the application of the young calculus to probability. His may have been the first treatment of the normal frequency ("bell-shaped") curve. There is an often-told story that, late in life, DeMoivre noticed that he seemed to need a quarter-hour more sleep each day. When the daily requirement reached 24 h, he died in his sleep.

6. Find $\left(-\frac{1}{2}+\frac{\sqrt{3}}{2}i\right)^3$.

$$\begin{aligned}\left(-\frac{1}{2}+\frac{\sqrt{3}}{2}i\right)^3 &= (1 \text{ cis } 120°)^3\\ &= 1^3 \text{ cis } 3(120°)\\ &= 1 \text{ cis } 360°\\ &= \text{cis } 0°\\ &= 1 + 0i\end{aligned}$$

$$\boldsymbol{\left(-\frac{1}{2}+\frac{\sqrt{3}}{2}i\right)^3 = 1}$$

Notice that

$$-\frac{1}{2}+\frac{\sqrt{3}}{2}$$

is a cube root of 1, because its cube is equal to 1. Obviously, 1 is also a cube root of itself. Are there more? Well, the nth roots of a complex number can be found using DeMoivre's theorem. Because, for example, $w^2 = z$, then z is the square root of w. That is, if $w = s \text{ cis } \phi$ and $z = r \text{ cis } \theta$,

$$(r \text{ cis } \theta)^{1/2} = s \text{ cis } \phi$$

or

$$\begin{aligned}r \text{ cis } \theta &= (s \text{ cis } \phi)^2\\ &= s^2 \text{ cis } 2\phi\end{aligned}$$

which means that

$$\begin{aligned}r &= s^2 &\quad \text{and} \quad 2\phi &= \theta + 2k\pi\\ r^{1/2} &= s & \phi &= \frac{\theta + 2k\pi}{2}\end{aligned}$$

so the expression can be rewritten as

$$(r \text{ cis } \theta)^{1/2} = r^{1/2} \text{ cis } \frac{\theta + 2k\pi}{2}$$

Again, it may be shown that the following theorem is true for all natural numbers n.

nTH ROOT THEOREM

$$(r \text{ cis } \theta)^{1/n} = r^{1/n} \text{ cis } \frac{\theta + 2\pi k}{n}$$

where n is a natural number.

This means that the n different values of $r^{1/n} \text{ cis}(\theta + 2\pi k)/n$ are the n distinct nth roots of r cis θ. Because there are n distinct roots, $k = 0, 1, 2, 3, \ldots, n - 1$ will generate them as shown in the following example.

EXAMPLES

7. Find the cube roots of 1 and write them in rectangular form.
First, rewrite in polar form.

$$1 = 1 + 0i = 1 \text{ cis } 0$$

Second, let z be the cube root and apply the nth root theorem.

$$z = (1 \text{ cis } 0)^{1/3} = 1^{1/3} \text{ cis } \frac{0 + 2\pi k}{3}$$

$1^{1/3}$ is the positive real cube root of 1.

$$= 1 \text{ cis } \frac{2\pi k}{3}$$

Finally, let $k = 0$, 1, and 2 to find the three distinct roots.

If $k = 0$ $\quad z_1 = 1 \text{ cis } \dfrac{2\pi \cdot 0}{3}$

$$= 1 \text{ cis } 0 = 1 + 0i$$

Notice that here we confirm the results of Example 6, and determine the third root.

If $k = 1$ $\quad z_2 = 1 \text{ cis } \dfrac{2\pi \cdot 1}{3}$

$$= 1 \text{ cis } \frac{2\pi}{3} = -\frac{1}{2} + \frac{\sqrt{3}}{2} i$$

If $k = 2$ $\quad z_3 = 1 \text{ cis } \dfrac{2\pi \cdot 2}{3}$

$$= 1 \text{ cis } \frac{4\pi}{3} = -\frac{1}{2} - \frac{\sqrt{3}}{2} i$$

$1 + 0i$, $-\frac{1}{2} + \frac{\sqrt{3}}{2} i$, and $-\frac{1}{2} - \frac{\sqrt{3}}{2} i$ are the three cube roots of 1.

8. Find the five fifth roots of $1 + i$ and plot them.

$$-1 + i = \sqrt{2} \text{ cis } 135° = 2^{1/2} \text{ cis } 135° \quad \text{or} \quad \frac{3\pi}{4}$$

$$z = (2^{1/2} \text{ cis } 135°)^{1/5}$$

$$= (2^{1/2})^{1/5} \text{ cis } \frac{135° + 360°k}{5}$$

$$\begin{array}{ll} \text{If } k = 0 & z_1 = 2^{1/10} \text{ cis } 27° \\ k = 1 & z_2 = 2^{1/10} \text{ cis } 99° \\ k = 2 & z_3 = 2^{1/10} \text{ cis } 171° \\ k = 3 & z_4 = 2^{1/10} \text{ cis } 243° \\ k = 4 & z_5 = 2^{1/10} \text{ cis } 315° \end{array}$$

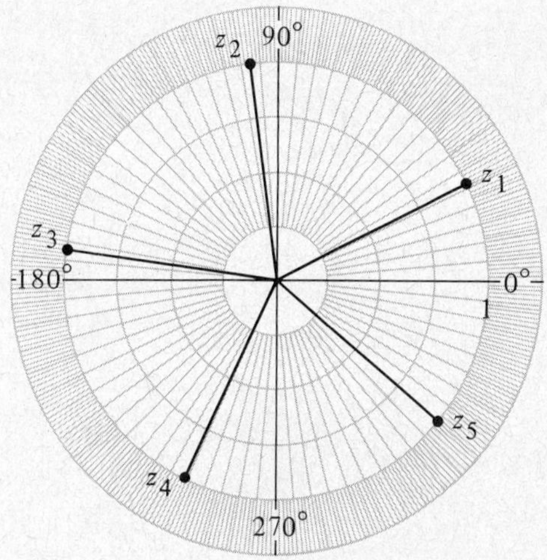

Notice the symmetry in Example 8, with the roots equally spaced about the circle with radius $2^{1/10}$ or approximately 1.07177.

exercise set 6.3

A. exercises

Perform the operations in Exercises 1–10 and simplify.

1. $[2(\cos 15° + i \sin 15°)][3(\cos 30° + i \sin 30°)]$ See Examples 1–4.

2. $[3(\cos 45° + i \sin 45°)][5(\cos 15° + i \sin 15°)]$

3. $(5 \text{ cis } 40°)(7 \text{ cis } 30°)$

4. $(3 \text{ cis } 50°)(11 \text{ cis } 80°)$

5. $(\sqrt{6} \text{ cis } 129°)(\sqrt{3} \text{ cis } 38°)$

6. $(\sqrt{10} \text{ cis } 77°)(\sqrt{2} \text{ cis } 222°)$

7. $\dfrac{6 \text{ cis } 125°}{3 \text{ cis } 55°}$

8. $\dfrac{15 \text{ cis } 95°}{3 \text{ cis } 75°}$

9. $\dfrac{\sqrt{2} \text{ cis } 244°}{\sqrt{6} \text{ cis } 157°}$

10. $\dfrac{\sqrt{10} \text{ cis } 187°}{\sqrt{5} \text{ cis } 132°}$

Use DeMoivre's theorem to find the indicated powers in Exercises 11–20.

See Examples 5 and 6.

11. $(2 - 2i)^4$
12. $(1 - \sqrt{3}i)^5$
13. $(\sqrt{3} + i)^3$
14. $(-1 + i)^6$
15. $(2 \text{ cis } 32°)^5$
16. $(3 \text{ cis } 28°)^4$
17. $(\text{cis } 12°)^{10}$
18. $(\text{cis } 18°)^{15}$
19. $(\sqrt{2} \text{ cis } 45°)^{10}$
20. $(\sqrt{3} \text{ cis } 30°)^{15}$

B. exercises

Find the roots indicated in Exercises 21–30 and plot the roots.

See Examples 7 and 8.

21. $(9 \text{ cis } 45°)^{1/2}$
22. $(25 \text{ cis } 30°)^{1/2}$
23. $\left(27 \text{ cis } \dfrac{\pi}{3}\right)^{1/3}$
24. $(8 \text{ cis } 135°)^{1/3}$
25. $(1 + i)^{1/5}$
26. $(1 - i)^{1/6}$
27. $(0 - i)^{1/3}$
28. $(-1 + 0i)^{1/4}$
29. Find the four fourth roots of 1.
30. Find the five fifth roots of 32.
31. Find the cube roots of i.
32. Find the fourth roots of $-16i$.
33. Find the cube roots of $4\sqrt{3} - 4i$.
34. Find the fourth roots of $2\sqrt{2} + 2\sqrt{2}i$.

Solve the equations in Exercises 35–40 using the nth root theorem.

35. $x^3 + 1 = 0$
36. $x^5 - 1 = 0$
37. $x^2 - i = 0$
38. $x^4 + i = 0$
39. $x^5 = 243 \text{ cis } 35°$
40. $x^6 = 64 \text{ cis } 36°$

C. exercises

41. Find the cube roots of 1 algebraically by solving $x^3 = 1$. (Recall difference of cubes.)

42. Prove:

$$\frac{r_1 \operatorname{cis} \theta_1}{r_2 \operatorname{cis} \theta_2} = \frac{r_1}{r_2} \operatorname{cis}(\theta_1 - \theta_2)$$

6.4 POLAR COORDINATES

Rectangular coordinates locate the position of a point in the plane by the distances from two given lines. The position of a point may also be specified by a distance from a given point and a direction from a given line. This coordinate system, called polar coordinates, is frequently used in mathematics and in the sciences. Moreover, it is a common notion in geography. For instance, it might be said that "Hawaii is some 2700 mi southwest of San Francisco." That is, the location is a certain distance from a point, San Francisco, in a certain direction relative to the north–south line. To form a polar coordinate system, a point is designated as the **origin** or **pole** and a ray extending to the right is called the **polar axis.** The polar axis provides a scale for r, and θ is measured counterclockwise from this axis. See Figure 6.3.

Figure 6.3 The polar coordinate system.

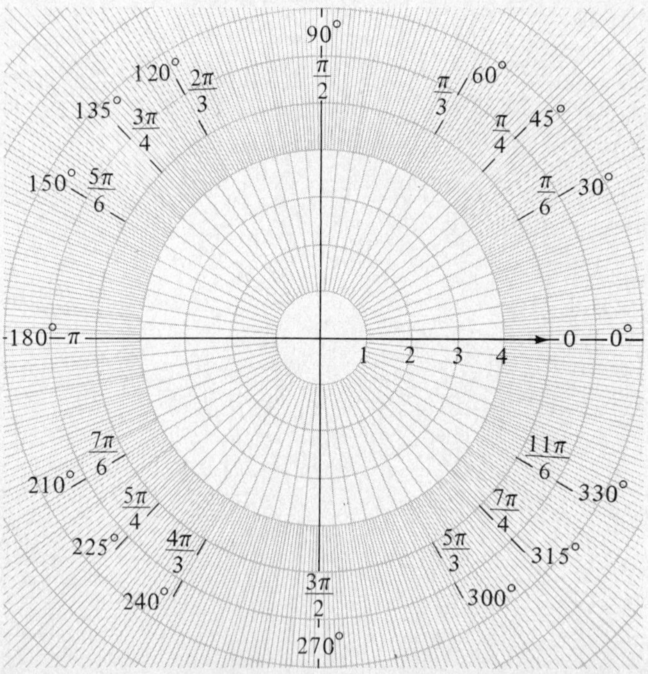

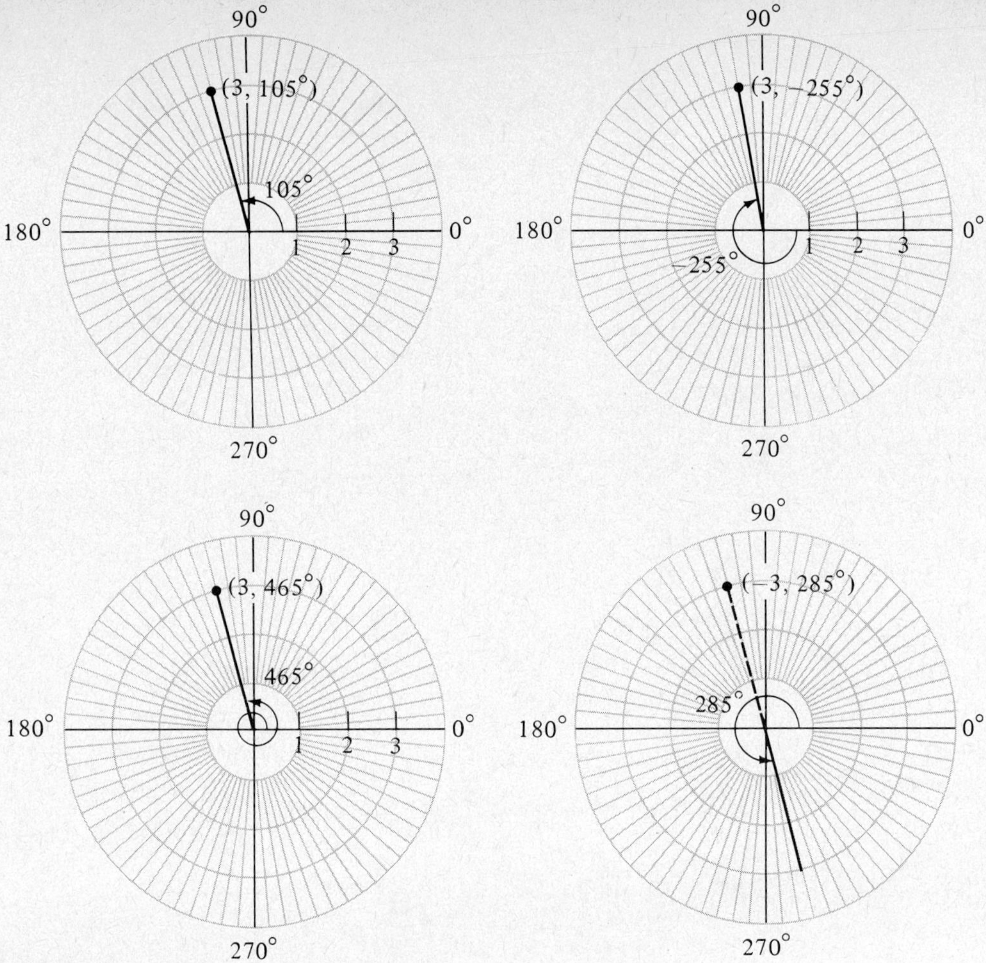

Figure 6.4 (3, 105°), (3, −255°), (3, 465°), and (−3, 285°).

A major difference between the rectangular and polar coordinate systems is seen immediately, if one plots (3, 105°), (3, −255°), (3, 465°), and (−3, 285°). Notice in Figure 6.4 that 105°, −255°, and 465° are coterminal, so that the first three pairs designate the same point. Further, 285° is a ray in the opposite direction from 105°, −255°, and 465°. But −3 is measured back in that direction, so (−3, 285°) designates the same point as the others.

In general, if (r, θ) is the coordinates of some point P, then

$$(r, \theta \pm 2k\pi), (-r, \theta \pm h\pi) \qquad \text{in radian measure}$$

and

$$(r, \theta \pm 360°k), (-r, \theta \pm 180°h) \qquad \text{in degrees}$$

all represent point P, where k is an integer and h is an odd integer.

EXAMPLES

Plot the points.

1. (2, 30°)
2. (1, − 135°)
3. (−3, 120°)

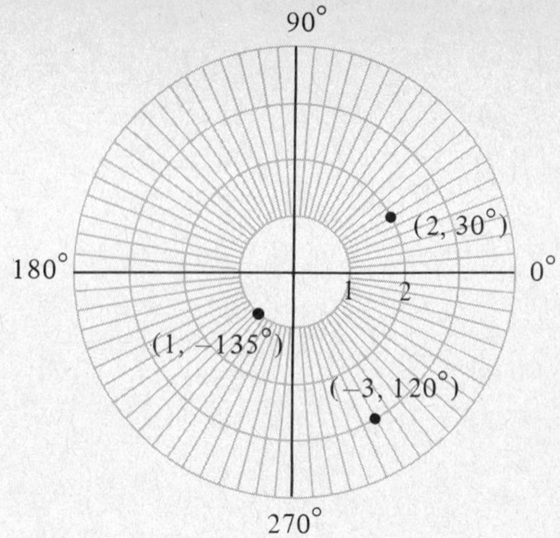

4. $\left(1, \frac{\pi}{3}\right)$
5. $\left(2.5, -\frac{5\pi}{4}\right)$
6. $\left(-2, \frac{\pi}{4}\right)$

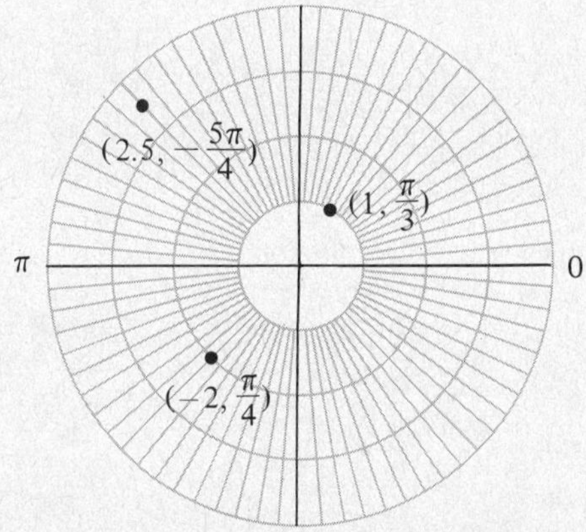

Changing from Polar to Rectangular Form

It is sometimes necessary to change the form of an equation from one coordinate system to another. As we have seen, the polar and rectangular systems are related by the following.

$$\mathbf{x = r \cos \theta \qquad y = r \sin \theta}$$

If specific values of θ are known, x and y can be found, changing from polar to rectangular coordinates.

EXAMPLE

Change to rectangular coordinates.

7. $\left(2\sqrt{3}, \frac{\pi}{3}\right)$

$$r = 2\sqrt{3} \quad \text{and} \quad \theta = \frac{\pi}{3}$$

so

$$x = 2\sqrt{3}\cos\frac{\pi}{3} \qquad y = 2\sqrt{3}\sin\frac{\pi}{3}$$

$$= 2\sqrt{3}\left(\frac{1}{2}\right) \qquad = 2\sqrt{3}\left(\frac{\sqrt{3}}{2}\right)$$

$$= \sqrt{3} \qquad = 3$$

$$\left(\mathbf{2\sqrt{3}, \frac{\pi}{3}}\right) \longleftrightarrow (\mathbf{\sqrt{3}, 3})$$

However, more importantly, an *equation* in r and θ can be converted to rectangular coordinates by the same substitution. The additional relationship

$$\boldsymbol{r^2 = x^2 + y^2} \quad \text{or} \quad \boldsymbol{r = \pm\sqrt{x^2 + y^2}}$$

is helpful.

EXAMPLES

Convert to rectangular coordinates and simplify.

8. $r = \dfrac{1}{\sin\theta + \cos\theta}$ Multiply both sides by $\sin\theta + \cos\theta$.

First simplify,

$$r(\sin\theta + \cos\theta) = 1$$

$$r\sin\theta + r\cos\theta = 1$$

then substitute, $y = r\sin\theta$ and $x = r\cos\theta$.

$$\mathbf{y + x = 1} \qquad \text{A line?}$$

9. $r = \dfrac{8}{2 - \cos\theta}$

$$2r - r\cos\theta = 8 \qquad \text{Clear the fraction.}$$

$$2(\boldsymbol{\sqrt{x^2 + y^2}}) - (\boldsymbol{x}) = 8$$ where $r = \sqrt{x^2 + y^2}$ because $r > 0$.

$$2\sqrt{x^2 + y^2} = x + 8 \qquad \text{Simplify.}$$

$$4(x^2 + y^2) = (x + 8)^2 \qquad \text{Square both sides.}$$

$$4x^2 + 4y^2 = x^2 + 16x + 64$$

$$\mathbf{3x^2 + 4y^2 - 16x - 64 = 0} \qquad \text{An ellipse?}$$

10. $r - 4 \sin \theta = 0$

$$r(r - 4 \sin \theta) = r(0)$$
$$r^2 - 4r \sin \theta = 0$$
$$(x^2 + y^2) - 4(y) = 0$$
$$\mathbf{x^2 + y^2 - 4y = 0}$$

Multiply both sides by r to obtain $r \sin \theta$.

A circle?

11. $r^2 = 9 \sin 2\theta$

$$r^2 = 9 \cdot 2 \sin \theta \cos \theta$$
$$= 18 \sin \theta \cos \theta$$
$$= 18 \left(\frac{y}{r}\right)\left(\frac{x}{r}\right)$$
$$= \frac{18xy}{r^2}$$
$$r^4 = 18xy$$
$$\mathbf{(x^2 + y^2)^2 = 18xy}$$

Notice that only the functions were substituted first.

Not a familiar rectangular curve.

Changing from Rectangular to Polar Form

Like the previous conversion, we make use of key relationships between the systems.

$$\mathbf{r = \sqrt{x^2 + y^2} \qquad \theta = \arctan \frac{y}{x}}$$

EXAMPLES

Change to polar form.

12. $(-\sqrt{6}, \sqrt{2})$

$$r = \sqrt{(-\sqrt{6})^2 + (\sqrt{2})^2} \qquad \theta = \arctan -\frac{\sqrt{2}}{\sqrt{6}}$$
$$= \sqrt{6 + 2} = \sqrt{8} \qquad = \arctan -\frac{1}{\sqrt{3}}$$
$$= 2\sqrt{2} \qquad = \frac{5\pi}{6}$$

The arctan relationship is convenient for changing coordinates.

$$\mathbf{(-\sqrt{6}, \sqrt{2}) \longleftrightarrow \left(2\sqrt{2}, \frac{5\pi}{6}\right)}$$

Why not $\frac{11\pi}{6}$ instead of $\frac{5\pi}{6}$?

13. $3x - 4y = 1$

$$3(x) - 4(y) = 1$$
$$3(r\cos\theta) - 4(r\sin\theta) = 1$$
$$3r\cos\theta - 4r\sin\theta = 1$$
$$\mathbf{r\,(3\cos\theta - 4\sin\theta) = 1}$$

14. $(x - 1)^2 + y^2 = 1$

$$x^2 - 2x + 1 + y^2 = 1$$
$$(x^2 + y^2) - 2x = 0$$
$$(r^2) - 2(r\cos\theta) = 0$$
$$r(r - 2\cos\theta) = 0$$

Notice that the arctan relationship is *not* useful here.

That is,

$$r = 0 \text{ or } r - 2\cos\theta = 0$$

Because $r = 0$ is simply the pole and it is included in $r - 2\cos\theta = 0$, that factor is redundant and can be dropped.

$$\mathbf{r - 2\cos\theta = 0}$$

15. $(x + y)^2 = x^2 - y^2$

$$(r\cos\theta + r\sin\theta)^2 = r^2\cos^2\theta - r^2\sin^2\theta \quad \text{Substitute.}$$
$$r^2\cos^2\theta + 2r^2\cos\theta\sin\theta + r^2\sin^2\theta = r^2\cos^2\theta - r^2\sin^2\theta \quad \text{Expand.}$$
$$r^2(\cos^2\theta + \sin^2\theta) + 2r^2\cos\theta\sin\theta = r^2(\cos^2\theta - \sin^2\theta) \quad \text{Factor.}$$
$$r^2(1) + r^2(\sin 2\theta) = r^2(\cos 2\theta) \quad \text{Use identities to simplify.}$$
$$r^2(1 + \sin 2\theta - \cos 2\theta) = 0$$
$$\mathbf{1 + \sin 2\theta - 2\theta = 0} \quad \text{After checking } r = 0 \text{ with the other factor.}$$

16. Consider Example 14 a little differently.

$$(x + y)^2 = x^2 - y^2 \quad \text{Expand first.}$$
$$x^2 + 2xy + y^2 = x^2 - y^2 \quad \text{Simplify.}$$
$$2xy - 2y^2 = 0$$
$$y(x + y) = 0 \quad \text{Factor.}$$
$$r\sin\theta(r\cos\theta + r\sin\theta) = 0 \quad \text{Substitute.}$$
$$\mathbf{\sin\theta(\cos\theta + \sin\theta) = 0} \quad r = 0?.$$

Because of the identities, there is a wide array of equivalent forms for most polar curves. Can you prove that the two results for Examples 15 and 16 are identical? A note of caution: An ordered pair of real numbers may now represent either rectangular or polar coordinates. One must be careful that the pair is identified properly.

exercise set 6.4

A. exercises

Plot the points in Exercises 1–10 and find three other sets of polar coordinates for each point such that θ is less than one rotation clockwise or counterclockwise.

See Examples 1–6.

1. $(2, 50°)$
2. $(4, 70°)$
3. $(1.5, 160°)$
4. $(-\frac{4}{3}, 230°)$
5. $(-3, 325°)$
6. $(2.5, -115°)$
7. $\left(\sqrt{3}, \frac{3\pi}{4}\right)$
8. $\left(\sqrt{2}, \frac{7\pi}{6}\right)$
9. $\left(-3, \frac{9\pi}{5}\right)$
10. $\left(2, -\frac{\pi}{3}\right)$

Find the rectangular coordinates of each point given in Exercises 11–20.

See Example 7.

11. $\left(4, \frac{\pi}{3}\right)$
12. $\left(2, \frac{\pi}{4}\right)$
13. $\left(-6, \frac{\pi}{6}\right)$
14. $\left(-1, \frac{2\pi}{3}\right)$
15. $(-\sqrt{2}, 135°)$
16. $(\sqrt{3}, -150°)$
17. $(0, 75°)$
18. $(\sqrt{5}, 0°)$
19. $(2, 600°)$
20. $(-6, 585°)$

Find the polar coordinates of each point given in Exercises 21–30 such that $r > 0$ and $0° \leq \theta < 360°$.

See Example 12.

21. $(2, 0)$
22. $(0, -1)$
23. $(3, 3)$
24. $(\sqrt{3}, 1)$
25. $\left(-\frac{1}{2}, \frac{\sqrt{3}}{2}\right)$
26. $\left(\frac{\sqrt{3}}{2}, -\frac{1}{2}\right)$
27. $(-3, 4)$
28. $(8, 15)$
29. $(-20, -21)$
30. $(12, -35)$

B. exercises

Convert each equation in Exercises 31–40 into rectangular form.

See Examples 8–11.

31. $r = 2 \csc \theta$
32. $r = 3 \sin 2\theta$
33. $r = 4$
34. $r = 2 \sin \theta$
35. $r^2 = 4 \cos 2\theta$
36. $r^2 = \tan^3 \theta + \tan \theta$
37. $r = \dfrac{9}{1 - \sin \theta}$
38. $r = \dfrac{1}{1 - 3 \cos \theta}$
39. $r = \dfrac{2}{\sin \theta - \cos \theta}$
40. $r^2 = \dfrac{1}{4 - 5 \sin^2 \theta}$

For Exercises 41–50, transform each equation into polar form.

See Examples 13–16.

41. $x = -2$
42. $y = 5$
43. $x^2 + y^2 = 25$
44. $x^2 = 4y$
45. $xy = x + y$
46. $x^2 + y^2 - 9x = 0$
47. $(x^2 + y^2)^3 = (x^2 - y^2)^2$
48. $y^2(x + 2) = x^2(x - 2)$
49. $9x^2 + y^2 = 9$
50. $x^2 - 4y^2 = 4$

C. exercises

51. **brain teaser** Prove that the polar equation $r = 2a \cos \theta$ is a circle passing through the origin, with center on the x axis, and radius a.

52. **brain teaser** Prove that the polar equation

$$r = \frac{ep}{1 - e \cos \theta}$$

is an ellipse for any real number p and $e < 1$.

6.5 POLAR GRAPHS

To sketch the graph of the solutions to an equation in polar form, the same methods employed to sketch rectangular graphs can be used. That is, a table of values can be computed, the points plotted, and the points connected with a smooth curve. It is also possible that a certain number of curves may have familiar equations. The necessary information may then be acquired by inspection as, for instance, the slope and y intercept of a line in rectangular form. First, let us consider polar equations for which a table of values can be generated. This process is aided considerably by a hand-held calculator and, of course, polar graph paper.

EXAMPLES

1. Sketch $r = 2(1 + \cos\theta)$.

r	θ
4	0°
3.73	30°
3	60°
2	90°
1	120°
0.27	150°
0	180°
0.27	210°
1	240°
2	270°
3	300°
3.73	330°

Plot the points from the table.

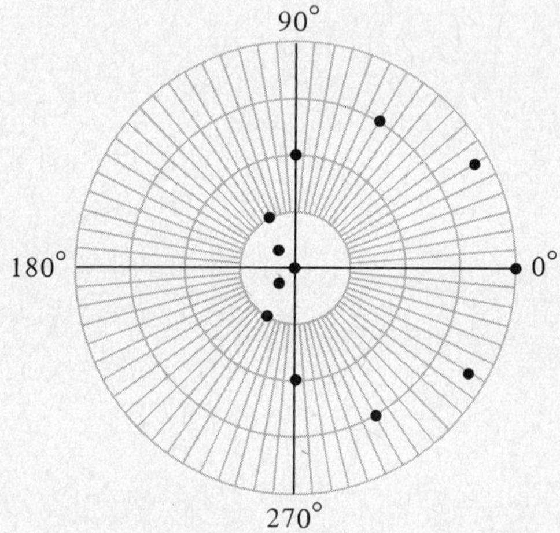

Notice how differently from a rectangular graph the polar curve develops as you sketch the curve through the points. The numbers in the figure indicate the order in which they are plotted.

Connect the points.

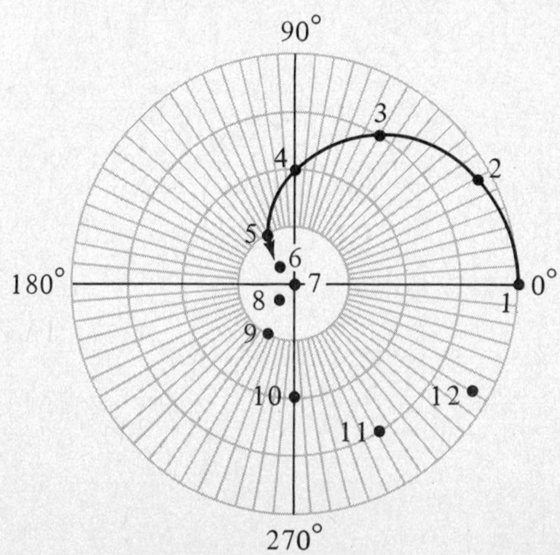

Complete the sketch.

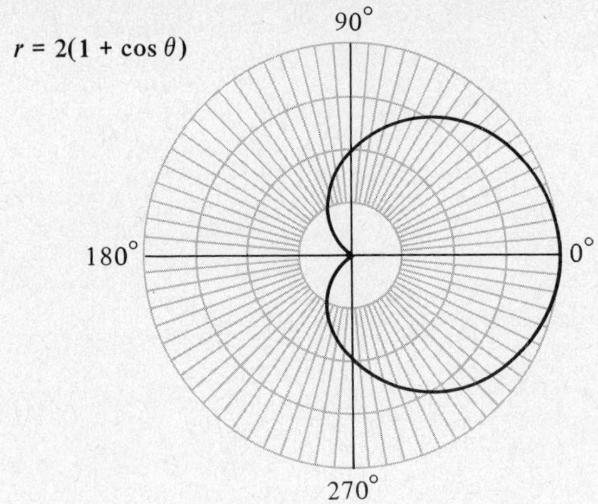

The finished graph is called a **cardioid,** as suggested by its heart shape.

2. Sketch $r = 3 \cos 2\theta$.

r	θ
3	0°
1.50	30°
−1.50	60°
−3	90°
.	.
.	.
.	.

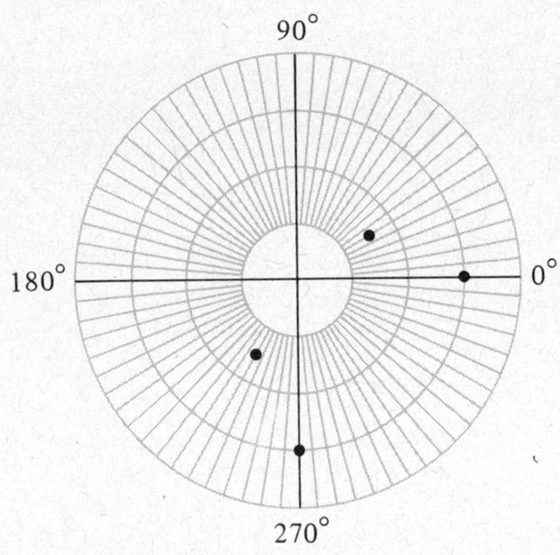

Note that taking θ every 30° produces rather scattered points.

To provide a better picture of the behavior of the graph, take θ every 15° (or every 10°). The closer interval may be necessary only until you understand that behavior.

r	θ
3	0°
2.6	15°
1.5	30°
0	45°
−1.5	60°
−2.6	75°
−3	90°
−2.6	105°
−1.5	120°
0	135°
.	.
.	.
.	.

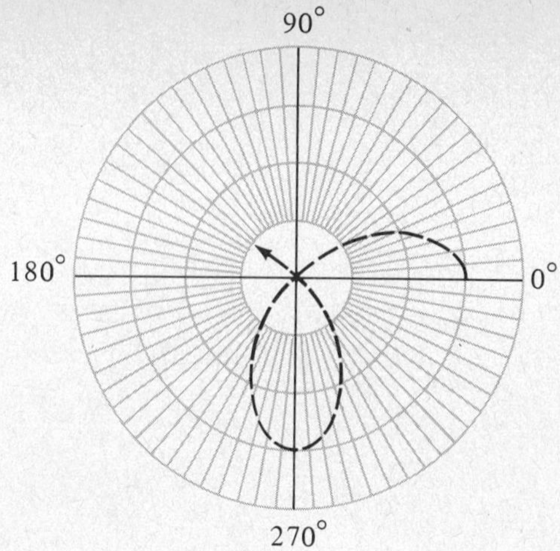

Plot the other points to see how the curve develops. Why can you be assured that the curve will be smooth and uninterrupted?

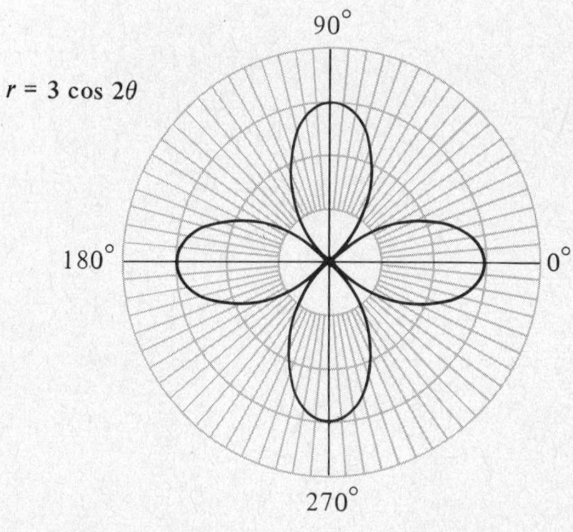

3. Sketch $\pi r = \theta$.

Note immediately that θ *must* be in radian measure so that r is a real number.

r	θ
0	0
$\frac{1}{6}$	$\frac{\pi}{6}$
$\frac{1}{3}$	$\frac{\pi}{3}$
$\frac{1}{2}$	$\frac{\pi}{2}$
$\frac{2}{3}$	$\frac{2\pi}{3}$
$\frac{5}{6}$	$\frac{5\pi}{6}$
1	π
.	.
.	.
.	.

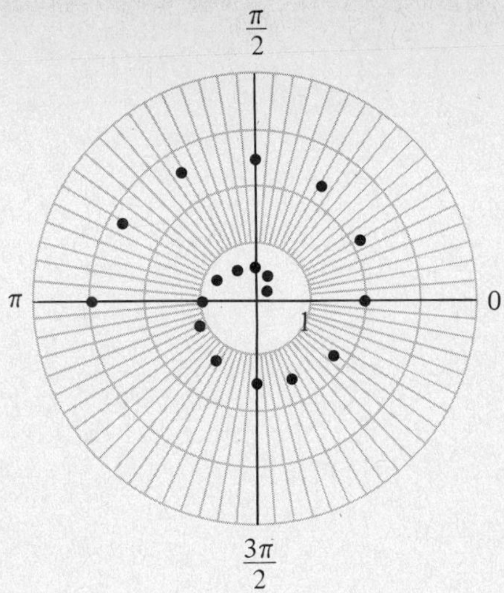

Compiling a table of values, we see that r will continue to increase without bound as θ rotates about the pole.

Thus,

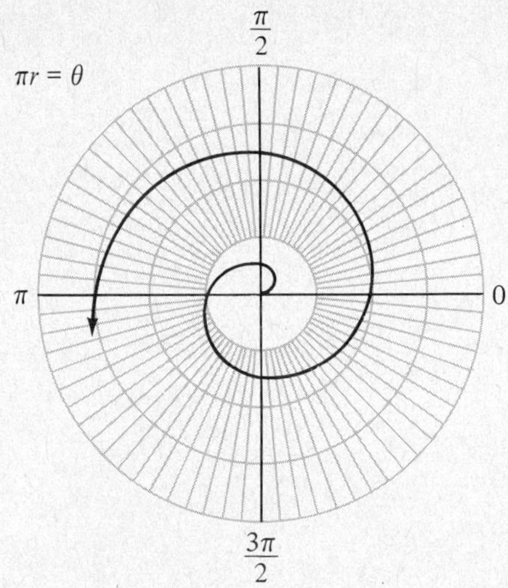

This curve is called the **spiral of Archimedes.**

Standard Forms of Polar Curves

Sketching polar curves by plotting points is time consuming, but there is perhaps no better way to observe the striking differences between polar and rectangular graphs. There are, however, a number of interesting

Figure 6.5 Standard forms of polar curves.

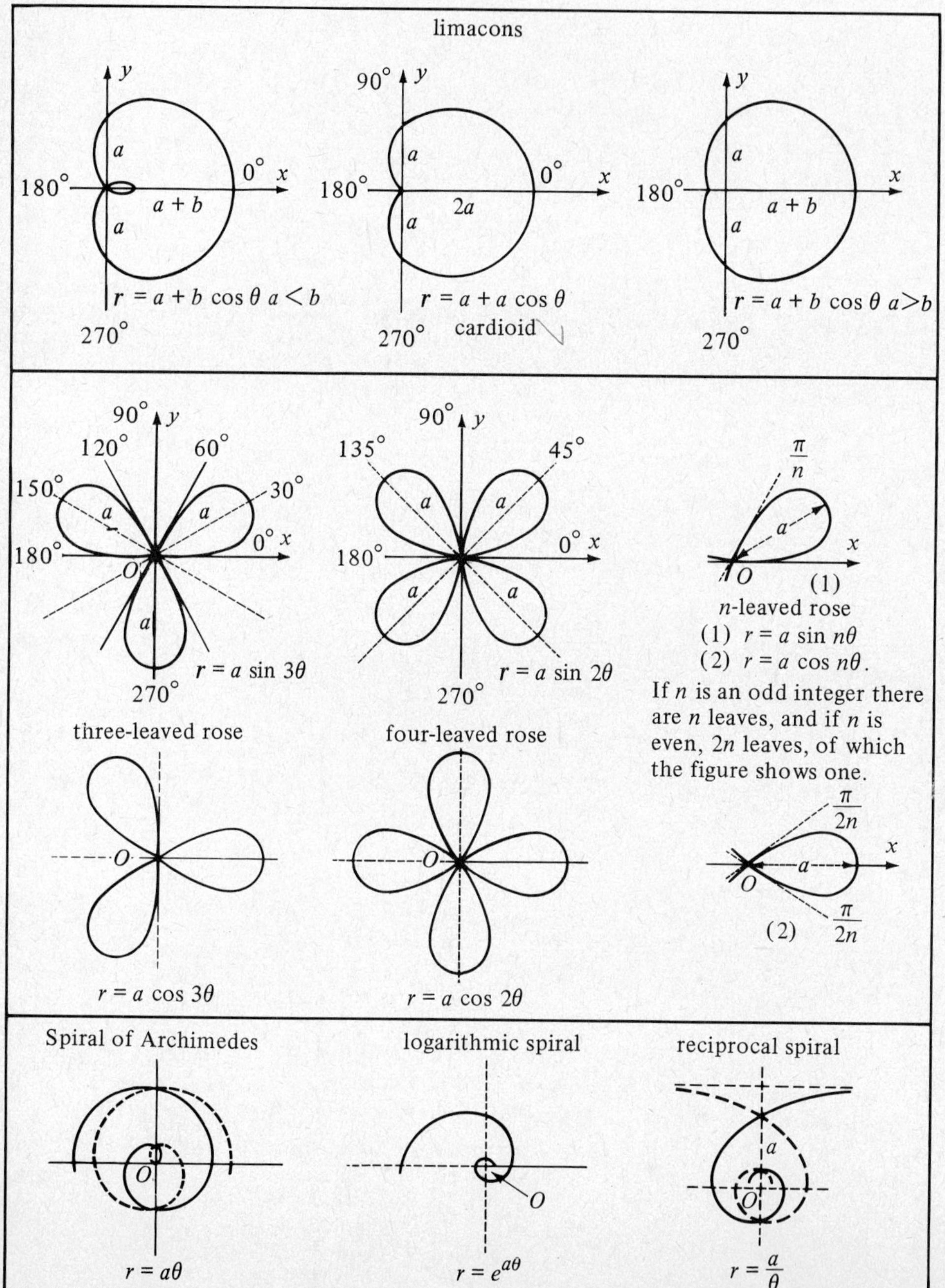

polar curves with distinctive shapes and easily recognizable equations. Notice that $r = a + b \cos \theta$ is in the form of Example 1, with $a = b = 2$. Such a curve is called a **limacon,** and when $a = b$ the limacon is a **cardioid.** A rough sketch can be rendered for a cardioid if the value of a is known. Consider the brief list of some of the more common polar curves in Figure 6.5. These and others are frequently included in handbooks of mathematical tables and formulas for easy reference.

EXAMPLES

4. Sketch $r = 2 + 3 \cos \theta$.

   The equation is in the form $r = a + b \cos \theta$, where $a = 2$, $b = 3$, and $a < b$. Thus the key points are $(a + b, 0°)$, $(a, 90°)$, $(a - b, 180°)$, and $(a, 270°)$.

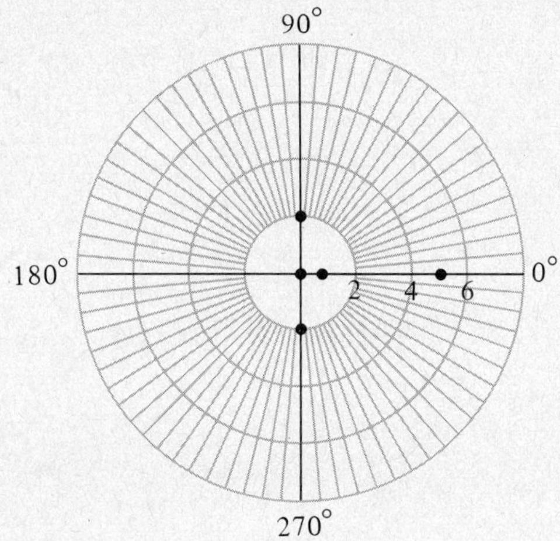

   A "rough" sketch of the curve can now be produced.

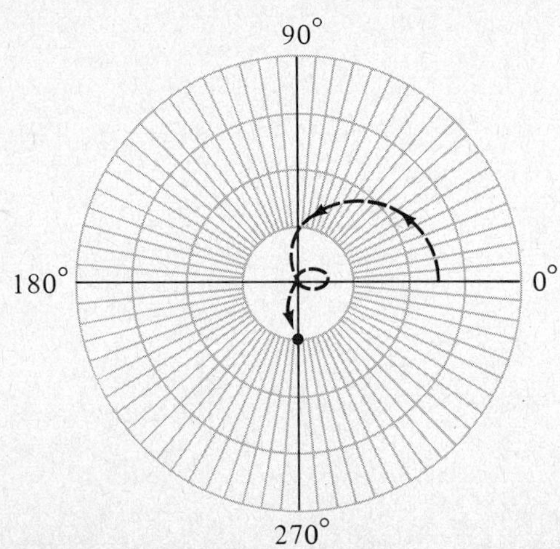

Now we can complete the sketch.

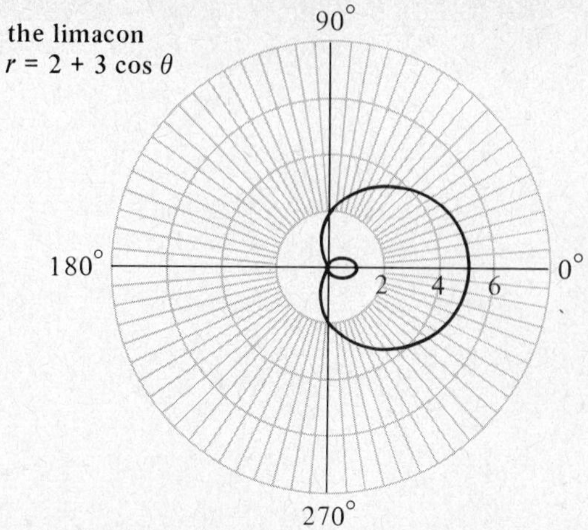

5. Sketch $r = 4 \cos 5\theta$.

The equation is in the form $r = a \cos n\theta$, with odd n, where $a = 4$ and $n = 5$. There are five leaves, $2\pi/5$ or 72° apart. We plot key points at 72°, 144°, 215°, and 288°.

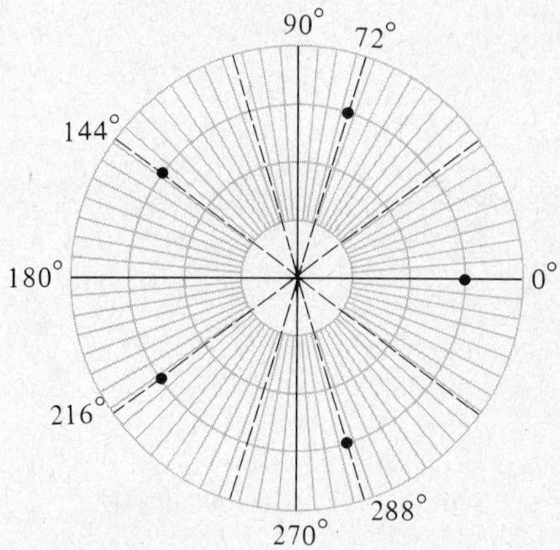

Now we use the guidelines and key points to sketch the curve.

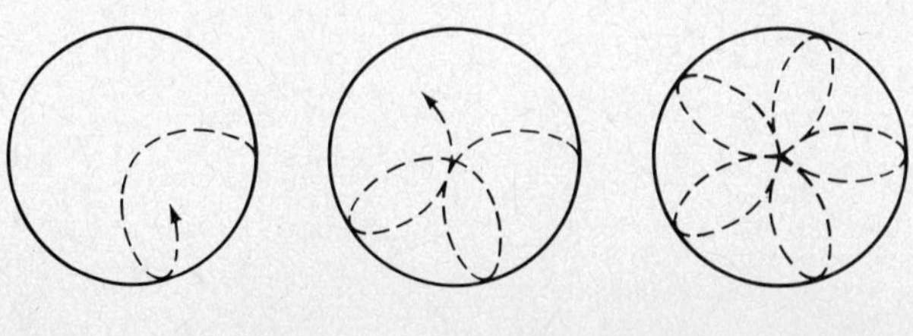

Thus,

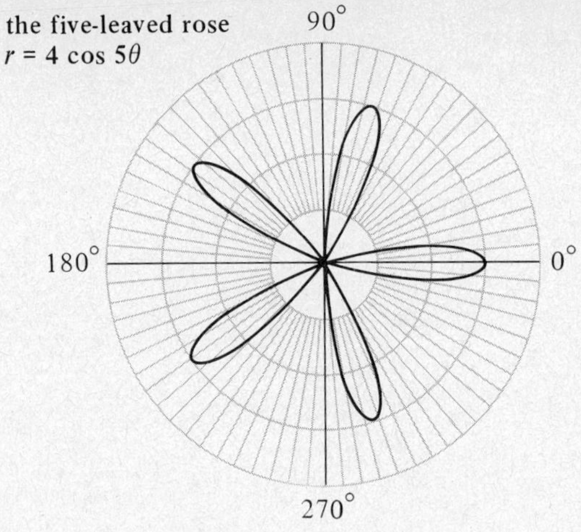

Sketching from Rectangular Graphs

Certain of the polar curves have familiar rectangular counterparts. That is, $y - 2 = 2 \sin x$ has been sketched in rectangular form. Can the graph of $r - 2 = 2 \sin \theta$ or $r = 2 + 2 \sin \theta$ be sketched by making use of its rectangular cousin? Yes, and it is a fairly fast and accurate means of obtaining a good representation without having to accumulate a table of many values.

To sketch $r = 2 + 2 \sin \theta$, first sketch the graph on a rectangular coordinate system (θ, r). That is, sketch $y = 2 + 2 \sin x$, where $x = \theta$ and $y = r$. Then the values for (r, θ) can be read by measuring r for each θ on the rectangular graph and marking that length on the terminal side of θ on the polar coordinate graph.

A small ruler or just a strip of paper can be used to measure and transfer r.

EXAMPLES

6. Sketch $r = 2 + 2 \sin \theta$.
   First sketch $y = 2 + 2 \sin x$ on rectangular axes.

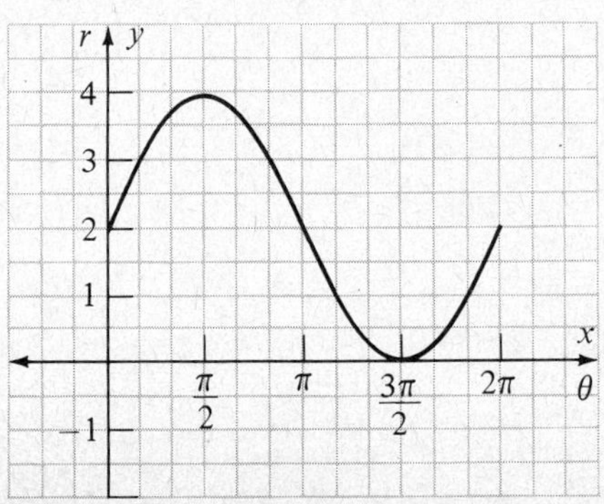

Now note values of r(or y) at selected values of θ (or x) and transfer these lengths to determine points on the polar grid.

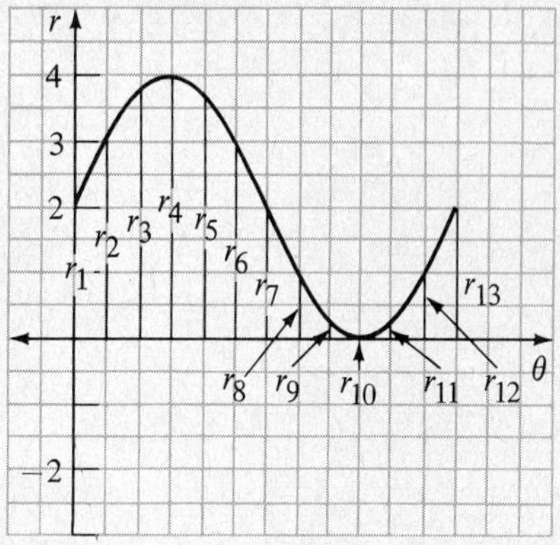

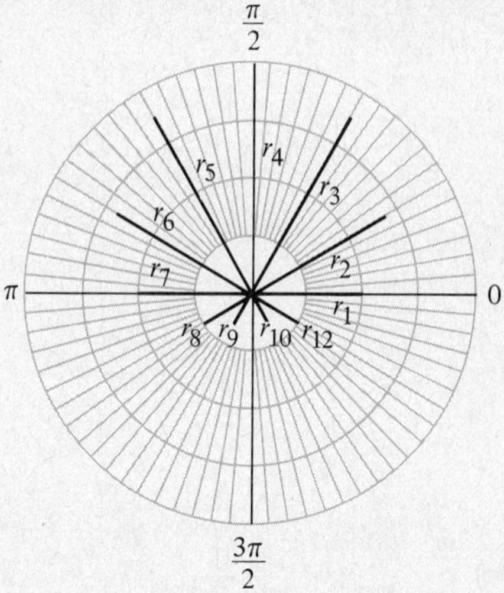

Finally, sketch the curve. If there are insufficient points, then any number of additional points may be plotted in the same way.

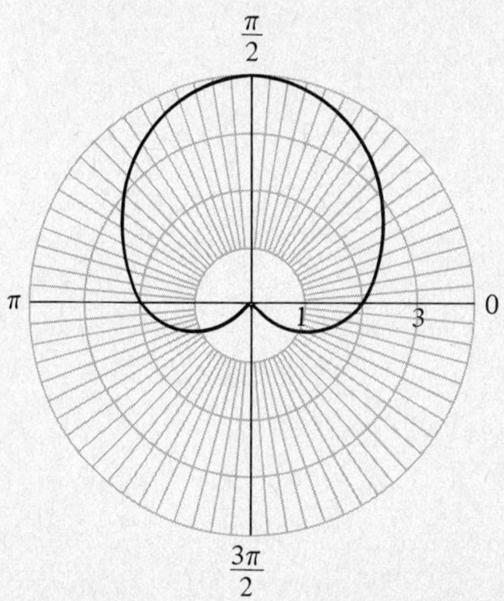

7. Sketch $r = 4 \sin 3\theta$.
Sketch $y = 4 \sin 3x$.

Transfer points to $r = 4 \sin 3\theta$.

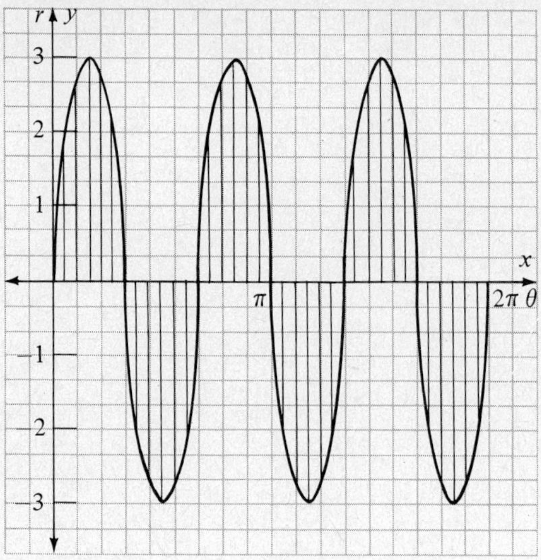

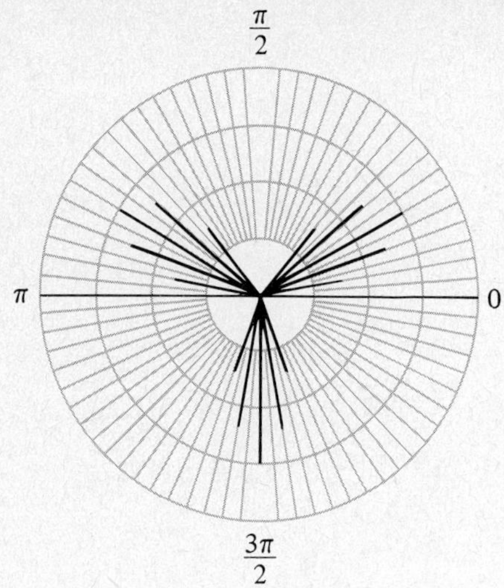

Thus,

$r = 4 \sin 3\theta$

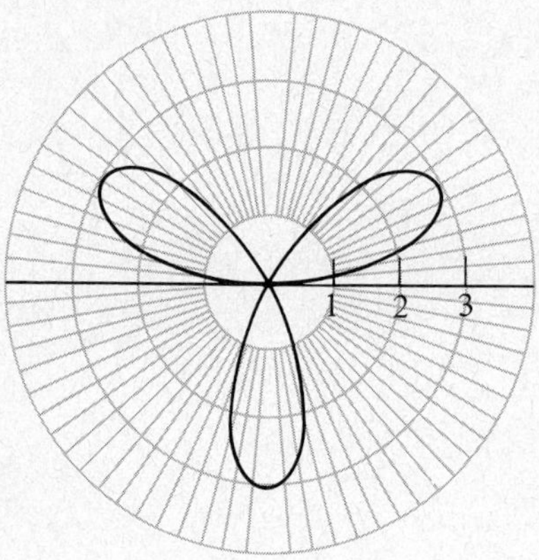

exercise set 6.5

A. exercises

Sketch each of the curves in Exercises 1–20.

See Examples 1–7.

1. $r = 4 \sec \theta$
2. $r = 4 \csc \theta$
3. $r = 3 \sin \theta$
4. $r = 3 \cos \theta$
5. $r = 2\theta$
6. $r\theta = 2$

7. $r = 4 \cos 3\theta$

8. $r = 4 \sin 5\theta$

9. $r \csc 4\theta = 3$

10. $r \sec 2\theta = 3$

11. $r = 3 + 3 \cos \theta$

12. $r = 3 + 3 \sin \theta$

13. $r = 3 + 2 \sin \theta$

14. $r = 3 + 2 \cos \theta$

15. $r = 2 + 3 \cos \theta$

16. $r = 2 + 3 \sin \theta$

17. $r = 2 - \sin \theta$

18. $r = 1 - 2 \cos \theta$

19. $r = 1 - \cos \theta$

20. $r = 3 - \sin \theta$

B. exercises

Compile a table of values and sketch the graphs of the curves in Exercises 21–34. (See Historical Aside.) a and b are arbitrary constants. Use the specified values.

21. **Gutschoven's curve**

    $r = a \cot \theta$

    Sketch for $a = 2$.

22. **Fermat's spiral (Fermat, 1636)**

    $r^2 = a^2\theta$

    Sketch for $a = 2$

23. **Limacon of Pascal**

    (Pascal, 1650)

    $r = 2a \cos \theta + b$

    Sketch for $a = b = 3$.

24. **Conchoid of Nicomedes**

    (Nicomedes, 225 B.C.)

    $r = a + b \sec \theta$

    Sketch for $a = 4$, $b = 2$.

25. **Lemniscate of Bernoulli**

    (Bernoulli, 1694)

    $r^2 = a^2 \cos 2\theta$

    Sketch for $a = 4$.

26. **Lemniscate of Gerono.**

    $r^2 = a^2 \sec^4 \theta \cos 2\theta$

    Sketch for $a = 4$.

27. **Tschirnhausen's cubic**

    (Tschirnhausen, 1679)

    $r \cos^3 \frac{\theta}{3} = a$

    Sketch for $a = \frac{1}{2}$.

28. **Trisectrix of Maclaurin**

    (Maclaurin, 1742)

    $r = a \sec \theta - 4a \cos \theta$

    Sketch for $a = 2$.

29. **Folium of Descartes**

    (Descartes, 1638)

    $r(\sin^3 \theta + \cos^3 \theta) = 3a \sin 2\theta$

    Sketch for $a = 3$.

30. **Cissoid of Diocles**

    (Diocles, 200 B.C.)

    $r = a \sin \theta \tan \theta$

    Sketch for $a = 4$.

HISTORICAL ASIDE

Discoveries in many fields are named for people, such as Haley's comet or the Richter scale. Many special curves are named for the people who studied them. One such curve is the so-called witch of Agnesi, named for Maria Gaetana Agnesi (1718–1799), a noted Italian mathematician at Bologna. The curve was actually first studied by Fermat (1666) and named by Grandi (1703). The name is apparently the result of a mixup. Agnesi seemingly confused Grandi's old Italian name vererio with the word versiera, which means goblin or witch. The more colorful name stuck, hence the "witch of Agnesi." See Exercise 33.

31. **Nephroid of Freeth**

$$r = a\left(1 + 2\sin\frac{\theta}{2}\right)$$

Sketch for $a = 3$.

32. **Cayley's sextic** (Maclaurin, 1718)

$$r = a\cos^3\frac{\theta}{3}$$

Sketch for $a = 4$.

C. exercises

33. Determine the point(s) of intersection of $r = \sin\theta$ and $r = \cos\theta$ by solving the equation $\sin\theta = \cos\theta$. Then graph the two curves on the same polar grid and determine the points of intersection by inspection. What do you observe? How do you explain the comparison?

34. Prove that $r = f(\theta)$ is symmetric with respect to the polar axis if it is unchanged when θ is replaced by $-\theta$.

35. Prove that $r = f(\theta)$ is symmetric with respect to the pole if it is unchanged when θ is replaced by $\theta + \pi$.

36. Prove that if a polar equation contains no function of θ other than $\cos n\theta$, then the replacement of the function by:
(a) $\sin n\theta$ rotates the curve through $90°/n$
(b) $-\cos n\theta$ rotates the curve through $180°/n$
(c) $-\sin n\theta$ rotates the curve through $270°/n$

37. Sketch the **witch of Agnesi** (See Historical Note),

$$r\sin\theta(r^2\cos^2\theta + 4a^2) = 8a^3$$

for $a = 1$.

6.6 CHAPTER REVIEW

Fill in the words or symbols that make the statements in Exercises 1–10 complete and correct.

1. (6.1) For the complex number $a + bi$, b is called the _________ _________.

2. (6.1) $i^{4n} =$ _________ for any integer n.

3. (6.1) $a + bi = c + di$ if and only if _________ and _________.

4. (6.1) $a + bi$ is the _________ of $a - bi$.

5. (6.2) $r(\cos\theta + i\sin\theta)$ is called the _________ form of a complex number and is abbreviated _________.

6. (6.2) If $a + bi = r(\cos\theta + i\sin\theta)$, then $r =$ _________ and $\theta =$ _________.

7. (6.3) $\dfrac{A \text{ cis } \alpha}{B \text{ cis } \beta} =$ ________.

8. (6.3) $(r \text{ cis } \theta)^n =$ ________, where n is a natural number.

9. (6.4) In a polar coordinate system, the ray extending to the right of the pole is called the ________.

10. (6.4) Many sets of polar coordinates may represent the ________ point.

11. (6.1) Simplify.

(a) $\sqrt{-27}$ (b) i^{73}
(c) $i(1 - 2i)$ (d) $i + (1 - 2i)$
(e) $(1 - 2i) + (2 + 3i)$ (f) $(1 - 2i) - (2 + 3i)$
(g) $(1 - 2i)(2 + 3i)$ (h) $(1 + 3i)(2 - 3i)$
(i) $\dfrac{1}{1 - 2i}$ (j) $\dfrac{1 - 2i}{2 + 3i}$

12. (6.2) Write the numbers in polar form.

(a) $2 - 2i$ (b) $-1 + \sqrt{3}i$

13. (6.2) Write the numbers in rectangular form.

(a) $3 \text{ cis } \dfrac{5\pi}{4}$ (b) $\sqrt{3} \text{ cis } 240°$

14. (6.3) Simplify.

(a) $\left(2 \text{ cis } \dfrac{\pi}{2}\right)\left(3 \text{ cis } \dfrac{\pi}{3}\right)$ (b) $\dfrac{\sqrt{2} \text{ cis } 250°}{\sqrt{6} \text{ cis } 105°}$

15. (6.3) Use DeMoivre's theorem to find:

(a) $(1 + i)^7$ (b) $(\sqrt{2} \text{ cis } 23°)^5$

16. (6.3) Find the roots and plot.

(a) $(25 \text{ cis } 60°)^{1/4}$ (b) $(0 - 8i)^{1/3}$
(c) The four fourth roots of i

17. (6.4) Plot the points, find three other sets of polar coordinates $(-360° < \theta < 360°)$ of the point, and find its rectangular coordinates.

(a) $(2, -30°)$ (b) $(\sqrt{3}, 225°)$

18. (6.4) Convert to polar form.

(a) $(\sqrt{2}, \sqrt{2})$ (b) $(-\sqrt{3}, 1)$

19. (6.4) Convert to rectangular form.

(a) $r = 3 \sec \theta$ (b) $r = \sin \theta - \cos \theta$

20. (6.5) Sketch the polar graphs.

(a) $r + 4 \sin \theta = 0$

(b) $r = 3 \sin 3\theta$

(c) $r = 1 + 3 \cos \theta$

(d) $r^2 = 9 \cos 2\theta$

(e) **Cote's spiral**
$r \sin 2\theta = 1$

(f) **Witch of Agnesi**
$r^2 = \tan^2 \theta + \cos^3 \theta$

appendix A
functions and graphs

A.1 COORDINATES AND GRAPHS

A **rectangular** or **Cartesian coordinate system** is, as you probably recall from previous study of algebra, the common means of representing an ordered pair as a point in a plane. Two number lines, called **axes,** are drawn perpendicular. The ordered pair (x, y) is located by measuring x on the horizontal or $\boldsymbol{x}$ **axis** and y on the vertical or $\boldsymbol{y}$ **axis.** The components of the ordered pair are called the **coordinates** of the point. The first component or x coordinate is called the **abscissa** and the second component or y coordinate is called the **ordinate.** The point is the **graph** of the ordered pair, which is often used as the name of the point. The point at which the two axes meet is called the **origin,** the point (0, 0). See Figure A.1.

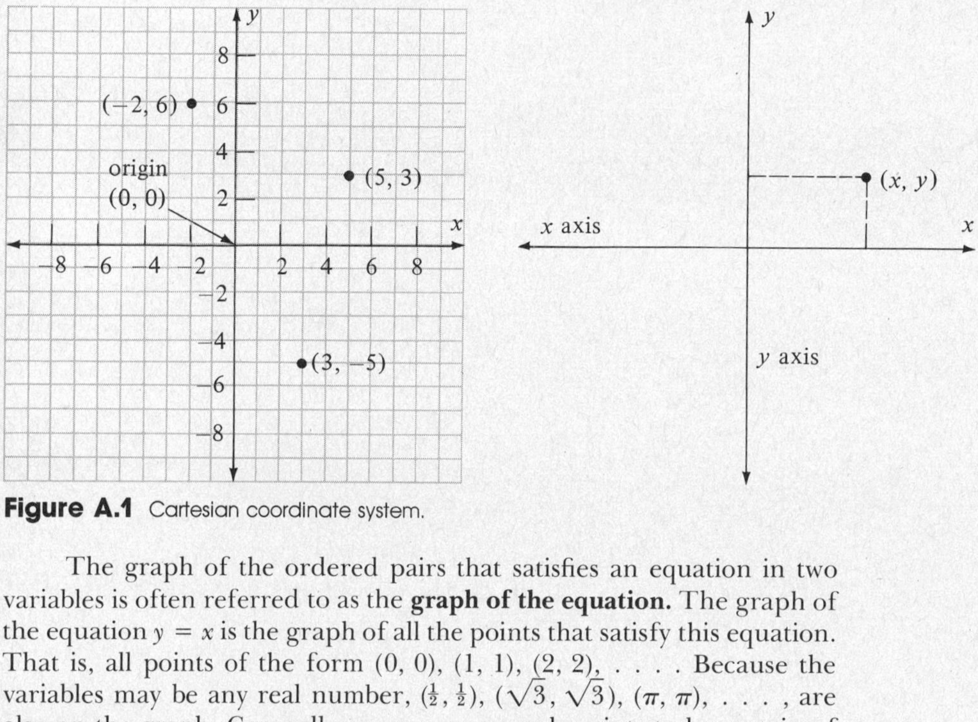

Figure A.1 Cartesian coordinate system.

The graph of the ordered pairs that satisfies an equation in two variables is often referred to as the **graph of the equation.** The graph of the equation $y = x$ is the graph of all the points that satisfy this equation. That is, all points of the form (0, 0), (1, 1), (2, 2), ... Because the variables may be any real number, $(\frac{1}{2}, \frac{1}{2})$, $(\sqrt{3}, \sqrt{3})$, (π, π), . . . , are also on the graph. Generally, we secure enough points to be certain of the behavior of the graph and then sketch a line or smooth curve through the points. This sketch represents the graph of the solutions of the equation or more simply the graph of the equation.

EXAMPLES

1. $y = 2x + 1$

   A table of values is compiled by finding the corresponding value of y for selected values of x.

x	y
−2	−3
−1	−1
0	1
1	3
2	5
3	7
.	.
.	.
.	.

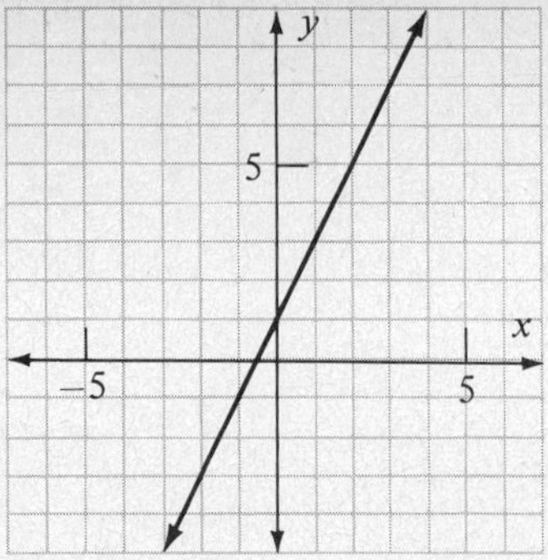

2. $y = x^2 - x + 2$

x	y
−2	8
−1	4
0	2
1	2
2	4
3	8

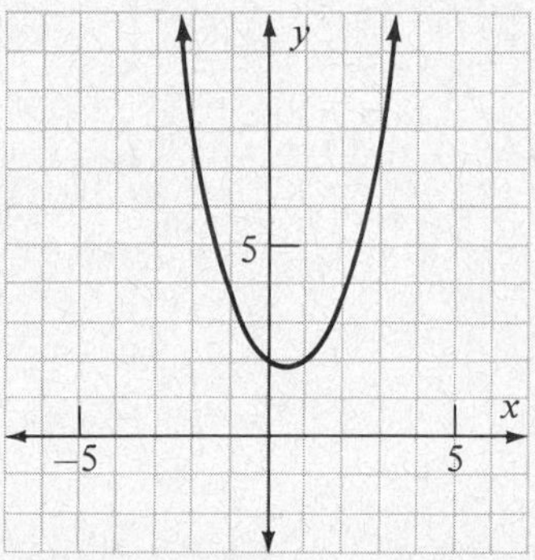

You may recognize this curve and the next as **parabolas.**

3. $2x = y^2 + y$

x	y
1	−2
0	−1
0	0
1	1
3	2
6	3

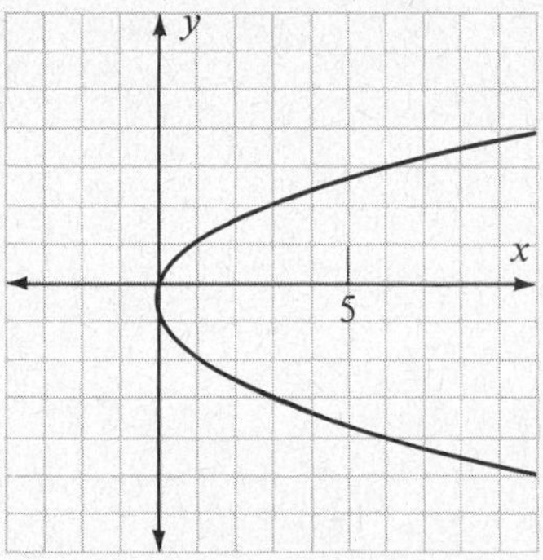

Notice here that it is more convenient to select values of y and evaluate x.

4. $y = x^3 + 4x^2 - 9$

x	y
−5	−34
−4	−9
−3	0
−2	−1
−1	−6
0	−9
1	−4
2	15
3	54

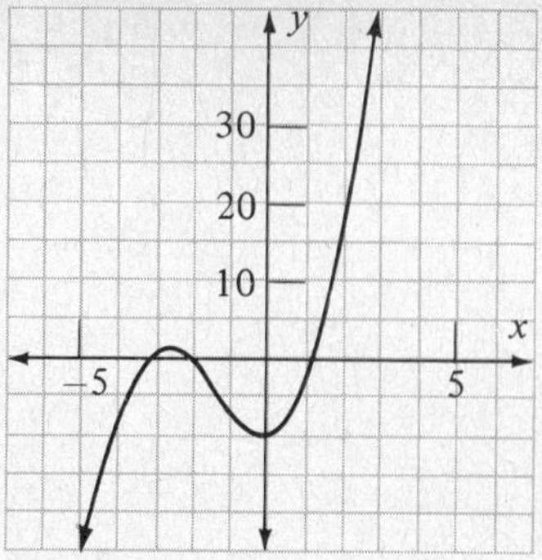

In some cases, more than just a few points will be necessary to obtain a representive sketch.

Not all graphs are smooth and unbroken curves. There are some very common processes that exhibit rather strange graphs. Consider, for example, the postal rates for first-class letters. The rate is 20¢ for the first ounce or fraction thereof and 17¢ for each additional ounce or fraction up to 12 oz. If x is the weight in ounces and y is the postage in cents, the graph is shown in Figure A.2. Notice that $y = 20$ for $0 < x \leq 1$, and $y = 37$ for $1 < x \leq 2$, and so on, and only positive values are used for x, which represents weight.

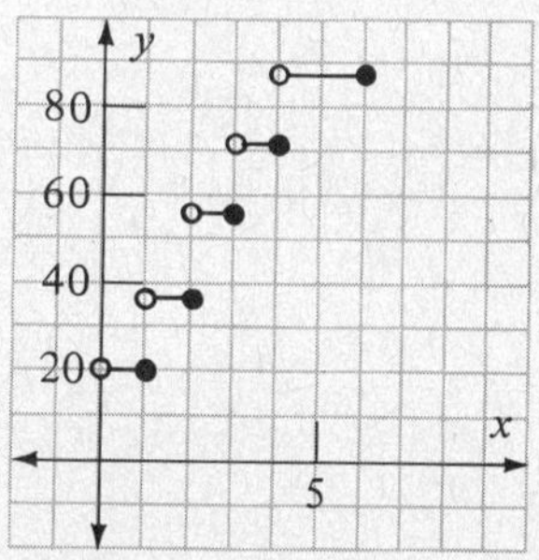

Figure A.2 First-class postage.

EXAMPLE

The equation $y = rx - 16x^2$ gives the height y that an object will reach in x seconds (s) if propelled upward at an initial rate of r ft/s.

5. A ball is tossed upward at 64 ft/s. Sketch the graph.

x	y
0	0
1	48
2	64
3	48
4	0

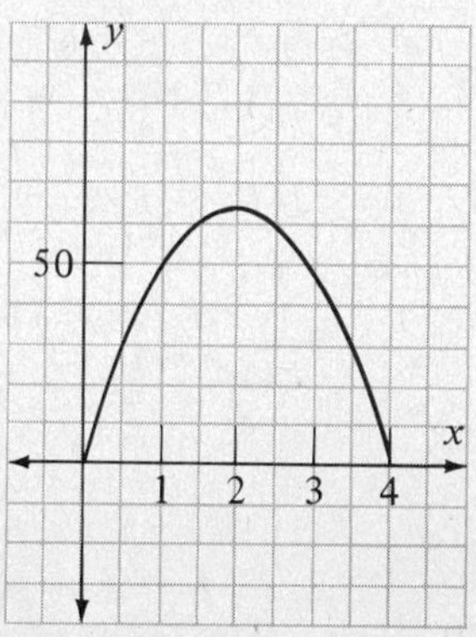

As in the postal rate illustration, there is an implicit restriction on one of the variables. Here y is always nonnegative because it represents height; thus $0 \leq x \leq 4$.

exercise set A.1

1. Plot the following points on the same coordinate axes: $(3, 7)$, $(-1, 1)$, $(-5, 7)$, $(-8, 7)$, $(-2.5, -1.25)$, $(-7, -8)$, $(-6, -8)$, $(-2, -2)$, $(2, -8)$, $(5, -8)$, $(-\frac{1}{2}, \frac{1}{4})$, $(4, 7)$, and $(3, 7)$. See Figure A.1.

2. **palatable plotting** Using the points in Exercise 1, connect consecutive points with line segments. The resulting figure will be identifiable if all points are plotted correctly.

In Exercises 3–10, sketch the graphs of the given pair of equations on the same set of coordinate axes. See Examples 1–4.

3. (a) $y = x + 1$ (b) $y = x - 1$

4. (a) $y = 2x - 3$ (b) $y = 2x + 3$

5. (a) $y = x^2 + 2$ (b) $x = y^2 + 2$

6. (a) $y = 3 - x^2$ (b) $x = 3 - y^2$

7. (a) $y = x^2$ (b) $y = (x - 3)^2$

8. (a) $y = \frac{1}{2}x^2$ (b) $y = \frac{1}{2}(x - 2)^2$

9. (a) $x = y^2$ (b) $x^2 = y$

10. (a) $y = 2x^2 - x^3$ (b) $y = x^3 + 2x^2$

11. The postal rate for single mailing of third-class matter is 20¢ for the first oz (or fraction) plus 17¢ for each additional oz up to 4 oz plus 14¢ for the next 2 oz (or fraction) plus 10¢ for each additional 2 oz (or fraction) up to (but not including) 16 oz. Sketch the graph of this rate structure using x for the weight in ounces and y for the charge in cents. See Figure A.2

12. The foreign airmail rate for delivery to most of Europe is 31¢ per $\frac{1}{2}$ oz (or fraction) through 2 oz and 26¢ per $\frac{1}{2}$ oz (or fraction) over 2 oz for letter mail. Sketch the graph of this rate schedule using x for the weight in ounces and y for the charge in cents.

The height y that an object will reach in x seconds is given by

$$y = rx - 16x^2$$

where r is the initial rate at which it is propelled. Air resistance is neglected.

13. Sketch the graph if a bullet is fired at 1600 ft/s. See Example 5.

14. A baseball is hurled upward at 96 ft/s. Sketch the graph.

15. The relationship between degrees Fahrenheit F, and degrees Celsius C is given by

$$F = \frac{9}{5}C + 32$$

Sketch the graph using the vertical scale for F and the horizontal scale for C.

16. Dry air rises and in doing so expands and cools. The temperature y, in degrees Fahrenheit, at height x ft is given by

$$y = T - 0.0055x$$

where T is the ground temperature. Sketch the graph of this equation where the ground temperature is 65°F. (Use 1 square = 1000 ft on the x scale and 1 square = 5 deg on the y scale.)

17. Hooke's law from physics states that each additional equal weight applied to a spring will stretch it an equal distance. The length y in centimeters for a given spring is given by

$$y = 20 + 0.001x$$

where x is the weight in grams. Sketch the graph of this equation. Choose convenient scales for the x and y axes.

A.2 RELATIONS AND FUNCTIONS

The deeper a diver goes, the greater is the pressure from the surrounding water. We say that the pressure is *related* to the depth. Relatedness is given a specific meaning in mathematics. A **relation** is a general association, correspondence, or pairing of elements from two sets. The first set, called the **domain,** contains the values of the **independent variable.** The second set is called the **range,** and its elements are the values of the **dependent variable.** A relation can be specified in a number of ways, including:

1. A set of ordered pairs. This specifies a correspondence where the first (or independent) variable is paired with the corresponding second (or dependent) variable.
2. A graph. This is the graphic representation of a set of ordered pairs.
3. An equation. The set of ordered pairs that satisfy the equation form a relation. The equation expresses the correspondence algebraically.

The Inverse of a Relation

Given a relation, another relation can be obtained by simply interchanging x and y, the numbers in the ordered pairs, or the variables in the equation. This new relation is called the **inverse** of the original relation. Graphically, this interchange produces a graph that is symmetric to the original with respect to the line $y = x$. Consider the relation $y = x^2 + 3$ with its graph shown in Figure A.3. When the variables are interchanged, the inverse of $y = x^2 + 3$ is $x = y^2 + 3$; both graphs are shown in Figure A.4. Notice that each point on $y = x^2 + 3$ is symmetric to some point on $x = y^2 + 3$ and conversely.

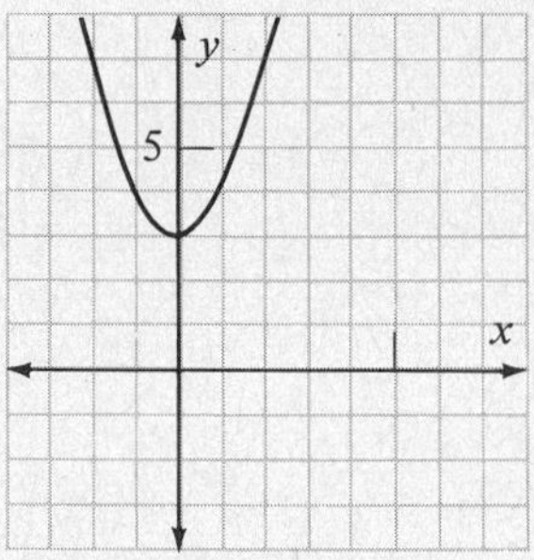

Figure A.3

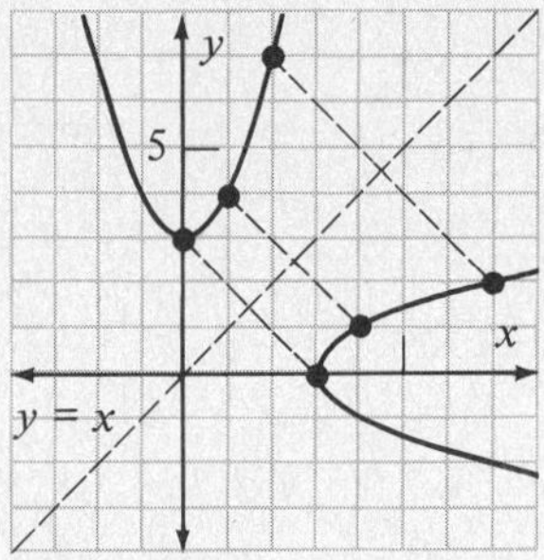

Figure A.4

It can be shown that the point (x, y) is symmetric to the point (y, x) with respect to the line $y = x$. We say that the symmetric graph is **reflected** in the line $x = y$ or is the **image** of the reflection.

EXAMPLES

Specify the inverse of each relation.

1. $R = \{(0, 1), (1, 3), (2, 5), (2, 7)\}$
   The inverse of R is $\{\mathbf{(1, 0), (3, 1), (5, 2), (7, 2)}\}$.
2. The relation S is specified by the graph in Figure A.5.

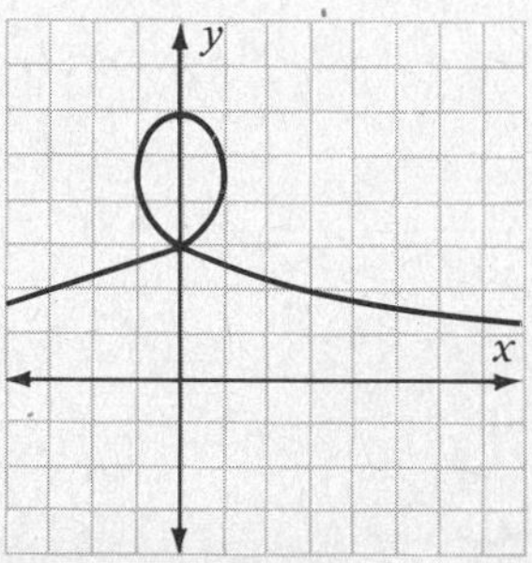

Figure A.5

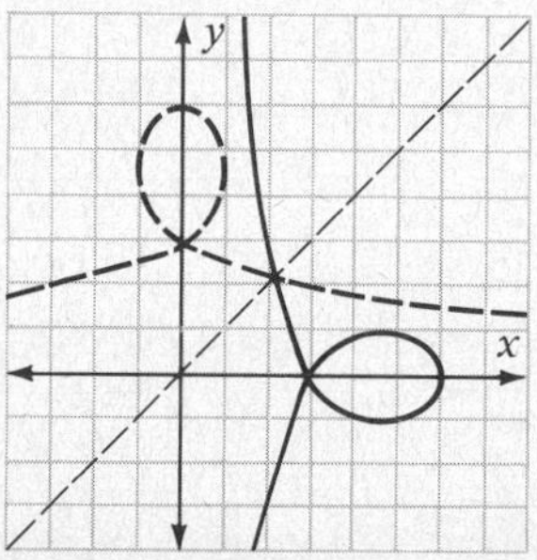

Figure A.6

The inverse of S is obtained by reflection in the line $y = x$ as shown in Figure A.6.

3. The relation T is specified by the equation $y = 2x - 6$.

   The inverse is obtained by interchanging x and y, so $x = 2y - 6$.

Functions

A **function** is a special relation. It is one of the most useful concepts of modern mathematics. Like a relation, a function may be thought of as a rule that describes how one quantity is determined by another or in a less intuitive sense, as an association, a correspondence, or a rule that associates each element of the domain with a *single* element of the range. The restriction to a single element of the range gives the function a special set of properties. Every function is a relation, but only certain relations are functions.

FUNCTION

A **function** f of a real number x is a rule that assigns one and only one real number y to each x in the domain.

The relationship between the two variables is usually given by an equation. For example, we say that pressure is a function of depth. The pressure y on a submerged body in pounds per square inch (lb/in.^{2}) can be approximated by

$$\mathbf{y = 0.43x + 14.7}$$

where x is the depth in feet.

We shall on occasion wish to indicate that y is a function of x without specifying a particular relationship. The notation

x is an element of the domain.

$$y = f(x)$$

$f(x)$ is the element of the range associated with x.

indicates that y is a function of x or simply "y equals f of x." Here $f(x)$ is the value assigned to x by the function f, and we speak of $f(x)$ as the value of f "at x." The example above may be written

$$\mathbf{f(x) = 0.43x + 14.7}$$

where the pressure $f(x)$ at depth x is given by $0.43x + 14.7$.

EXAMPLES

Given $f(x) = 2x - 1$ and $g(x) = 3 - x^2$ in Examples 4–8. Find the indicated values.

4. $f(3)$ $\qquad f(x) = 2x - 1$

$f(\mathbf{3}) = 2(\mathbf{3}) - 1 = 6 - 1$

$\mathbf{f(3) = 5}$

5. $g(-1)$ $\qquad g(x) = 3 - x^2$

$g(\mathbf{-1}) = 3 - (\mathbf{-1})^2 = 3 - 1$

$\mathbf{g(-1) = 2}$

6. $f(\frac{1}{2})$ $\qquad f(\frac{1}{2}) = 2(\frac{1}{2}) - 1 = 1 - 1$

$\mathbf{f(\frac{1}{2}) = 0}$

7. $g(5)$ $\qquad g(\mathbf{5}) = 3 - (\mathbf{5})^2 = 3 - 25$

$\mathbf{g(5) = -22}$

8. $f(-1) + g(2) = [2(-1) - 1] + [3 - (2)^2]$

$\qquad = [-3] + [-1]$

$\mathbf{f(-1) + g(2) = -4}$

9. If $f(x) = 0.43x + 14.7$, find $f(100)$.

$$f(\mathbf{100}) = 0.43(\mathbf{100}) + 14.7$$
$$f(100) = 43 + 14.7$$
$$\mathbf{f(100) = 57.7}$$

10. What is the pressure on a submerged body 75 ft below the surface?

$$f(x) = 0.43x + 14.7$$
$$f(\mathbf{75}) = 0.43(\mathbf{75}) + 14.7$$
$$= 46.95$$

The pressure is approximately 47 lb/in.^{2}.

exercise set A.2

Given that $f(x) = 2 - 3x$ and $g(x) = x^2 - 9$, find the specified values in Exercises 1–10.

1. $f(1)$
2. $f(-1)$
3. $g(-2)$
4. $g(2)$
5. $f(3) + g(-1)$
6. $f(-2) - g(5)$
7. $f(0) + g(0)$
8. $g(4) + f(3)$
9. $g(-3) - f(\frac{1}{3})$
10. $f(\frac{2}{3}) - g(3)$

Exercises 11–15 refer to the following table. If x is the year, {1930, 1940, 1950, 1960, 1965, 1970, 1975} is the domain, then let $p(x)$ = price per pound for pork, $b(x)$ = price per pound for beef, $L(x)$ = price per pound for lamb, $m(x)$ = price per pound for milk, $E(x)$ = price per dozen for eggs, $c(x)$ = price per bushel for corn, and $w(x)$ = price per bushel for wheat.

AVERAGE PRICES RECEIVED BY U.S. FARMERS*

YEAR	PORK	BEEF	LAMB	MILK	EGGS	CORN	WHEAT
1930	0.088	0.077	0.078	0.022	23.7	0.66	0.55
1940	0.054	0.076	0.081	0.018	18.0	0.67	0.60
1950	0.180	0.233	0.251	0.039	36.3	2.00	1.52
1960	0.153	0.204	0.179	0.042	36.1	1.74	1.00
1965	0.196	0.199	0.228	0.042	33.7	1.35	1.16
1970	0.227	0.271	0.264	0.057	39.1	1.33	1.33
1975	0.461	0.323	0.421	0.088	52.5	2.54	3.56
x	$p(x)$	$b(x)$	$L(x)$	$m(x)$	$E(x)$	$c(x)$	$w(x)$

* *Source:* U.S. Department of Agriculture.

Find each of the following.

11. (a) $p(1960)$ (b) $b(1940)$
    (c) $m(1970)$ (d) $w(1930)$

12. (a) $b(1975)$ (b) $L(1930)$
    (c) $E(1965)$ (d) $c(1940)$

13. (a) $E(1975) - E(1970)$ (b) $c(1965) - c(1960)$
    (c) $w(1975) - c(1975)$ (d) $b(1960) - L(1960)$

14. (a) $\dfrac{p(1975) + b(1975) + L(1975)}{3}$

(b) What is the meaning (in words) of part (a).

15. (a) $b(1975) - b(1960)$
(b) What is the meaning (in words) of part (a).

16. Biologists have determined that the rate at which a cricket chirps is a function of temperature. The number of chirps per minutes is approximated by

$$r(T) = \tfrac{1}{4}T + 40$$

where T is in degrees Fahrenheit.
(a) Find $r(80)$, $r(64)$, and $r(74)$.
(b) What is the rate at 100°F?

17. Normal blood pressure for an adult is a function of age. Normal blood pressure for age a years is approximated by

$$p(a) = \tfrac{1}{2}a + 110$$

(a) Find $p(24)$, $p(35)$, and $p(52)$.
(b) What is the normal blood pressure for a 50-year-old?

18. The distance in miles that can be seen to the horizon at sea is approximated by

$$d(h) = \sqrt{\tfrac{8}{5}h}$$

where h is the height in feet above the surface of the water.
(a) Find $d(8)$, $d(10)$, $d(40)$, $d(90)$, and $d(200)$.
(b) How far can one see from a height of 40 ft above the water?

19. calculator problem The distance in feet necessary to stop a car traveling x mi/h is approximated by

$$d(x) = 0.071x^2 + 0.73x$$

(a) Find $d(15)$, $d(25)$, $d(40)$, $d(55)$, and $d(70)$.
(b) What distance is necessary to stop a car moving at 35 mi/h?

20. The volume of a sphere is a function of the length of its radius. The volume is given by

$$V(r) = \tfrac{4}{3}\pi r^3$$

where r is the radius of the sphere.
(a) Find $V(1)$, $V(2)$, $V(3)$, and $V(\pi)$
(b) What is the volume of a large hot-air balloon 12 ft in radius?

A.3 THE GRAPH OF A FUNCTION

$f(x) = 2$ is a function because, regardless of the choice of x, the corresponding value of y is 2. The definition of function does not require that y be different for each x, only that y have a single value for each x. Thus $y^2 = x$ or $y = \pm\sqrt{x}$ does not specify a function because it produces two distinct values of y for each positive value of x.

THE GRAPH OF A FUNCTION
The **graph of a function** f is the set of points (x, f(x)) in the plane.

The graph of $y = f(x)$ will be peculiar because there will be a unique $f(x)$ for each x in the domain. That is, we may interpret this graphically by saying that it is a function if and only if a vertical line meets the graph in at most one point.

EXAMPLES

Which of the graphs shown is the graph of a function?

1. Each vertical line will meet the graph in exactly one point, so this is the graph of a function.

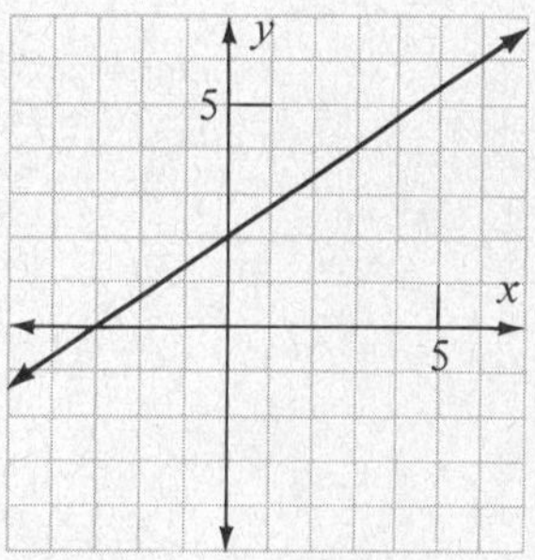

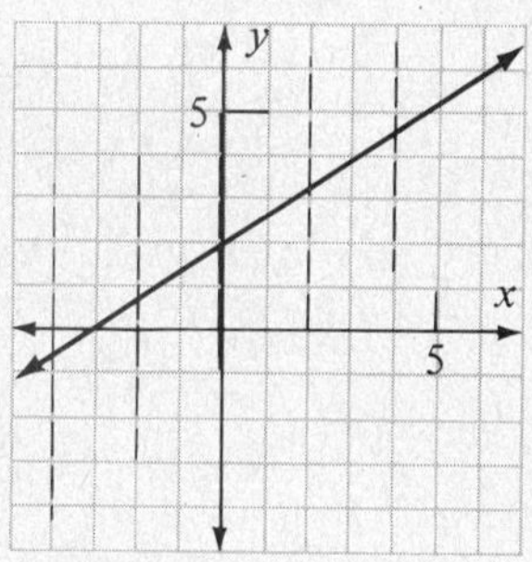

2. Certain vertical lines will meet the graph in more than one point, so the vertical line test fails and this is not the graph of a function.

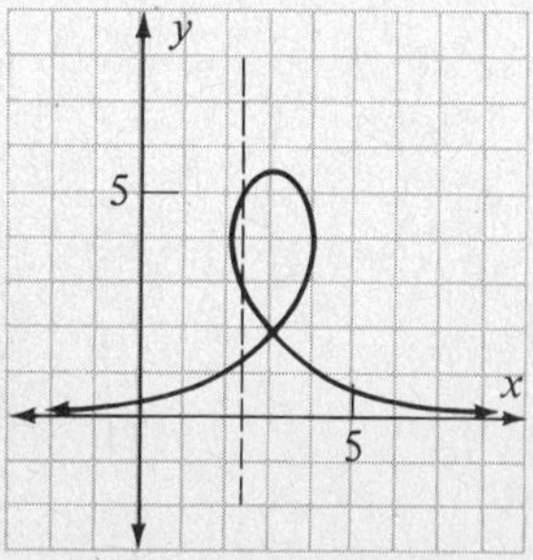

3. The vertical line test succeeds, so this is the graph of a function.

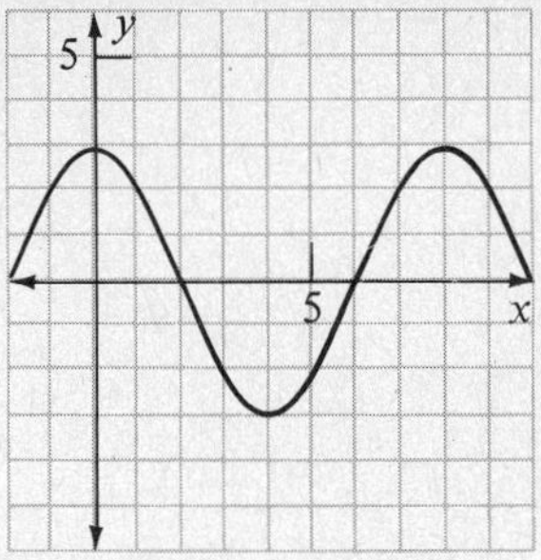

The Inverse of a Function

The function f assigns a value y from the range to each value x from the domain. The inverse of f reverses the correspondence and assigns a value x to each y. The inverse of f is indicated by f^{-1} if it is itself a function.

> **INVERSE FUNCTIONS**
> For f and f^{-1} both functions, if $y = f(x)$, then $x = f^{-1}(y)$.

The inverse is found by interchanging x and y and then testing to see that the resulting equation defines y as a function of x.

EXAMPLES

Find the inverse function, if it exists.

4. $y = 2x - 1$

The inverse is

interchanged

$$\mathbf{x} = 2\mathbf{y} - 1$$

and solving for y,

$$2y = x + 1$$

$$y = \tfrac{1}{2}x + \tfrac{1}{2}$$

So if $f(x) = 2x - 1$, then $\boldsymbol{f^{-1}(x) = \tfrac{1}{2}x + \tfrac{1}{2}}$.

5. $g(x) = x^2$

The inverse of $y = x^2$ is $x = y^2$. Solving for y, $y = \pm\sqrt{x}$, which is *not* a function. Thus g does not have an inverse function.

Symmetry

A function and its inverse are symmetric with respect to the line $y = x$. Having the graph of a function, its inverse is easily sketched as a reflection or image of the function in the line $y = x$. Other symmetries provide information useful when working with the graphs of certain functions.

Consider a point (x, y) relative to the coordinate axes. The point $(-x, y)$ is an equal distance on the opposite side of the y axis from (x, y). The points are symmetric with respect to the y axis. Figure A.7 illustrates this as well as (x, y) and $(x, -y)$ symmetric with respect to the x axis, and (x, y) and $(-x, -y)$ symmetric to the origin.

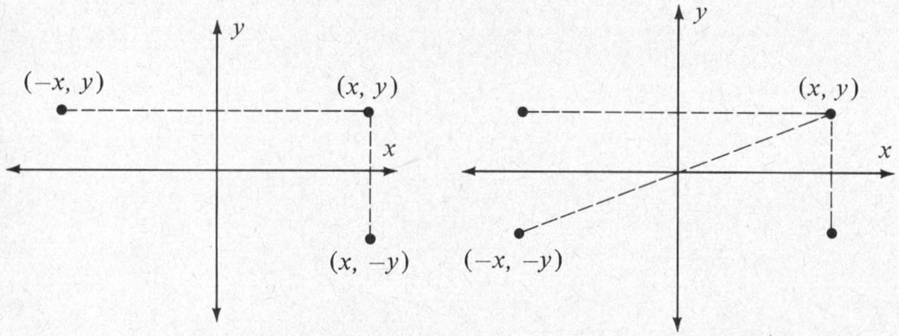

Figure A.7 Symmetry with respect to the axes.

> Given $y = f(x)$, the graphs of:
>
> $y = f(x)$ and $-y = f(x)$ are symmetric with respect to the x axis.
> $y = f(x)$ and $y = f(-x)$ are symmetric with respect to the y axis.
> $y = f(x)$ and $-y = f(-x)$ are symmetric with respect to the origin.

The graphs of two functions are symmetric if every point of one has an image in the other and conversely. The graphs of $y = (x + 3)^2$ and $y = (x - 3)^2$ are shown in Figure A.8. Notice that they are symmetric with respect to the y axis. Similarly, $y = \frac{1}{2}x + 2$ and $y = -\frac{1}{2}x - 2$ are symmetric with respect to the x axis.

We may test the equations for symmetry using a method similar to the procedure for finding inverses that were symmetric with respect to $y = x$.

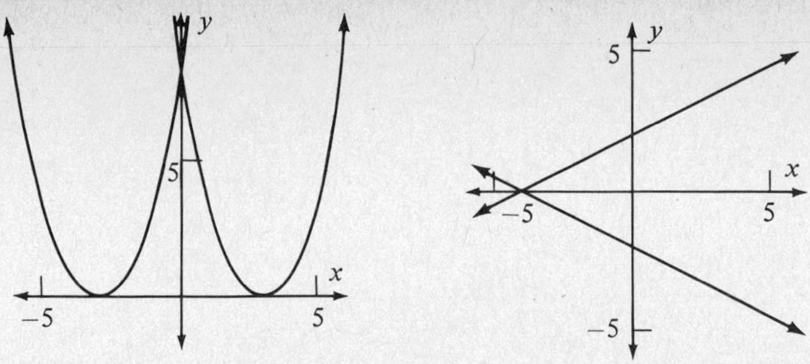

Figure A.8

> **TESTS FOR SYMMETRY**
>
> Two graphs are symmetric with respect to:
>
> The x axis if replacing (x, y) with (x, −y) in one equation produces the other equation.
>
> The y axis if replacing (x, y) with (−x, y) in the one equation produces the other equation.
>
> The origin if replacing (x, y) with (−x, −y) in one equation produces the other equation.

EXAMPLES

6. Show that $y = \frac{1}{2}x + 2$ and $y = -\frac{1}{2}x - 2$ are symmetric with respect to the x axis.

   Replace x and y in $y = \frac{1}{2}x + 2$ by x and $-y$.

$$\mathbf{-y} = \tfrac{1}{2}\mathbf{x} + 2$$

$$y = -\tfrac{1}{2}x - 2$$

   Thus $y = \frac{1}{2}x + 2$ and $y = -\frac{1}{2}x - 2$ are symmetric with respect to the x axis.

7. Show that $y = (x + 3)^2$ and $y = (x - 3)^2$ are symmetric with respect to the y axis.

   Replace x and y in $y = (x + 3)^2$ by $-x$ and y

$$\mathbf{y} = (\mathbf{-x} + 3)^2 = (x - 3)^2$$

$$y = (x - 3)^2$$

   [Notice that $(a - b)^2 = (b - a)^2$.] Thus the graphs are symmetric with respect to the y axis.

8. Show that $y = x^2$ and $y = -x^2$ are symmetric with respect to the origin.

Replace (x, y) by $(-x, -y)$.

$$y = x^2$$
$$(-y) = (-x)^2$$
$$-y = x^2$$
$$y = -x^2$$

Thus the graphs are symmetric with respect to the origin.

Translation

Points may be shifted or **translated** as well as reflected in the plane. Thus $y - k = f(x - h)$ is the graph of $y = f(x)$ shifted h units horizontally and k units vertically. That is, if x and y are replaced by $x - h$ and $y - k$, the graph is translated over h units and up k units.

TRANSLATION

The graph of $y - k = f(x - h)$ is the graph of $y = f(x)$ translated h units horizontally and k units vertically.

First sketch the graph of $y = x^2$ as shown in Figure A.9. Then sketch the graph of $y - 3 = (x - 5)^2$ or $y = x^2 - 10x + 28$ as shown in Figure A.10. Notice that each point on the graph of $y = x^2$ has been

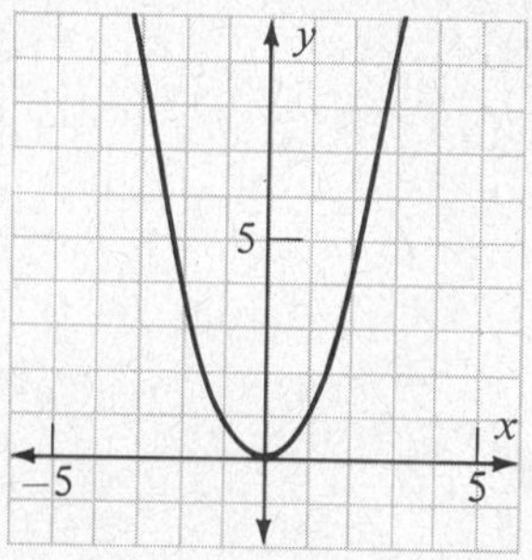

Figure A.9

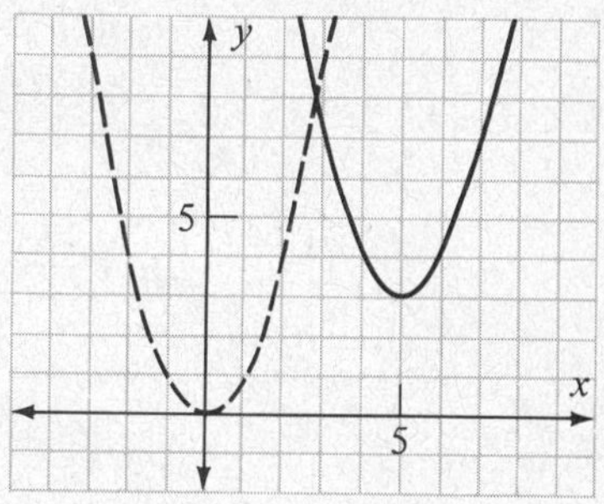

Figure A.10

moved 5 units to the right and up 3 units to form the graph of $y - 3 = (x - 5)^2$. This can be very helpful, because compiling a table of points for $y = x^2$ is much simpler than for that of $y = x^2 - 10x + 28$. Essentially, the graph of $y = f(x)$ could be drawn by treating the point (h, k) as if it were the origin to produce the graph of $y - k = f(x - h)$.

EXAMPLES

Let $y = f(x)$ be the function shown in Figure A.11. Graph each indicated function.

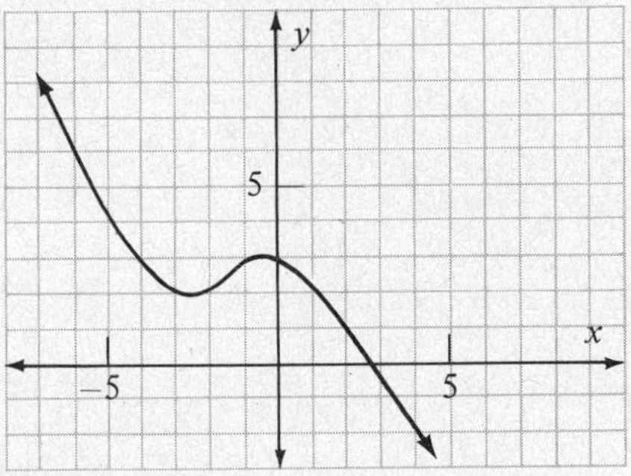

Figure A.11 $y = f(x)$.

9. $y - 2 = f(x - 4)$

This is the graph of $y = f(x)$ translated 4 units over and 2 units up. First translate several key points, then sketch the graph of the translated function.

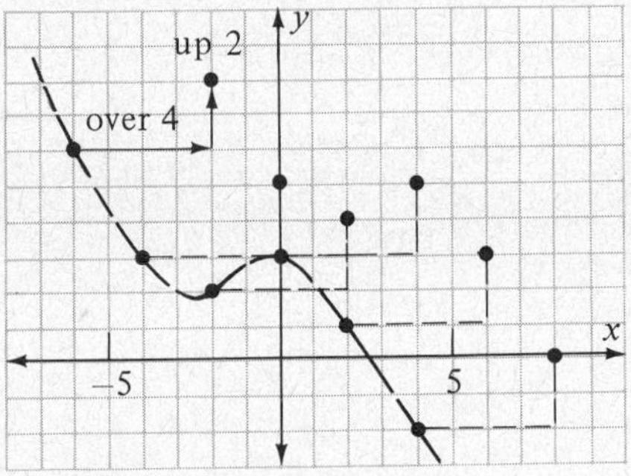

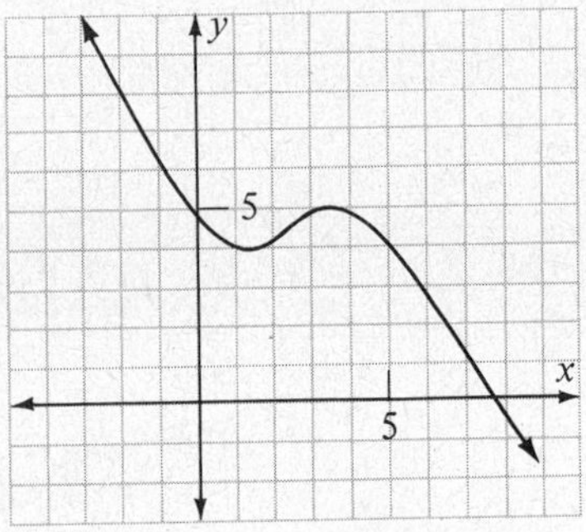

10. $y + 3 = f(x - 1)$

This is the graph of $y = f(x)$ translated 1 unit over and 3 units *down* $[-3$ units up because $y + 3 = y - (-3)]$.

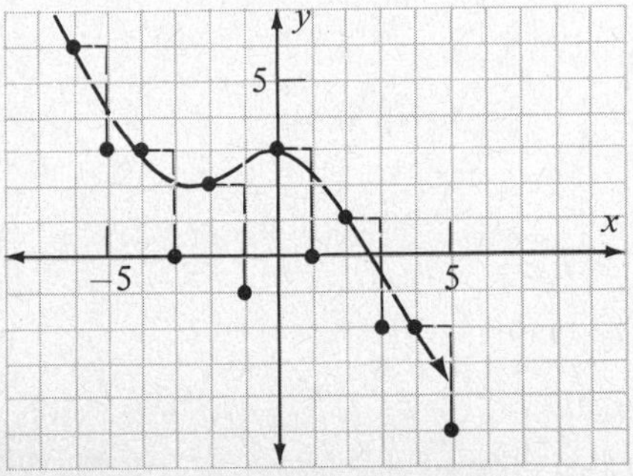

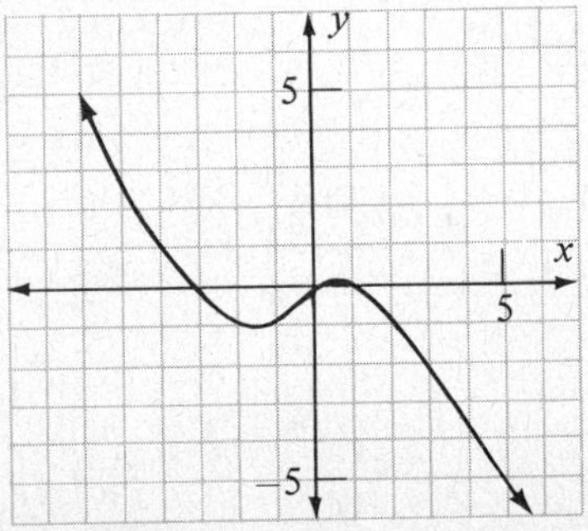

exercise set A.3

In Exercises 1 and 2, determine if the graph is that of a function and if so if its inverse also a function.

1. (a)

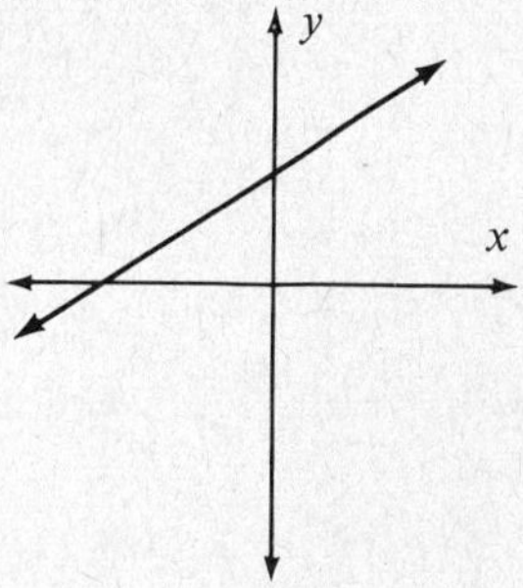

(b)

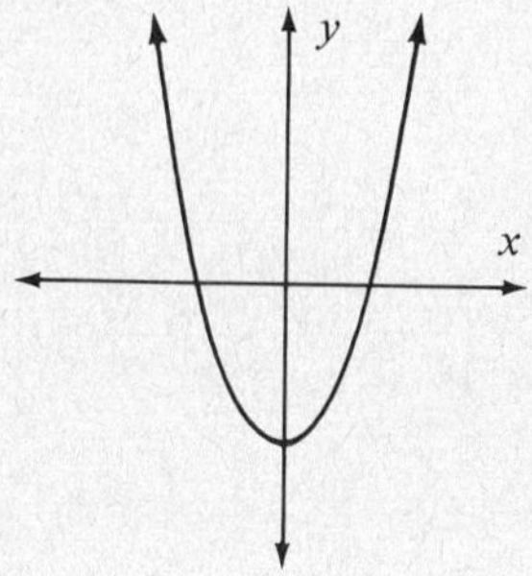

(c)

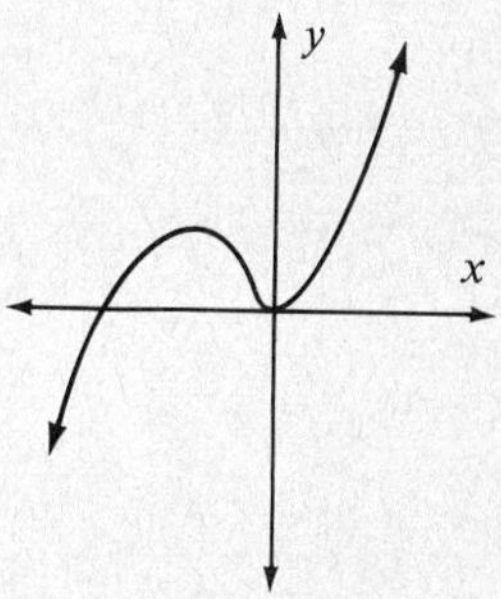

(d)

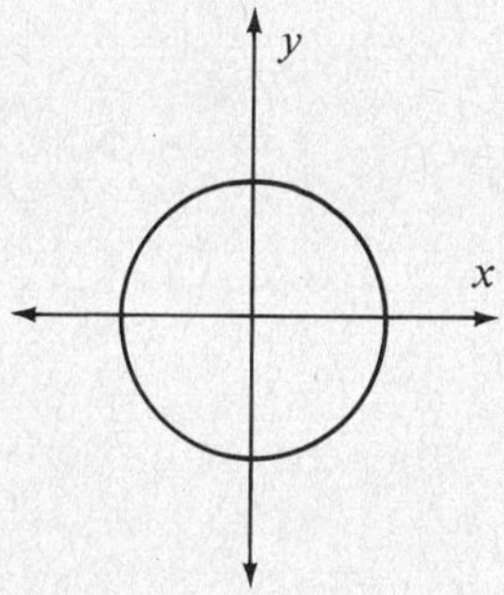

2. (a)

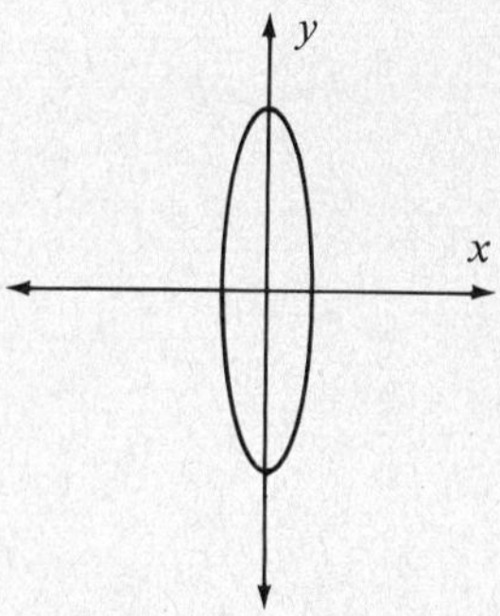

(b)

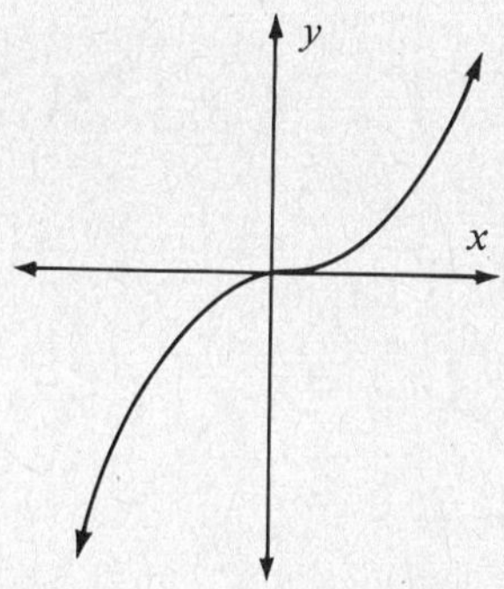

(c)

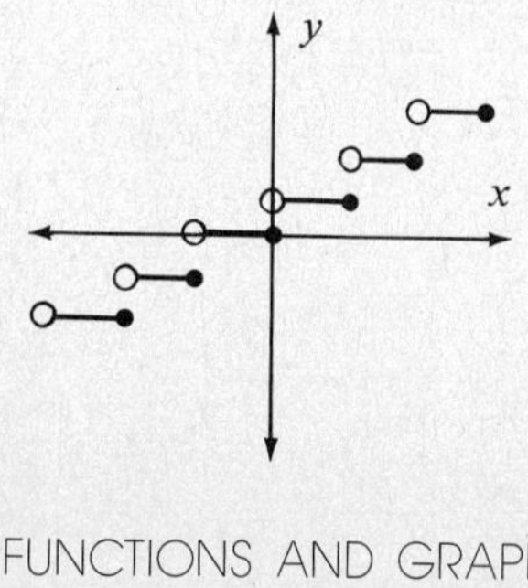

(d)

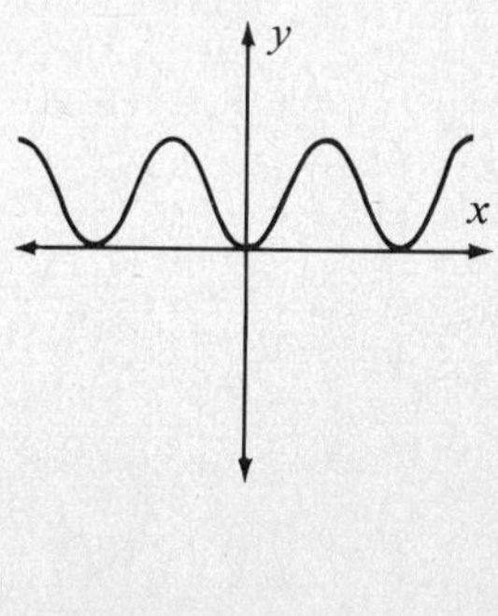

In Exercises 3 and 4, find the inverse function if it exists. See Examples 4 and 5.

3. (a) $y = 3x + 1$ (b) $y = 3x^2 + 1$

4. (a) $y = x^2 + 1$ (b) $y = \dfrac{x + 1}{x - 1} \quad (x \neq -1)$

In Exercises 5–7, use the graphs in Exercises 1 and 2 to sketch the image of the graph in the stated axis. See Figure A.8.

5. (a) 1(a), y axis (b) 1(b), x axis

6. (a) 1(c), y axis (b) 2(b), x axis

7. (a) 2(c), x axis (b) 2(d), y axis

Find the equation of the graph symmetric to the given function in the given axis in Exercises 8–10. See Examples 6–8.

8. (a) $y = 4x - 3$, x axis
   (b) $y = \frac{1}{3}x + 2$, y axis

9. (a) $y = x^2 - 8x + 9$, y axis
   (b) $y = (x - 2)^2$, x axis

10. (a) $y = x^3 - 2x^2 + x - 1$, x axis
    (b) $y = 5 - x - x^3$, y axis

Sketch the graph of the given function in Exercises 11–15, given $y = f(x)$ and $y = g(x)$ in Figure A.12.

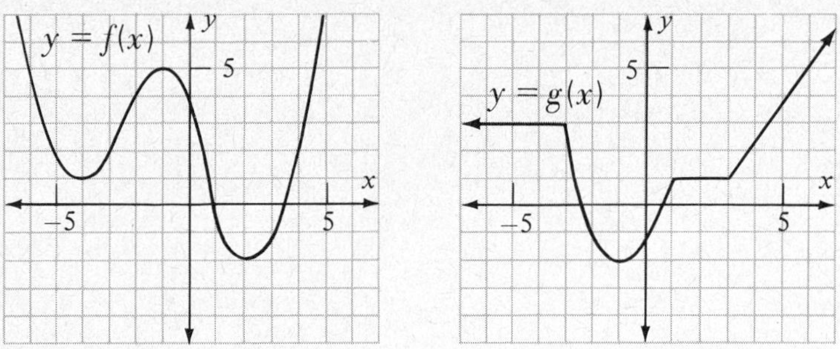

Figure A.12

11. (a) $y = f(x - 2)$ (b) $y = f(x + 3)$

12. (a) $y - 2 = f(x)$ (b) $y + 3 = f(x)$

13. (a) $y - 1 = f(x - 3)$ (b) $y - 3 = f(x - 1)$

14. (a) $y - 2 = g(x + 1)$ (b) $y + 4 = g(x - 2)$

15. (a) $y + 3 = g(x + 5)$ (b) $y + 5 = g(x + 3)$

16. The number of degrees Fahrenheit is a function of the number of degrees Celsius, as given by

See Exercise 15 in Section A.1.

$$f(C) = \tfrac{9}{5}C + 32$$

or

$$F = \tfrac{9}{5}C + 32$$

Find f^{-1}.

17. The temperature y is a function of the height x above ground:

See Exercise 16 in Section A.1.

$$f(x) = T - 0.0055x$$

or

$$y = T - 0.0055x$$

where T is the ground temperature. Find f^{-1} if the ground temperature is 65°.

18. The rate at which a cricket chirps is a function of the temperature:

See Exercise 16 in Section A.2.

$$r(t) = \tfrac{1}{4}t + 40$$

Find r^{-1}.

19. Blood pressure is a function of age:

See Exercise 17 in Section A.2.

$$p(a) = \tfrac{1}{2}a + 110$$

Find p^{-1}.

20. The distance visible to the horizon is a function of height above the sea:

See Exercise 18 in Section A.2.

$$d(h) = \sqrt{\tfrac{8}{5}h}$$

(a) What is the implicit restriction on h?
(b) Find d^{-1}, if it exists.
(c) d gives "the distance one can see to the horizon at sea given one's height above the water." Explain d^{-1} in words.

A.4 EXPONENTIAL FUNCTIONS

Exponential functions are of particular importance because of their wide application. Growth and decay are described by these functions: population growth, radioactive decay, growth of money at compound interest, or decrease of atmospheric pressure with increased altitude.

EXPONENTIAL FUNCTION

The function $f(x) = b^x$ is called an **exponential function,** where the base b is positive and not equal to 1.

First, we shall sketch the graphs of several exponential functions to observe their behavior.

EXAMPLE

1. Sketch the graph of $f(x) = 2^x$ from a table of values.

x	y
-2	$2^{-2} = \frac{1}{2^2} = \frac{1}{4}$
-1	$2^{-1} = \frac{1}{2}$
0	$2^0 = 1$
1	$2^1 = 2$
2	$2^2 = 4$
3	$2^3 = 8$
4	$2^4 = 16$

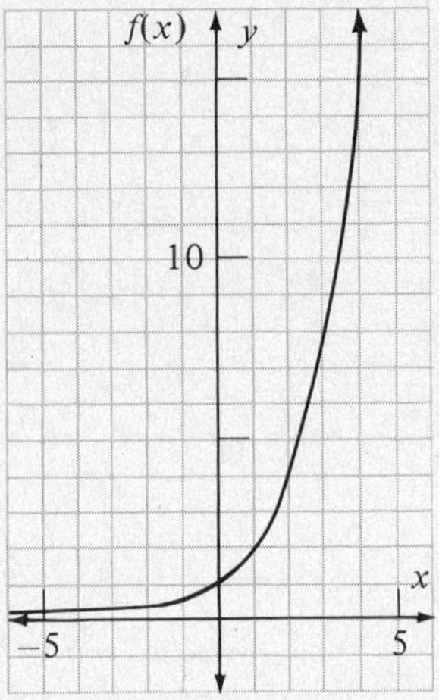

As x decreases, the graph of $y = 2^x$ approaches $y = 0$ as an asymptote; and as x increases, the function increases rapidly and without bound. It is everywhere increasing as x increases. This behavior is called **monotonic increasing.** If (x, y) is replaced by $(-x, y)$, then $y = 2^{-x} = (\frac{1}{2})^x$ is obtained, which is the reflection of $y = 2^x$ in the y axis.

EXAMPLE

2. Sketch $y = 2^{-x}$.

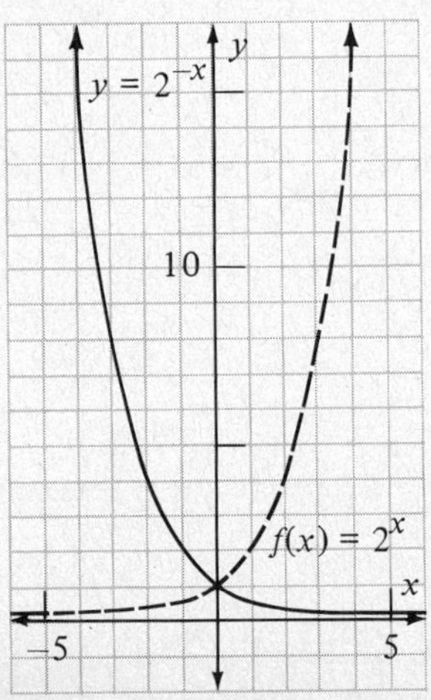

In general, $f(x) = b^x$ is an increasing function for $b > 1$ and decreasing for $0 < b < 1$. The use of these functions for growth and decay can now be seen to be more reasonable. It is more difficult to accumulate a table of values if b is not an integer. However, the hand-held calculator can be of great assistance. Consider $f(x) = 2(3.1)^{x/10}$; what is $f(7)$?

[3.1] [y^x] [0.7] [=] [×] [2] [=] 4.4155334

$$f(7) = 2(3.1)^{0.7}$$
$$\approx 4.4155334$$

The [y^x] key allows the quick computation of many values in the following example.

EXAMPLE

3. Sketch $y = 2(3.1)^{x/10}$.

x	y
−6	1.01
−4	1.27
−2	1.59
0	2
2	2.51
4	3.14
6	3.94
8	4.94
10	6.2

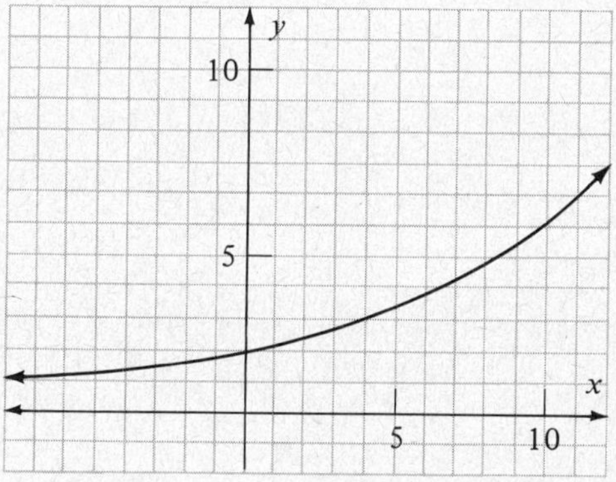

Radioactive substances decompose or decay. This decay is measured by its **half-life,** the time required for one-half of the substance to decompose. The amount A remaining after t years is given by

$$\mathbf{A = A_0(\tfrac{1}{2})^{t/h}}$$

where A_0 is the original amount of a substance with a half-life of h.

In archaeology, carbon-14 dating is a standard method of determining the age of certain artifacts. There is a constant percentage of the total carbon present in all living matter that is in the radioactive form

^{14}C. When the organic substance dies, the ^{14}C begins to decay. The amount remaining at any given time is a measure of the artifact's age.

EXAMPLE

4. Carbon-14 has a half-life of 5570 years. How much ^{14}C remains after 1000 years, if 100 mg was present originally?

$$\begin{aligned} A &= A_0(\tfrac{1}{2})^{t/h} \\ &= 100(\tfrac{1}{2})^{1000/5570} \\ &\approx 88.298864 \end{aligned}$$

Approximately 88.3 mg of the ^{14}C remains.

A very practical application of the exponential function utilizes the following formula to determine the amount of money accumulated in an account earning compound interest.

COMPOUND INTEREST

The amount A accumulated on P dollars invested for t years at annual rate r compounded n times a year is given by

$$A = P\left(1 + \frac{r}{n}\right)^{nt}$$

If the principal, rate, time, and number of periods are known (constant), then the amount A is a function of time t.

EXAMPLE

5. What is the final balance on an account earning $11\frac{3}{4}$ percent compounded quarterly for 6 years, if \$2750 was the initial balance?

Note that $P = 2750$, $r = 0.1175$ (11.75 percent), $n = 4$ (quarterly), and $t = 6$.

$$A = 2750\left(1 + \frac{0.1175}{4}\right)^{4\cdot 6}$$

Key	Display
0.1175	0.1175
÷	0.1175
4	4
+	0.029375
1	1
=	1.029375
y^x	1.029375
24	24
=	2.0033959
×	2.0033959
2750	2750
=	5509.3389

$$A \approx 5509.34$$

The final balance will be $5509.34.

To experience what is meant by money growing exponentially, consider $1000 at 8 percent compounded semiannually. The amount in the account is a function of time.

$$A = P\left(1 + \frac{r}{n}\right)^{nt}$$

$$= 1000\left(1 + \frac{0.08}{2}\right)^{2 \cdot t}$$

$$f(t) = 1000(1.04)^{2t}$$

The function is graphed in the following example.

EXAMPLE

6. Sketch $A = 1000(1.04)^{2t}$.

t	A
0	1000
1	1081.60
2	1169.86
3	1265.32
4	1368.57
5	1480.24
6	1601.03
7	1731.68
8	1872.98
9	2025.82
10	2191.12

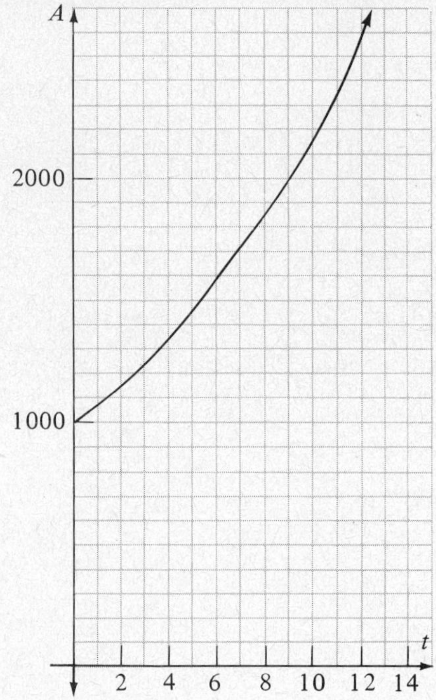

Because the function is exponential (with $b > 1$), the amount increases as time increases. Of course, an increase in the rate would also cause an increase in amount. However, what about the number of times the interest is compounded? Will a greater number of interest periods each year mean a corresponding gain in interest? Some financial institutions advertise "interest compounded daily" or mention "continuous interest." What actually happens as n increases? Suppose that we simplify the quantities involved and take \$1.00 for 1 year at 100 percent, so that $P = t = r = 1$.

$$\begin{aligned} A &= P\left(1 + \frac{r}{n}\right)^{nt} \\ &= 1\left(1 + \frac{1}{n}\right)^{n \cdot 1} \\ &= \left(1 + \frac{1}{n}\right)^{n} \end{aligned}$$

Now a table can be compiled for the amount earned if interest is compounded annually, semiannually, quarterly, monthly, and so on, by letting $n = 1, 2, 4, 12, 52, \ldots$.

n	$1 + \frac{1}{n}$	$\left(1 + \frac{1}{n}\right)^n$	
1	2	2	Annually
2	1.5	2.25	Semiannually
4	1.25	2.44	Quarterly
12	1.083	2.61	Monthly
52	1.019	2.69	Weekly
360	1.0027	2.715	Daily
8,640	1.0001157	2.7181	Hourly
518,400	1.000001929	2.71826	Every minute
31,104,000	1.0000000321	2.7182818	Every second

There are substantial increases until weekly or daily computation, but beyond that the increases flatten out. In fact, no matter how often the interest is compounded, the amount accumulated does not exceed a number that is approximately 2.7182818. This number is significant enough that it is denoted by its own symbol, the letter ***e***, and is called the **natural base.** It is not unlike π, an irrational nonrepeating decimal representation. It has a great many applications, because it describes continuous growth. If interest were compounded continuously, we could rewrite the compound interest formula as

$$A = Pe^{tr}$$

However, whereas interest is compounded a finite number of times (discretely), many natural phenomena grow continuously. Many kinds of growth and decay are described by an exponential function of the form

$$y = y_0 e^{kt}$$

where t is time, y_0 is the number or amount present at $t = 0$, and k is a constant.

For instance, human population growth is described by

$$\boldsymbol{P = P_0 e^{rt}}$$

where an initial population P_0 at $t = 0$ grows to P in t years at rate r.

EXAMPLES

7. Honolulu, Hawaii, had a 1960 population of 294,194. If during the next 10 years it grew at 7.63 percent, what was Honolulu's population in 1970 (to the nearest thousand)?

$$P = P_0 e^{rt}$$
$$= 294{,}194e^{(0.0763)(10)}$$
$$= 294{,}194e^{0.763}$$

[0.763] [e^x] [×] [294194] [=] 630958.07

$$P \approx 630{,}958.07$$
$$\approx 631{,}000$$

The population of Honolulu in 1970 was approximately 631,000.

(Note that if you do not have an [e^x] key, you can enter the approximation 2.7182818 and use the [y^x] key. We consider another way to handle this in the next section.)

It should be noted that such phenomena are described or modeled by the exponential functions. Actually, they ignore factors such as immigration and migration.

8. The atmospheric pressure P in pounds per square inch approximated by

$$\mathbf{P = 14.7e^{-0.21a}}$$

where a is the altitude above sea level in miles. What is the atmospheric pressure at Denver, Colorado, the so-called mile-high city?

$$P = 14.7e^{-0.21(1)}$$
$$\approx 11.915588$$

The pressure is approximately 11.9 lb/in.^{2}.

exercise set A.4

Sketch the graphs of the exponential functions in each of Exercises 1–6 on a single set of axes.

1. (a) $y = 2^x$ (b) $y = 3^x$ (c) $y = 4^x$ See Examples 1 and 2.

2. (a) $y = 2^{-x}$ (b) $y = 3^{-x}$ (c) $y = 4^{-x}$

3. (a) $y = (1.5)^x$ (b) $y = (1.5)^{2x}$ (c) $y = (1.5)^{x/2}$

4. (a) $y = 0.5^x$ (b) $y = 0.5^{3x}$ (c) $y = 0.5^{x/3}$

5. (a) $y = \pi^x$ (b) $y = \pi^{2x}$ (c) $y = \pi^{x/2}$ See Example 3.

6. (a) $y = e^x$ (b) $y = e^{2x}$ (c) $y = e^{x/2}$

7. Uranium-234 has a half-life of 252,000 yr. How much ^{234}U remains of 10 mg after 1900 years? See Example 4.

8. Radium-226 has a half-life of 1622 years. How much of 25 mg of ^{226}Ra remains after a century?

9. Find the amount accumulated in an amount if \$1500 is left for 5 years at $12\frac{1}{4}$ percent compounded quarterly. See Example 5.

10. A \$600 balance is allowed to accumulate interest for 10 years. What is the final balance if the rate is $11\frac{1}{2}$ percent compound quarterly?

11. Houston, Texas, had a population of 938,219 in 1960. Over the next 15 years it grew at 2.66 percent. What was its 1975 population (to the nearest hundred thousand)? See Example 7.

12. The 1960 population of Reno, Nevada, was 51,470. If Reno grew at 3.48 percent during the following decade, what was its population in 1970 (to the nearest hundred)?

13. Mount McKinley in Alaska is the highest point in North America. The altitude is 20,320 ft (or 3.85 mi). What is the atmospheric pressure on top? See Example 8.

14. The lowest land point on the surface of the earth is the Dead Sea (Israel-Jordan), where the altitude is 1299 ft below sea level (−0.25 mi). What is the atmospheric pressure at this point?

15. Members of a discussion group tend to rank themselves exponentially by the number of times they enter the discussion. The number of times the nth ranked individual in a group of 10 enters the discussion is P_n and is approximated by

$$P_n = P_1 e^{0.11(1-n)}$$

where P_1 is the number of times the first-ranked person participates. For each 1000 times the top person enters the group's discussion, how many times should the bottom-ranked person be expected to participate?

16. A law of light absorption of a medium for a beam of light passing through is given by

$$I = I_0 e^{-rt}$$

where I_0 is the original intensity of the light in lumens, and I is the intensity after passing through t cm of a medium whose absorption coefficient is r. Find the intensity of a 100-lumen beam after it passes through 3.14 cm of a medium with an absorption coefficient of 0.089.

17. A healing law for skin wounds states that

$$A = A_0 e^{-0.1t}$$

where A is the approximate area (in square centimeters) of unhealed skin after t days when the original area of the wound was A_0. Sketch healing curve for a wound that is 10 cm² initially.

18. The yearly sales S of a given product will decrease at a constant yearly rate without the benefit of promotion. After t years, S is approximated by

$$S = S_0 e^{-kt}$$

where S_0 is the sales in the initial year, and k is called the *sales decay constant.* Sketch the graph of sales where S_0 is a million dollars and k is 0.21.

19. The rate at which a person learns certain specific tasks can be described by a "learning curve." The number of words typed per minute by a student with n days of instruction is approximated by

$$N = 80(1 - e^{-0.018n})$$

Sketch the learning curve by computing the rate achieved every two weeks, that is, after every 10 days of instruction.

20. The current I, in amperes, in an electrical circuit is given by

$$I = \frac{E}{R}(1 - e^{-Rt/L})$$

for t s after the circuit is closed. E is the voltage, R is the resistance in ohms, and L is the inductance in henrys. Sketch the graph of this behavior for $E = 12$, $R = 5$, and $L = 2.05$, by computing I for every tenth of a second after the switch is flipped.

A.5 LOGARITHMIC FUNCTIONS

The inverse of the exponential function is important in itself and has an equally broad range of applications. Consider

$$y = f(x) = b^x, \qquad b > 0, b \neq 1$$

The inverse is

$$x = b^y \qquad \text{where } y = f^{-1}(x)$$

but, at this point, we have no means to solve explicitly for y. Toward this end, we might first express the meaning of y in words.

y is the exponent of b that yields x

which is very wordy. Abbreviation may simplify it further.

$$y = \text{ex. of } b \text{ that yields } x$$

or

$$y = \text{ex}_b\, x$$

with a subscript for the base. The last notation would be satisfactory, but historically this function has been named the **logarithm function** and the standard notation is

$$y = \log_b x$$

> If $f(x) = b^x$, then $f^{-1}(x) = \log_b x$ for $b > 0$, $b \neq 1$, and
>
> $$\underbrace{\mathbf{x = b^y}}_{\text{exponential form}} \text{ is equivalent to } \underbrace{\mathbf{y = \log_b x}}_{\text{logarithmic form}}$$
>
> The logarithm base b of x is the exponent y of b that yields x.

The equivalence of the exponential and logarithmic forms allow any exponential expression to be written in logarithmic form and conversely, as illustrated in the following examples.

EXAMPLES

Change from exponential to logarithmic form.

	EXPONENTIAL FORM	LOGARITHMIC FORM
1.	$8 = 2^3$	$3 = \log_2 8$
2.	$49 = 7^2$	$2 = \log_7 49$
3.	$0.1 = 10^{-1}$	$-1 = \log_{10} 0.1$
4.	$9 = 27^{2/3}$	$\frac{2}{3} = \log_{27} 9$
5.	$r = s^t$	$t = \log_2 r$

Change from logarithmic form to exponential form.

	LOGARITHMIC FORM	EXPONENTIAL FORM
6.	$\log_3 \frac{1}{9} = -2$	$\frac{1}{9} = 3^{-2}$
7.	$\log_{10} 10{,}000 = 4$	$10{,}000 = 10^4$
8.	$\log_8 16 = \frac{4}{3}$	$16 = 8^{4/3}$
9.	$\log_{10} 0.001 = -3$	$0.001 = 10^{-3}$
10.	$\log_m n = p$	$n = m^p$

Natural and Common Logarithms

Although the base may be *any* positive number other than 1, just two bases are generally employed when applying the logarithm function. If the base is 10, $y = \log_{10} x$ is written $y = \log x$ without the subscript and is called the **common logarithm function.** Values of this function are called **common logarithms.** Your calculator will have a LOG or log x key to compute common logarithms. The **natural logarithm function** is $y = \log_e x$ or $y = \ln x$, where e is the natural base. $\ln x$ is read "lon x" and is the **natural logarithm** of x. Scientific calculators will have an LN or ln x key to compute natural logarithms.

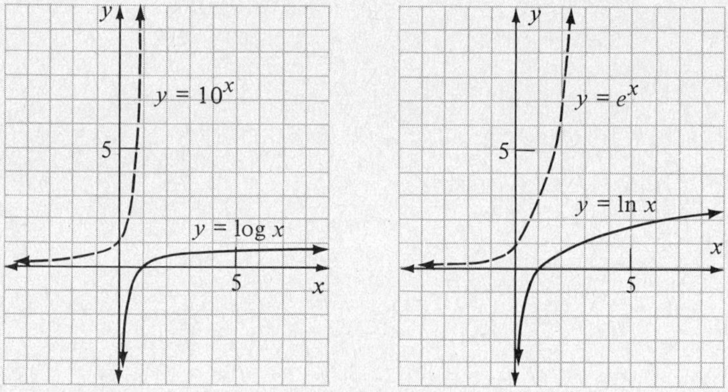

Figure A.13 The graphs of y = log x and y = ln x.

The graphs of the common and natural logarithm functions are shown in Figure A.13. The graphs are easily obtained from $y = 10^x$ and $y = e^x$ as reflections in the line $y = x$. The graph of a function such as $y = \frac{1}{2}\ln(3x - 2)$ can be constructed by compiling a table of values. To compute the values of $\frac{1}{2}\ln(3x - 2)$, $3x - 2$ must be positive, so $x > \frac{2}{3}$. Letting $x = 0.7$,

$$\begin{aligned} y &= \tfrac{1}{2}\ln[3(0.7) - 2] \\ &= \tfrac{1}{2}\ln(2.1 - 2) \\ &= \tfrac{1}{2}\ln 0.1 \end{aligned}$$

EXAMPLE

11. Compute $y = \frac{1}{2}\ln 0.1$.

[0.1] [LN] [÷] [2] [=] -1.1512925

$y \approx -1.1512925$

$\boldsymbol{y \approx -1.15}$

The calculator may now be used to compile the table in the following example.

EXAMPLE

12. Sketch the graph of $y = \frac{1}{2}\ln(3x - 2)$.

x	y
0.7	−1.15
0.8	−0.46
0.9	−0.18
1	0
2	0.69
3	0.97
4	1.15
5	1.28
6	1.39
7	1.47
8	1.55
9	1.61
10	1.67

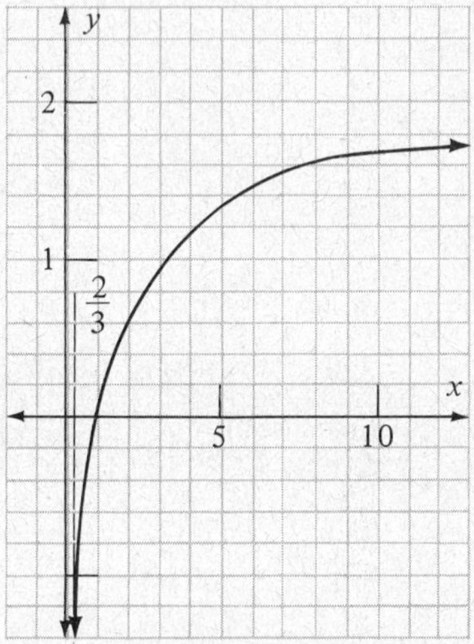

An immediate consequence of our definitions is the ability to explore more fully the applications of the previous section. Recall that population growth can be described by

$$\boldsymbol{P = P_0 e^{rt}}$$

See Example 7 of Section A.4.

EXAMPLES

13. Honolulu, Hawaii, grew during the 1960s at 7.63 percent. At that rate, how long would it take Honolulu to double its population?

Here we want the time t, so

$$\boldsymbol{P = P_0 e^{rt}}$$

$$\frac{P}{P_0} = e^{rt}$$

and we can rewrite this in logarithmic form.

$$rt = \ln \frac{P}{P_0}$$

$$\mathbf{t = \frac{1}{r} \ln \frac{P}{P_0}}$$

If the population "doubles," P will be twice P_0, or $P = 2P_0$, and r is given as 0.0763.

$$t = \frac{1}{0.0763} \ln \frac{2P_0}{P_0}$$

$$= \frac{1}{0.0763} \ln 2 \quad \text{or} \quad \frac{\ln 2}{.0763}$$

[2] [LN] [÷] [.0763] [=] 9.0844978

$$t \approx 9.0844978$$

$$\approx 9.1$$

Honolulu will double its population in 9.1 years.

14. Honolulu's population was 630,528 in 1970 and 705,381 in 1975. What was its rate of growth over this period?

From Example 11,

$$rt = \ln \frac{P}{P_0}$$

$$r = \frac{1}{t} \ln \frac{P}{P_0}$$

$$= \frac{1}{5} \ln \frac{705{,}381}{630{,}528}$$

[705381] [÷] [630528] [=] [LN] [÷] [5] [=]

$$r \approx 0.022436085$$

$$\approx 0.0224$$

From 1970 to 1975, Honolulu grew at approximately 2.24 percent annually.

As can now be seen, with the aid of logarithms an exponential equation can be solved explicitly for any of its variables. The hand-held calculator can then be used to evaluate the expression. Some of these calculator models have an [LN] key but no [e^x] key. Because $y = \ln x$ and $y = e^x$ are inverses functions, the (inverse) INV key can be used to evaluate e^x on these models. Even fewer calculators have a $\boxed{10^x}$ key, so that the $\boxed{\text{INV}}$ $\boxed{\text{LOG}}$ is often necessary if the calculator has an INV key. It is used the same as e^x, because $y = 10^x$ and $y = \log x$ are also inverses. The use of the $\boxed{\text{INV}}$ key is illustrated in the following examples.

EXAMPLES

15. Find $e^{3.01}$.

$\boxed{3.01}$ $\boxed{\text{INV}}$ $\boxed{\text{LN}}$ 20.287400

$$e^{3.01} \approx \mathbf{20.2874}$$

16. Find $10^{-2.47}$.

$\boxed{2.47}$ $\boxed{+/-}$ $\boxed{\text{INV}}$ $\boxed{\text{LOG}}$ 0.0033884416

$$\mathbf{10^{-2.47} \approx 0.0033884416}$$

The concentration of hydrogen ions in aqueous solution, indicated by $[H^+]$, is given by

$$\mathbf{[H^+] = \frac{1}{10^{pH}}}$$

where pH is the hydrogen potential of the solution and $[H^+]$ is in moles per liter. The pH is a measure of the acidity or alkalinity of a solution.

$$[H^+] = 10^{-pH}$$

$$-pH = \log[H^+]$$

$$\mathbf{pH = -\log[H^+]}$$

Pure water has a $[H^+]$ of 10^{-7} moles per liter at standard conditions.

$$pH = -\log 10^{-7}$$

$$-pH = \log 10^{-7}$$

$$10^{-7} = 10^{-pH}$$

so

$$pH = 7$$

A substance of pH of less than 7 is considered acid and above 7 is alkaline.

EXAMPLES

17. Lemon juice has a pH of approximately 2.1. Find the $[H^+]$ for lemon juice.

$$[H^+] = 10^{-pH}$$
$$= 10^{-2.1}$$
$$\approx 0.007943282$$

$[H^+]$ for lemon juice is about 0.0079 moles per liter.

18. A sample of fruit jelly has a hydrogen concentration of 3.25 moles per liter. Find its pH.

$$pH = -\log[H^+]$$
$$= -\log 3.25$$
$$\approx 0.51188336$$

The fruit jelly has a pH of about 0.512 (very acid).

19. Ammonia (10 percent NH_3) has a $[H^+]$ of 1.5849×10^{-12} mole per liter. What is its pH?

$$pH = -\log(1.5849 \times 10^{-12})$$

A number in scientific notation can be entered directly into the calculator with the EXP or EEX or similar key.

	Key	Display	
	[1.5849]	1.5849	
[EXP] or	[EEX]	1.5849	00
	[12]	1.5849	12
[+/−] or	[CHS]	1.5849	−12
	[LOG]	−11.799998	
[+/−] or	[CHS]	11.799998	

$$pH \approx 11.799998$$

The ammonia has a pH of approximately 11.8 (very alkaline).

exercise set A.5

Change from exponential to logarithmic form in Exercises 1–6.

1. (a) $125 = 5^3$ (b) $243 = 3^5$ See Examples 1–5.

2. (a) $4096 = 4^6$ (b) $1296 = 6^4$

3. (a) $100 = 10^2$ (b) $0.01 = 10^{-2}$

4. (a) $10^5 = 100{,}000$ (b) $0.00001 = 10^{-5}$

5. (a) $64 = 16^{3/2}$ (b) $27^{4/3} = 81$

6. (a) $32^{-2/5} = \frac{1}{4}$ (b) $25 = 125^{2/3}$

Change from logarithmic to exponential form in Exercises 7–12.

See Examples 6–10.

7. (a) $4 = \log_5 625$ (b) $5 = \log_4 1024$

8. (a) $3 = \log_2 8$ (b) $2 = \log_{11} 121$

9. (a) $3 = \log 1000$ (b) $\log 0.001 = -3$

10. (a) $\log 1{,}000{,}000 = 6$ (b) $-5 = \log 0.00001$

11. (a) $\log_{81} 27 = \frac{3}{4}$ (b) $\log_8 \frac{1}{4} = -\frac{2}{3}$

12. (a) $\frac{3}{2} = \log_{25} 125$ (b) $\log_{343} 49 = \frac{2}{3}$

Compute each quantity in Exercises 13–16.

See Example 11.

13. (a) $\log 4.72$ (b) $\ln 4.72$

14. (a) $\ln 0.0194$ (b) $\log 0.0194$

15. (a) $\frac{1}{8} \log \frac{710}{650}$ (b) $\frac{1}{0.12} \log 8$

16. (a) $0.875 \ln 4.32$ (b) $0.25 \log \frac{842}{750}$

Sketch the graph of each function in Exercise 17–20.

See Example 12.

17. (a) $y = \frac{3}{2} \ln x$ (b) $y = \frac{1}{5} \log 2x$

18. (a) $y = \frac{1}{3} \log x$ (b) $y = \frac{3}{4} \ln 3x$

19. (a) $y = 0.3 \log(4x - 1)$ (b) $y = 0.72 \ln(2x + 5)$

20. (a) $y = 0.6 \ln(3x + 1)$ (b) $y = 0.41 \ln(3x - 7)$

Solve for the variable indicated in underlined type in Exercises 21–24.

See Example 13.

21. (a) $P = 14.7e^{-0.21\underline{a}}$ (b) $H = 3.21 \times 10^{-\underline{p}}$

22. (a) $P_n = P_1 e^{0.11(1-\underline{n})}$ (b) $I = I_0 e^{-\underline{r}t}$

23. (a) $A = A_0e^{-0.1t}$ (b) $S = S_0e^{-kt}$

24. (a) $N = 80(1 - e^{-0.018n})$ (b) $T = \frac{E}{R}(1 - e^{-Rt/L})$

25. The state of Texas grew from 1960 to 1970 at the rate of 1.56 percent. How long will it take the state to double its population at this rate? See Example 13.

26. The state of New York grew at the rate of 0.834 percent between 1960 and 1970. How long will it take the state to double its population at this rate?

27. San Antonio, Texas, had a population of 708,582 in 1970 and 773,248 in 1975. What was the rate of growth over the period? See Example 14.

28. San Jose, California, grew from 204,196 in 1960 to 555,707 in 1975. What was the rate of growth over that period?

29. Find the $[H^+]$ of each substance with the given pH. See Example 17.
   (a) Saturated limewater, 12.4
   (b) Trisodium phosphate (2%), 11.95
   (c) Milk, 6.7
   (d) Blood, 7.4
   (e) Plant tissue, 5.2

30. *Find the pH of each substance with the given* $[H^+]$. See Examples 18–19.
   (a) Vinegar, 1.26×10^{-3}
   (b) Tomatoes, 6.31×10^{-5}
   (c) Milk of magnesia, 3.16×10^{-11}
   (d) Bread, 3.16×10^{-3}
   (e) Soft drink, 1.0×10^{-3}

A.6 PROPERTIES OF LOGARITHMS

It is the properties of the logarithm function that give it its unique character. These properties make logarithms useful in solving equations, especially in the work of the calculus.

First, recall that if $x = b^y$, then $y = \log_b x$; but if the later expression for y is substituted into the first, then $x = b^{\log_b x}$. Similarly, if $y = b^x$, then $x = \log_b x$ and, by substitution, $x = \log_b b^x$. These are our first two properties.

$$\mathbf{b}^{\log_b x} = \mathbf{x} \tag{1}$$

$$\log_b \mathbf{b}^x = \mathbf{x} \tag{2}$$

These two identities can now be put to work to develop a number of other identities that describe important properties of logarithms. Consider the logarithm of the product of two numbers, $\log_b xy$.

$$b^{\log_b xy} = xy \qquad \text{by (1)}$$

and

$$x = b^{\log_b x} \quad \text{and} \quad y = b^{\log_b y}$$

so

$$\begin{aligned} xy &= b^{\log_b x}\, b^{\log_b y} \\ &= b^{\log_b x + \log_b y} \qquad \text{adding exponents} \\ b^{\log_b xy} &= b^{\log_b x + \log_b y} \end{aligned}$$

Thus

$$\log_b xy = \log_b x + \log_b y$$

Similarly, properties for $\log_b (x/y)$ and $\log_b x^n$ can be developed from the first two properties.

$$\log_b xy = \log_b x + \log_b y \qquad (3)$$

The logarithm of the product of two numbers is the sum of their logarithms.

$$\log_b \frac{x}{y} = \log_b x - \log_b y \qquad (4)$$

The logarithm of the quotients of two numbers is the difference of their logarithms.

$$\log_b x^n = n \log_b x \qquad (5)$$

The logarithm of the power of a number is the power times the logarithm of the number.

Three other useful properties follow easily as special cases.

$$\log_b b^n = n \qquad (6)$$

$$\log_b b = 1 \qquad (7)$$

$$\log_b 1 = 0 \qquad (8)$$

This set of properties gives sufficient power to work with most logarithmic expressions and to solve explicitly for their variables. The following examples illustrate the properties in use.

EXAMPLES

Expand, using the properties of logarithms.

1. $\log \pi r^2$

$\log \pi r^2 = \log \pi + \log r^2$ (Prop. 3)

$\mathbf{\log \pi r^2 = \log \pi + 2 \log r}$ (Prop. 5)

2. $\log_b \frac{2x^3}{y}$

$\log_b \frac{2x^3}{y} = \log_b 2x^3 - \log_b y$ (Prop. 4)

$= \log_b 2 + \log_b x^3 - \log_b y$ (Prop. 3)

$\mathbf{\log_b \frac{2x^3}{y} = \log_b 2 + 3 \log_b x - \log_b y}$ (Prop. 5)

3. $\log_b \sqrt{\frac{x}{yz}}$

$\log_b \sqrt{\frac{x}{yz}} = \log_b \left(\frac{x}{yz}\right)^{1/2}$

$= \frac{1}{2} \log_b \frac{x}{yz}$ (Prop. 5)

$= \frac{1}{2} (\log_b x - \log_b yz)$ (Prop. 4)

$= \frac{1}{2} \log_b x - (\log_b y + \log_b z)$ (Prop. 3)

$\mathbf{\log_b \sqrt{\frac{x}{yz}} = \frac{1}{2} (\log_b x - \log_b y - \log_b z)}$

EXAMPLES

Write as a single logarithm.

4. $\log b + \log h - \log 2$

$\log b + \log h - \log 2 = \log bh - \log 2$ (Prop. 3)

$\mathbf{\log b + \log h - \log 2 = \log \frac{bh}{2}}$ (Prop. 4)

5. Alternative solution for Example 4.

$$\log b + \log h - \log 2 = \log b + \log h + (-1)\log 2$$

$$= \log b + \log h + \log 2^{-1} \qquad \text{(Prop. 5)}$$

$$= \log b + \log h + \log \tfrac{1}{2}$$

$$\mathbf{\log b + \log h - \log 2 = \log \tfrac{1}{2}bh} \qquad \text{(Prop. 3)}$$

6. $\ln P_0 + rt$

$$\ln P_0 + rt = \ln P_0 + \ln e^{rt} \qquad \text{(Prop. 2)}$$

$$\mathbf{\ln P_0 + rt = \ln P_0 e^{rt}} \qquad \text{(Prop. 3)}$$

7. $\log_b 2 + \log_b \pi + \frac{1}{2}\log_b 1 - \frac{1}{2}\log_b g$

$$= \log_b 2 + \log_b \pi + \frac{1}{2}(\log_b 1 - \log_b g)$$

$$= \log_b 2 + \log_b \pi + \frac{1}{2}\log_b \frac{1}{g} \qquad \text{(Prop. 4)}$$

$$= \log_b 2 + \log_b \pi + \log_b \left(\frac{1}{g}\right)^{1/2} \qquad \text{(Prop. 5)}$$

$$= \log_b 2 + \log_b \pi + \log_b \sqrt{\frac{1}{g}}$$

$$= \mathbf{\log_b 2\pi \sqrt{\frac{1}{g}}} \qquad \text{(Prop. 3)}$$

As the examples illustrate, logarithmic expressions can be expanded or simplified by the use of fundamental properties of logarithms. These properties may also be employed to solve logarithmic equations, if we have a few additional properties.

PROPERTIES OF EXPONENTIALS

Given that x and y are nonzero real numbers and a and b are positive real numbers other than 1.

$$\textbf{If } a^x = a^y\textbf{, then } x = y \qquad \textbf{(9)}$$

If the bases of two equal numbers are equal, then their exponents are also equal.

$$\textbf{If } a^x = b^x\textbf{, then } a = b \qquad \textbf{(10)}$$

If the exponents of two equal numbers are equal, then their bases are also equal.

If you examine the deriviation of Property 3, you will see that the last step uses Property 8 tacitly. There are analogous properties for logarithms, but first some examples of the properties of exponentials.

EXAMPLES

Solve for the variable.

8. $7^x = 343$

$7^x = 7^3$

$\mathbf{x = 3}$ (Prop. 9)

9. $25^y = 125$

$(5^2)^x = 5^3$

$5^{2x} = 5^3$

$2x = 3$ (Prop.9)

$\mathbf{x = \frac{3}{2}}$

10. $27^z = \frac{1}{9}$

$(3^3)^z = 3^{-2}$

$3^{3z} = 3^{-2}$

$3z = -2$ (Prop.9)

$\mathbf{z = -\frac{2}{3}}$

11. $b^4 = 16$

$b^4 = 2^4$

$\mathbf{b = 2}$ (Prop. 10)

Given that m and n are positive real numbers and b is a positive real number other than 1.

If $\log_b m = \log_b n$, then $m = n$ **(11)**

If the logarithms are equal then the numbers are equal, and conversely:

If $m = n$, then $\log_b m = \log_b n$ **(12)**

The logarithms of equal numbers are equal.

EXAMPLES

Solve for the variables.

12. $\frac{1}{2}\log 3 - \log x = 3\log 2$

$\log 3^{1/2} - \log x = \log 2^3$ (Prop. 5)

$\log \sqrt{3} - \log x = \log 8$

$\log \frac{\sqrt{3}}{x} = \log 8$ (Prop. 4)

$\frac{\sqrt{3}}{x} = 8$ (Prop. 11)

$\sqrt{3} = 8x$

$\mathbf{x = \frac{\sqrt{3}}{8}}$

$x \approx 0.21650635$

13. $\ln 3 = \ln A - 2$

$\ln 3 = \ln A - 2\ln e$ (Prop. 6)

$= \ln A + \ln e^{-2}$ (Prop. 5)

$= \ln Ae^{-2}$ (Prop. 3)

$3 = Ae^{-2}$ (Prop. 11)

$\mathbf{A = 3e^2}$

$A \approx 22.167168$

14. $10^M = e^{1.73}$

$\ln 10^M = \ln e^{1.73}$ (Prop. 12)

$= 1.73$ (Prop. 2)

$M \ln 10 = 1.73$ (Prop. 5)

$\mathbf{M = \frac{1.73}{\ln 10}}$

$M \approx 0.75132945$

15. $2.41^x = 19.9$

$\log 2.41^x = \log 19.9$ (Prop. 12)

$x \log 2.41 = \log 19.9$ (Prop. 5)

$\mathbf{x = \frac{\log 19.9}{\log 2.41}}$

$x \approx 3.3999872$

16. $\log x + \log(x - 3) = 1$

$$\log x(x - 3) = 1$$

$$10^1 = x(x - 3)$$

$$x^2 - 3x - 10 = 0$$

$$(x + 2)(x - 5) = 0$$

$$x = -2 \quad \text{or} \quad x = 5$$

But $x = -2$ gives $\log(-2)$ and $\log(-5)$ in the original equation, which has no meaning. Thus $x = -2$ is extraneous and $x = 5$ is the only solution.

$$\mathbf{x = 5}$$

The properties developed thus far can be used to evaluate a logarithm in any base despite the fact that a calculator or tables will have only natural or common bases. Consider $L = \log_b N$, where b is any valid base:

$$L = \log_b N$$

$$b^L = N$$

$$\log_x b^L = \log_x N \qquad \text{(Prop. 11)}$$

$$L \log_x b = \log_x N \qquad \text{(Prop. 5)}$$

$$L = \frac{\log_x N}{\log_x b}$$

Thus, the logarithm for a number can be found in any base, if the logarithm of the number and of the base can be found in some other base.

CHANGE-OF-BASE THEOREM

Given any positive real number N and positive real numbers b and x other than 1.

$$\log_b N = \frac{\log_x N}{\log_x b} \qquad (13)$$

The logarithm of a number in any base is the quotient of the logs (or "lons") of the number and of the base. Note that

$$\log_b N = \frac{\log N}{\log b} = \frac{\ln N}{\ln b}$$

EXAMPLES

17. Find $\log_\pi 5$.

$$\log_\pi 5 = \frac{\log 5}{\log \pi}$$

$$\approx \frac{0.69897000}{0.49714987}$$

$$\mathbf{\log_\pi 5 \approx 1.4059543}$$

18. Alternative solution for Example 17.

$$\log_\pi 5 = \frac{\ln 5}{\ln \pi}$$

$$\approx \frac{1.6094379}{1.1447298}$$

$$\mathbf{\log_\pi \approx 1.4059543}$$

exercise set A.6

Use the properties of logarithms to expand the expressions in Exercises 1–10.

See Examples 1–3.

1. $\log_b 2\pi r$
2. $\log_b Prt$
3. $\log \frac{4}{3}\pi r^3$
4. $\log 8\sqrt{h}$
5. $\ln \sqrt[3]{\frac{c}{s^2}}$
6. $\ln ar^{n-1}$
7. $\log_b \frac{s^2}{r\sqrt{t}}$
8. $\log_b \frac{s\sqrt{r}}{t^2}$
9. $\ln S_0 e^{-kt}$
10. $\ln Ae^{rt}$

Write each of the expressions in Exercises 11–20 as a single logarithm, if possible.

See Examples 4–7.

11. $2 \log_b A + \log_b B$
12. $2(\log_b A - \log_b B)$
13. $\log_b A_0 - \frac{t}{h} \log_b \frac{1}{2}$
14. $\log_b k + \frac{3}{2} \log_b n$
15. $\log M + \frac{1}{2} \log 3 - 3 \log N$
16. $\frac{1}{3} \ln M - \ln 2 - 2 \ln N$
17. $\ln\sqrt{2e} - \ln \pi$
18. $\log e + \log\sqrt{10}$
19. $\log P + 1$
20. $2 - \ln Q$

Solve for the variables in Exercises 21–40.

See Examples 8–11.

21. $11^x = 1331$
22. $5^x = 3125$
23. $8^y = 16$
24. $81^y = 27$
25. $b^3 = 64$
26. $b^5 = 32$
27. $(\frac{2}{3})^n = \frac{81}{16}$
28. $(\frac{10}{3})^n = 0.09$
29. $8^{x-3} = \frac{1}{16}$
30. $27^{x+2} = \frac{1}{81}$
31. $\log_2 16 = x$
32. $\log_5 \frac{1}{25} = x$
33. $\log_7 N = -2$
34. $\log_3 N = -3$
35. $\log_b 8 = \frac{3}{2}$
36. $\log_b 9 = \frac{2}{3}$
37. $\log 2 + \log x = \log 6$
38. $\log x - \log 3 = \log 9$ See Example 11.
39. $3 \log 2 + \log x = 2 \log 3$
40. $\frac{1}{2} \log 3 - \log x = 3 \log \frac{1}{2}$

In Exercises 41–50, solve for the variable indicated in bold type.

41. $\ln \mathbf{N} = \ln 80 + \ln(1 - e^{-0.018n})$ See Example 12.
42. $\ln \mathbf{I} = \ln E - \ln R + \ln(1 - e^{-Rt/L})$
43. $\ln \mathbf{P} = \ln 14.7 - 0.21a$
44. $\ln \mathbf{A} = -0.1t + \ln A_0$
45. $\log \mathbf{x} - \log 3 = 1$
46. $\ln 3 - \ln \mathbf{x} = 1$ See Example 16.
47. $\log \mathbf{x} + \log(\mathbf{x} + 9) = 1$
48. $\log \mathbf{x} + \log(7 - \mathbf{x}) = 1$
49. $\log(25 - \mathbf{x}) - 2 = \log \frac{1}{\mathbf{x}}$
50. $2 - \log \mathbf{x} = \log(\mathbf{x} + 21)$

Evaluate the variables in Exercies 51–60.

51. $10^N = e^{2.04}$
52. $e^N = 10^{1.48}$ See Examples 14 and 15.
53. $3.78^x = 6.52$
54. $6.52^x = 3.32$
55. $\pi = 0.59^n$
56. $1.59 = \pi^n$
57. $y = \log_2 11$
58. $y = \log_{17} 34$ See Examples 17 and 18.
59. $\log_{3.12} 1454 = L$
60. $\log_{12.3} 15.81 = L$

appendix B
trigonometric tables

TABLE I
TRIGONOMETRIC FUNCTIONS, VALUES IN DEGREES

θ degrees	θ radians	sin θ	csc θ	tan θ	cot θ	sec θ	cos θ		
0° 00′	.0000	.0000	undefined	.0000	undefined	1.000	1.0000	1.5708	**90° 00′**
10	.0029	.0029	343.8	.0029	343.8	1.000	1.0000	1.5679	50
20	.0058	.0058	171.9	.0058	171.9	1.000	1.0000	1.5650	40
30	.0087	.0087	114.6	.0087	114.6	1.000	1.0000	1.5621	30
40	.0116	.0116	85.95	.0116	85.94	1.000	.9999	1.5592	20
50	.0145	.0145	68.76	.0145	68.75	1.000	.9999	1.5563	10
1° 00′	.0175	.0175	57.30	.0175	57.29	1.000	.9998	1.5533	**89° 00′**
10	.0204	.0204	49.11	.0204	49.10	1.000	.9998	1.5504	50
20	.0233	.0233	42.98	.0233	42.96	1.000	.9997	1.5475	40
30	.0262	.0262	38.20	.0262	38.19	1.000	.9997	1.5446	30
40	.0291	.0291	34.38	.0291	34.37	1.000	.9996	1.5417	20
50	.0320	.0320	31.26	.0320	31.24	1.001	.9995	1.5388	10
2° 00′	.0349	.0349	28.65	.0349	28.64	1.001	.9994	1.5359	**88° 00′**
10	.0378	.0378	26.45	.0378	26.43	1.001	.9993	1.5330	50
20	.0407	.0407	24.56	.0407	24.54	1.001	.9992	1.5301	40
30	.0436	.0436	22.93	.0437	22.90	1.001	.9990	1.5272	30
40	.0465	.0465	21.49	.0466	21.47	1.001	.9989	1.5243	20
50	.0495	.0494	20.23	.0495	20.21	1.001	.9988	1.5213	10
3° 00′	.0524	.0523	19.11	.0524	19.08	1.001	.9986	1.5184	**87° 00′**
10	.0553	.0552	18.10	.0553	18.07	1.002	.9985	1.5155	50
20	.0582	.0581	17.20	.0582	17.17	1.002	.9983	1.5126	40
30	.0611	.0610	16.38	.0612	16.35	1.002	.9981	1.5097	30
40	.0640	.0640	15.64	.0641	15.60	1.002	.9980	1.5068	20
50	.0669	.0669	14.96	.0670	14.92	1.002	.9978	1.5039	10
4° 00′	.0698	.0698	14.34	.0699	14.30	1.002	.9976	1.5010	**86° 00′**
10	.0727	.0727	13.76	.0729	13.73	1.003	.9974	1.4981	50
20	.0756	.0756	13.23	.0758	13.20	1.003	.9971	1.4952	40
30	.0785	.0785	12.75	.0787	12.71	1.003	.9969	1.4923	30
40	.0814	.0814	12.29	.0816	12.25	1.003	.9967	1.4893	20
50	.0844	.0843	11.87	.0846	11.83	1.004	.9964	1.4864	10
5° 00′	.0873	.0872	11.47	.0875	11.43	1.004	.9962	1.4835	**85° 00′**
10	.0902	.0901	11.10	.0904	11.06	1.004	.9959	1.4806	50
20	.0931	.0929	10.76	.0934	10.71	1.004	.9957	1.4777	40
30	.0960	.0958	10.43	.0963	10.39	1.005	.9954	1.4748	30
40	.0989	.0987	10.13	.0992	10.08	1.005	.9951	1.4719	20
50	.1018	.1016	9.839	.1022	9.788	1.005	.9948	1.4690	10
6° 00′	.1047	.1045	9.567	.1051	9.514	1.006	.9945	1.4661	**84° 00′**
		cos θ	**sec θ**	**cot θ**	**tan θ**	**csc θ**	**sin θ**	**θ radians**	**θ degrees**

Table I, Continued
Trigonometric Functions, Values in Degrees

θ degrees	θ radians	sin θ	csc θ	tan θ	cot θ	sec θ	cos θ		
6° 00′	.1047	.1045	9.567	.1051	9.514	1.006	.9945	1.4661	**84° 00′**
10	.1076	.1074	9.309	.1080	9.255	1.006	.9942	1.4632	50
20	.1105	.1103	9.065	.1110	9.010	1.006	.9939	1.4603	40
30	.1134	.1132	8.834	.1139	8.777	1.006	.9936	1.4573	30
40	.1164	.1161	8.614	.1169	8.556	1.007	.9932	1.4544	20
50	.1193	.1190	8.405	.1198	8.345	1.007	.9929	1.4515	10
7° 00′	.1222	.1219	8.206	.1228	8.144	1.008	.9925	1.4486	**83° 00′**
10	.1251	.1248	8.016	.1257	7.953	1.008	.9922	1.4457	50
20	.1280	.1276	7.834	.1287	7.770	1.008	.9918	1.4428	40
30	.1309	.1305	7.661	.1317	7.596	1.009	.9914	1.4399	30
40	.1338	.1334	7.496	.1346	7.429	1.009	.9911	1.4370	20
50	.1367	.1363	7.337	.1376	7.269	1.009	.9907	1.4341	10
8° 00′	.1396	.1392	7.185	.1405	7.115	1.010	.9903	1.4312	**82° 00′**
10	.1425	.1421	7.040	.1435	6.968	1.010	.9899	1.4283	50
20	.1454	.1449	6.900	.1465	6.827	1.011	.9894	1.4254	40
30	.1484	.1478	6.765	.1495	6.691	1.011	.9890	1.4224	30
40	.1513	.1507	6.636	.1524	6.561	1.012	.9886	1.4195	20
50	.1542	.1536	6.512	.1554	6.435	1.012	.9881	1.4166	10
9° 00′	.1571	.1564	6.392	.1584	6.314	1.012	.9877	1.4137	**81° 00′**
10	.1600	.1593	277	.1614	6.197	1.013	.9872	1.4108	50
20	.1629	.1622	166	.1644	6.084	1.013	.9868	1.4079	40
30	.1658	.1650	6.059	.1673	5.976	1.014	.9863	1.4050	30
40	.1687	.1679	5.955	.1703	5.871	1.014	.9858	1.4021	20
50	.1716	.1708	855	.1733	5.769	1.015	.9853	1.3992	10
10° 00′	.1745	.1736	5.759	1.763	5.671	1.015	.9848	1.3963	**80° 00′**
10	.1774	.1765	665	.1793	5.576	1.016	.9843	1.3934	50
20	.1804	.1794	575	.1823	5.485	1.016	.9838	1.3904	40
30	.1833	.1822	5.487	.1853	5.396	1.017	.9833	1.3875	30
40	.1862	.1851	403	.1883	5.309	1.018	.9827	1.3846	20
50	.1891	.1880	320	.1914	5.226	1.018	.9822	1.3817	10
11° 00′	.1920	.1908	5.241	.1944	5.145	1.019	.9816	1.3788	**79° 00′**
10	.1949	.1937	164	.1974	5.066	1.019	.9811	1.3759	50
20	.1978	.1965	089	.2004	4.989	1.020	.9805	1.3730	40
30	.2007	.1994	5.016	.2035	4.915	1.020	.9799	1.3701	30
40	.2036	.2022	4.945	.2065	4.843	1.021	.9793	1.3672	20
50	.2065	.2051	876	.2095	4.773	1.022	.9787	1.3643	10
12° 00′	.2094	.2079	4.810	.2126	4.705	1.022	.9781	1.3614	**78° 00′**
		cos θ	**sec θ**	**cot θ**	**tan θ**	**csc θ**	**sin θ**	**θ radians**	**θ degrees**

Table I, Continued
Trigonometric Functions, Values in Degrees

θ degrees	θ radians	sin θ	csc θ	tan θ	cot θ	sec θ	cos θ		
12° 00′	.2094	.2079	4.810	.2126	4.705	1.022	.9781	1.3614	**78° 00′**
10	.2123	.2108	4.745	.2156	4.638	1.023	.9775	1.3584	50
20	.2153	.2136	4.682	.2186	4.574	1.024	.9769	1.3555	40
30	.2182	.2164	4.620	.2217	4.511	1.024	.9763	1.3526	30
40	.2211	.2193	4.560	.2247	4.449	1.025	.9757	1.3497	20
50	.2240	.2221	4.502	.2278	4.390	1.026	.9750	1.3468	10
13° 00′	.2269	.2250	4.445	.2309	4.331	1.026	.9744	1.3439	**77° 00′**
10	.2298	.2278	4.390	.2339	4.275	1.027	.9737	1.3410	50
20	.2327	.2306	4.336	.2370	4.219	1.028	.9730	1.3381	40
30	.2356	.2334	4.284	.2401	4.165	1.028	.9724	1.3352	30
40	.2385	.2363	4.232	.2432	4.113	1.029	.9717	1.3323	20
50	.2414	.2391	4.182	.2462	4.061	1.030	.9710	1.3294	10
14° 00′	.2443	.2419	4.134	.2493	4.011	1.031	.9703	1.3265	**76° 00′**
10	.2473	.2447	4.086	.2524	3.962	1.031	.9696	1.3235	50
20	.2502	.2476	4.039	.2555	3.914	1.032	.9689	1.3206	40
30	.2531	.2504	3.994	.2586	3.867	1.033	.9681	1.3177	30
40	.2560	.2532	3.950	.2617	3.821	1.034	.9674	1.3148	20
50	.2589	.2560	3.906	.2648	3.776	1.034	.9667	1.3119	10
15° 00′	.2618	.2588	3.864	.2679	3.732	1.035	.9659	1.3090	**75° 00′**
10	.2647	.2616	3.822	.2711	3.689	1.036	.9652	1.3061	50
20	.2676	.2644	3.782	.2742	3.647	1.037	.9644	1.3032	40
30	.2705	.2672	3.742	.2773	3.606	1.038	.9636	1.3003	30
40	.2734	.2700	3.703	.2805	3.566	1.039	.9628	1.2974	20
50	.2763	.2728	3.665	.2836	3.526	1.039	.9621	1.2945	10
16° 00′	.2793	.2756	3.628	.2867	3.487	1.040	.9613	1.2915	**74° 00′**
10	.2822	.2784	3.592	.2899	3.450	1.041	.9605	1.2886	50
20	.2851	.2812	3.556	.2931	3.412	1.042	.9596	1.2857	40
30	.2880	.2840	3.521	.2962	3.376	1.043	.9588	1.2828	30
40	.2909	.2868	3.487	.2994	3.340	1.044	.9580	1.2799	20
50	.2938	.2896	3.453	.3026	3.305	1.045	.9572	1.2770	10
17° 00′	.2967	.2924	3.420	.3057	3.271	1.046	.9563	1.2741	**73° 00′**
10	.2996	.2952	3.388	.3089	3.237	1.047	.9555	1.2712	50
20	.3025	.2979	3.357	.3121	3.204	1.048	.9546	1.2683	40
30	.3054	.3007	3.326	.3153	3.172	1.048	.9537	1.2654	30
40	.3083	.3035	3.295	.3185	3.140	1.049	.9528	1.2625	20
50	.3113	.3062	3.265	.3217	3.108	1.050	.9520	1.2595	10
18° 00′	.3142	.3090	3.236	.3249	3.078	1.051	.9511	1.2566	**72° 00′**
		cos θ	**sec θ**	**cot θ**	**tan θ**	**csc θ**	**sin θ**	**θ radians**	**θ degrees**

Table I, Continued
Trigonometric Functions, Values in Degrees

θ degrees	θ radians	sin θ	csc θ	tan θ	cot θ	sec θ	cos θ		
18° 00′	.3142	.3090	3.236	.3249	3.078	1.051	.9511	1.2566	**72° 00′**
10	.3171	.3118	3.207	.3281	3.047	1.052	.9502	1.2537	50
20	.3200	.3145	3.179	.3314	3.018	1.053	.9492	1.2508	40
30	.3229	.3173	3.152	.3346	2.989	1.054	.9483	1.2479	30
40	.3258	.3201	3.124	.3378	2.960	1.056	.9474	1.2450	20
50	.3287	.3228	3.098	.3411	2.932	1.057	.9465	1.2421	10
19° 00′	.3316	.3256	3.072	.3443	2.904	1.058	.9455	1.2392	**71° 00′**
10	.3345	.3283	3.046	.3476	2.877	1.059	.9446	1.2363	50
20	.3374	.3311	3.021	.3508	2.850	1.060	.9436	1.2334	40
30	.3403	.3338	2.996	.3541	2.824	1.061	.9426	1.2305	30
40	.3432	.3365	2.971	.3574	2.798	1.062	.9417	1.2275	20
50	.3462	.3393	2.947	.3607	2.773	1.063	.9407	1.2246	10
20° 00′	.3491	.3420	2.924	.3640	2.747	1.064	.9397	1.2217	**70° 00′**
10	.3520	.3448	2.901	.3673	2.723	1.065	.9387	1.2188	50
20	.3549	.3475	2.878	.3706	2.699	1.066	.9377	1.2159	40
30	.3578	.3502	2.855	.3739	2.675	1.068	.9367	1.2130	30
40	.3607	.3529	2.833	.3772	2.651	1.069	.9356	1.2101	20
50	.3636	.3557	2.812	.3805	2.628	1.070	.9346	1.2072	10
21° 00′	.3665	.3584	2.790	.3839	2.605	1.071	.9336	1.2043	**69° 00′**
10	.3694	.3611	2.769	.3872	2.583	1.072	.9325	1.2014	50
20	.3723	.3638	2.749	.3906	2.560	1.074	.9315	1.2985	40
30	.3752	.3665	2.729	.3939	2.539	1.075	.9304	1.1956	30
40	.3782	.3692	2.709	.3973	2.517	1.076	.9293	1.1926	20
50	.3811	.3719	2.689	.4006	2.496	1.077	.9283	1.1897	10
22° 00′	.3840	.3746	2.669	.4040	2.475	1.079	.9272	1.1868	**68° 00′**
10	.3869	.3773	2.650	.4074	2.455	1.080	.9261	1.1839	50
20	.3898	.3800	2.632	.4108	2.434	1.081	.9250	1.1810	40
30	.3927	.3827	2.613	.4142	2.414	1.082	.9239	1.1781	30
40	.3956	.3854	2.595	.4176	2.394	1.084	.9228	1.1752	20
50	.3985	.3881	2.577	.4210	2.375	1.085	.9216	1.1723	10
23° 00′	.4014	.3907	2.559	.4245	2.356	1.086	.9205	1.1694	**67° 00′**
10	.4043	.3934	2.542	.4279	2.337	1.088	.9194	1.1665	50
20	.4072	.3961	2.525	.4314	2.318	1.089	.9182	1.1636	40
30	.4102	.3987	2.508	.4348	2.300	1.090	.9171	1.1606	30
40	.4131	.4014	2.491	.4383	2.282	1.092	.9159	1.1577	20
50	.4160	.4041	2.475	.4417	2.264	1.093	.9147	1.1548	10
24° 00′	.4189	.4067	2.459	.4452	2.246	1.095	.9135	1.1519	**66° 00′**
		cos θ	sec θ	cot θ	tan θ	csc θ	sin θ	θ radians	θ degrees

Table I, Continued
Trigonometric Functions, Values in Degrees

θ degrees	θ radians	sin θ	csc θ	tan θ	cot θ	sec θ	cos θ		
24° 00′	.4189	.4067	2.459	.4452	2.246	1.095	.9135	1.1519	**66° 00′**
10	.4218	.4094	2.443	.4487	2.229	1.096	.9124	1.1490	50
20	.4247	.4120	2.427	.4522	2.211	1.097	.9112	1.1461	40
30	.4276	.4147	2.411	.4557	2.194	1.099	.9100	1.1432	30
40	.4305	.4173	2.396	.4592	2.177	1.100	.9088	1.1403	20
50	.4334	.4200	2.381	.4628	2.161	1.102	.9075	1.1374	10
25° 00′	.4363	.4226	2.366	.4663	2.145	1.103	.9063	1.1345	**65° 00′**
10	.4392	.4253	2.352	.4699	2.128	1.105	.9051	1.1316	50
20	.4422	.4279	2.337	.4734	2.112	1.106	.9038	1.1286	40
30	.4451	.4305	2.323	.4770	2.097	1.108	.9026	1.1257	30
40	.4480	.4331	2.309	.4806	2.081	1.109	.9013	1.1228	20
50	.4509	.4358	2.295	.4841	2.066	1.111	.9001	1.1199	10
26° 00′	.4538	.4384	2.281	.4877	2.050	1.113	.8988	1.1170	**64° 00′**
10	.4567	.4410	2.268	.4913	2.035	1.114	.8975	1.1141	50
20	.4596	.4436	2.254	.4950	2.020	1.116	.8962	1.1112	40
30	.4625	.4462	2.241	.4986	2.006	1.117	.8949	1.1083	30
40	.4654	.4488	2.228	.5022	1.991	1.119	.8936	1.1054	20
50	.4683	.4514	2.215	.5059	1.977	1.121	.8923	1.1025	10
27° 00′	.4712	.4540	2.203	.5095	1.963	1.122	.8910	1.0996	**63° 00′**
10	.4741	.4566	2.190	.5132	949	1.124	.8897	1.0966	50
20	.4771	.4592	2.178	.5169	935	1.126	.8884	1.0937	40
30	.4800	.4617	2.166	.5206	1.921	1.127	.8870	1.0908	30
40	.4829	.4643	2.154	.5243	907	1.129	.8857	1.0879	20
50	.4858	.4669	2.142	.5280	894	1.131	.8843	1.0850	10
28° 00′	.4887	.4695	2.130	.5317	1.881	1.133	.8829	1.0821	**62° 00′**
10	.4916	.4720	2.118	.5354	868	1.134	.8816	1.0792	50
20	.4945	.4746	2.107	.5392	855	1.136	.8802	1.0763	40
30	.4974	.4772	2.096	.5430	1.842	1.138	.8788	1.0734	30
40	.5003	.4797	2.085	.5467	829	1.140	.8774	1.0705	20
50	.5032	.4823	2.074	.5505	816	1.142	.8760	1.0676	10
29° 00′	.5061	.4848	2.063	.5543	1.804	1.143	.8746	1.0647	**61° 00′**
10	.5091	.4874	2.052	.5581	792	1.145	.8732	1.0617	50
20	.5120	.4899	2.041	.5619	780	1.147	.8718	1.0588	40
30	.5149	.4924	2.031	.5658	1.767	1.149	.8704	1.0559	30
40	.5178	.4950	2.020	.5696	756	1.151	.8689	1.0530	20
50	.5207	.4975	2.010	.5735	744	1.153	.8675	1.0501	10
30° 00′	.5236	.5000	2.000	.5774	1.732	1.155	.8660	1.0472	**60° 00′**
		cos θ	**sec θ**	**cot θ**	**tan θ**	**csc θ**	**sin θ**	**θ radians**	**θ degrees**

Table I, Continued
Trigonometric Functions, Values in Degrees

θ degrees	θ radians	sin θ	csc θ	tan θ	cot θ	sec θ	cos θ		
30° 00′	.5236	.5000	2.000	.5774	1.732	1.155	.8660	1.0472	**60° 00′**
10	.5265	.5025	1.990	.5812	1.720	1.157	.8646	1.0443	50
20	.5294	.5050	1.980	.5851	1.709	1.159	.8631	1.0414	40
30	.5323	.5075	1.970	.5890	1.698	1.161	.8616	1.0385	30
40	.5352	.5100	1.961	.5930	1.686	1.163	.8601	1.0356	20
50	.5381	.5125	1.951	.5969	1.675	1.165	.8587	1.0327	10
31° 00′	.5411	.5150	1.942	.6009	1.664	1.167	.8572	1.0297	**59° 00′**
10	.5440	.5175	1.932	.6048	1.653	1.169	.8557	1.0268	50
20	.5469	.5200	1.923	.6088	1.643	1.171	.8542	1.0239	40
30	.5498	.5225	1.914	.6128	1.632	1.173	.8526	1.0210	30
40	.5527	.5250	1.905	.6168	1.621	1.175	.8511	1.0181	20
50	.5556	.5275	1.896	.6208	1.611	1.177	.8496	1.0152	10
32° 00′	.5585	.5299	1.887	.6249	1.600	1.179	.8480	1.0123	**58° 00′**
10	.5614	.5324	1.878	.6289	1.590	1.181	.8465	1.0094	50
20	.5643	.5348	1.870	.6330	1.580	1.184	.8450	1.0065	40
30	.5672	.5373	1.861	.6371	1.570	1.186	.8434	1.0036	30
40	.5701	.5398	1.853	.6412	1.560	1.188	.8418	1.0007	20
50	.5730	.5422	1.844	.6453	1.550	1.190	.8403	.9977	10
33° 00′	.5760	.5446	1.836	.6494	1.540	1.192	.8387	.9948	**57° 00′**
10	.5789	.5471	1.828	.6536	1.530	1.195	.8371	.9919	50
20	.5818	.5495	1.820	.6577	1.520	1.197	.8355	.9890	40
30	.5847	.5519	1.812	.6619	1.511	1.199	.8339	.9861	30
40	.5876	.5544	1.804	.6661	1.501	1.202	.8323	.9832	20
50	.5905	.5568	1.796	.6703	1.492	1.204	.8307	.9803	10
34° 00′	.5934	.5592	1.788	.6745	1.483	1.206	.8290	.9774	**56° 00′**
10	.5963	.5616	1.781	.6787	1.473	1.209	.8274	.9745	50
20	.5992	.5640	1.773	.6830	1.464	1.211	.8258	.9716	40
30	.6021	.5664	1.766	.6873	1.455	1.213	.8241	.9687	30
40	.6050	.5688	1.758	.6916	1.446	1.216	.8225	.9657	20
50	.6080	.5712	1.751	.6959	1.437	1.218	.8208	.9628	10
35° 00′	.6109	.5736	1.743	.7002	1.428	1.221	.8192	.9599	**55° 00′**
10	.6138	.5760	1.736	.7046	1.419	1.223	.8175	.9570	50
20	.6167	.5783	1.729	.7089	1.411	1.226	.8158	.9541	40
30	.6196	.5807	1.722	.7133	1.402	1.228	.8141	.9512	30
40	.6225	.5831	1.715	.7177	1.393	1.231	.8124	.9483	20
50	.6254	.5854	1.708	.7221	1.385	1.233	.8107	.9454	10
36° 00′	.6283	.5878	1.701	.7265	1.376	1.236	.8090	.9425	**54° 00′**
		cos θ	**sec θ**	**cot θ**	**tan θ**	**csc θ**	**sin θ**	**θ radians**	**θ degrees**

Table I, Continued
Trigonometric Functions, Values in Degrees

θ degrees	θ radians	sin θ	csc θ	tan θ	cot θ	sec θ	cos θ		
36° 00′	.6283	.5878	1.701	.7265	1.376	1.236	.8090	.9425	**54° 00′**
10	.6312	.5901	1.695	.7310	1.368	1.239	.8073	.9396	50
20	.6341	.5925	1.688	.7355	1.360	1.241	.8056	.9367	40
30	.6370	.5948	1.681	.7400	1.351	1.244	.8039	.9338	30
40	.6400	.5972	1.675	.7445	1.343	1.247	.8021	.9308	20
50	.6429	.5995	1.668	.7490	1.335	1.249	.8004	.9279	10
37° 00′	.6458	.6018	1.662	.7536	1.327	1.252	.7986	.9250	**53° 00′**
10	.6487	.6041	1.655	.7581	1.319	1.255	.7969	.9221	50
20	.6516	.6065	1.649	.7627	1.311	1.258	.7951	.9192	40
30	.6545	.6088	1.643	.7673	1.303	1.260	.7934	.9163	30
40	.6574	.6111	1.636	.7720	1.295	1.263	.7916	.9134	20
50	.6603	.6134	1.630	.7766	1.288	1.266	.7898	.9105	10
38° 00′	.6632	.6157	1.624	.7813	1.280	1.269	.7880	.9076	**52° 00′**
10	.6661	.6180	1.618	.7860	1.272	1.272	.7862	.9047	50
20	.6690	.6202	1.612	.7907	1.265	1.275	.7844	.9018	40
30	.6720	.6225	1.606	.7954	1.257	1.278	.7826	.8988	30
40	.6749	.6248	1.601	.8002	1.250	1.281	.7808	.8959	20
50	.6778	.6271	1.595	.8050	1.242	1.284	.7790	.8930	10
39° 00′	.6807	.6293	1.589	.8098	1.235	1.287	.7771	.8901	**51° 00′**
10	.6836	.6316	1.583	.8146	1.228	1.290	.7753	.8872	50
20	.6865	.6338	1.578	.8195	1.220	1.293	.7735	.8843	40
30	.6894	.6361	1.572	.8243	1.213	1.296	.7716	.8814	30
40	.6923	.6383	1.567	.8292	1.206	1.299	.7698	.8785	20
50	.6952	.6406	1.561	.8342	1.199	1.302	.7679	.8756	10
40° 00′	.6981	.6428	1.556	.8391	1.192	1.305	.7660	.8727	**50° 00′**
10	.7010	.6450	1.550	.8441	1.185	1.309	.7642	.8698	50
20	.7039	.6472	1.545	.8491	1.178	1.312	.7623	.8668	40
30	.7069	.6494	1.540	.8541	1.171	1.315	.7604	.8639	30
40	.7098	.6517	1.535	.8591	1.164	1.318	.7585	.8610	20
50	.7127	.6539	1.529	.8642	1.157	1.322	.7566	.8581	10
41° 00′	.7156	.6561	1.524	.8693	1.150	1.325	.7547	.8552	**49° 00′**
10	.7185	.6583	1.519	.8744	1.144	1.328	.7528	.8523	50
20	.7214	.6604	1.514	.8796	1.137	1.332	.7509	.8494	40
30	.7243	.6626	1.509	.8847	1.130	1.335	.7490	.8465	30
40	.7272	.6648	1.504	.8899	1.124	1.339	.7470	.8436	20
50	.7301	.6670	1.499	.8952	1.117	1.342	.7451	.8407	10
42° 00′	.7330	.6691	1.494	.9004	1.111	1.346	.7431	.8378	**48° 00′**
		cos θ	sec θ	cot θ	tan θ	csc θ	sin θ	θ radians	θ degrees

TABLE I, CONTINUED

TRIGONOMETRIC FUNCTIONS, VALUES IN DEGREES

θ degrees	θ radians	sin θ	csc θ	tan θ	cot θ	sec θ	cos θ		
42° 00′	.7330	.6691	1.494	.9004	1.111	1.346	.7431	.8378	**48° 00′**
10	.7359	.6713	1.490	.9057	1.104	1.349	.7412	.8348	50
20	.7389	.6734	1.485	.9110	1.098	1.353	.7392	.8319	40
30	.7418	.6756	1.480	.9163	1.091	1.356	.7373	.8290	30
40	.7447	.6777	1.476	.9217	1.085	1.360	.7353	.8261	20
50	.7476	.6799	1.471	.9271	1.079	1.364	.7333	.8232	10
43° 00′	.7505	.6820	1.466	.9325	1.072	1.367	.7314	.8203	**47° 00′**
10	.7534	.6841	1.462	.9380	1.066	1.371	.7294	.8174	50
20	.7563	.6862	1.457	.9435	1.060	1.375	.7274	.8145	40
30	.7592	.6884	1.453	.9490	1.054	1.379	.7254	.8116	30
40	.7621	.6905	1.448	.9545	1.048	1.382	.7234	.8087	20
50	.7650	.6926	1.444	.9601	1.042	1.386	.7214	.8058	10
44° 00′	.7679	.6947	1.440	.9657	1.036	1.390	.7193	.8029	**46° 00′**
10	.7709	.6967	1.435	.9713	1.030	1.394	.7173	.7999	50
20	.7738	.6988	1.431	.9770	1.024	1.398	.7153	.7970	40
30	.7767	.7009	1.427	.9827	1.018	1.402	.7133	.7941	30
40	.7796	.7030	1.423	.9884	1.012	1.406	.7112	.7912	20
50	.7825	.7050	1.418	.9942	1.006	1.410	.7092	.7883	10
45° 00′	.7854	.7071	1.414	1.000	1.000	1.414	.7071	.7854	**45° 00′**
		cos θ	sec θ	cot θ	tan θ	csc θ	sin θ	θ radians	θ degrees

appendix C
answers to odd-numbered exercises

CHAPTER ONE. The Trigonometric Ratios

Exercise Set 1.1, page 9

1. $b = 8$ 3. $AB = 65$ 5. $C = 61$ 7. $AC = 8$ 9. $c = \sqrt{34}$ 11. Right triangle 13. Not a right triangle 15. Right triangle 17. Not a right triangle 19. Right triangle 21. 5 23. 17 25. 25 27. 5 29. $\sqrt{58} \approx 7.616$ 31. $\alpha = \angle DQA = \angle AQD$ 33. $\angle QAD = \delta = \angle DAQ$ 35. $\angle U = \angle QUA = \angle AUQ$ 37. $\angle AOR = \angle 2 = \angle ROA$ 39. $\angle RAC = \angle 1 = \angle CAR$ 41. 90° 43. 120° 45. 4° 47. 270° 49. 540° 51. $\frac{1}{4}$ of a rotation, 90° 53. $\frac{1}{3}$ of a rotation, 120° 55. $\frac{1}{15}$ of a rotation, 24° 57. $\frac{13}{12}$ rotations, 390° 59. Two rotations, 720° 61. 140° 63. 280° 65. 173° 67. 169°25′ 69. 316°42′40″ 71. 37° 73. 143° 75. 120° 77. 52° 79. 19° 81. 57° 83. 86° 85. 94° 87. 94° 89. 86° 91. 59° 93. 64° 95. 57° 97. 121° 99. 59°

Exercise Set 1.2, page 19

1. 6 3. 8 5. 32 7. 8.5 9. 8 11. $\sqrt{14}$ 13. $6\sqrt{3}$ 15. $2\sqrt{3}$ 17. $2\sqrt{10}$ 19. $4\sqrt{5}$ 21. 5 23. 13 25. $\sqrt{2}$ 27. $2\sqrt{3}$ 29. $4\sqrt{2}$ 31. The building is 63 ft tall. 33. The pole is 24 ft tall. 35. The area is $9\sqrt{3}$ yd².

Exercise Set 1.3, page 29

1. $\sin B = \frac{b}{a}$ 3. $\sin N = \frac{n}{t}$ 5. $\sin \alpha = \frac{y}{z}$

$\cos B = \frac{y}{a}$ $\cos N = \frac{m}{t}$ $\cos \alpha = \frac{x}{z}$

$\tan B = \frac{b}{y}$ $\tan N = \frac{n}{m}$ $\tan \alpha = \frac{y}{x}$

$\cot B = \frac{y}{b}$ $\cot N = \frac{m}{n}$ $\cot \alpha = \frac{x}{y}$

$\sec B = \frac{a}{y}$ $\sec N = \frac{t}{m}$ $\sec \alpha = \frac{z}{x}$

$\csc B = \frac{a}{b}$ $\csc N = \frac{t}{n}$ $\csc \alpha = \frac{z}{y}$

7. $\frac{20}{29}$ 9. $\frac{73}{55}$ 11. $\frac{29}{20}$ 13. $\frac{73}{55}$ 15. $\cot T$ or $\tan R$ 17. $\sec T$ or $\csc R$ 19. $\sec X$ or $\csc Z$ 21. $\frac{7}{24}$ 23. $\frac{35}{12}$ 25. $\frac{2}{3}$ 27. $\sqrt{10}$ 29. $\frac{65}{63}$ 31. $1/\sqrt{2}$ or $\sqrt{2}/2$ 33. 1 35. $\sqrt{2}$ 37. $\sqrt{3}$ 39. $1/\sqrt{2}$ or $\sqrt{2}/2$ 41. True 43. True 45. True 47. True 49. True 51. 24 53. $12\sqrt{6}$ 55. $16\sqrt{3}$ 57. $8\sqrt{6}$ 59. $18\sqrt{3}$ 61. $12\sqrt{3}$ 63. $24\sqrt{3}$ 65. 32 67. $32\sqrt{2}$ 69. 64

71. The total span is $(12 + 20\sqrt{3})$ ft or approximately 46.6 ft.

73. The height is $(2\sqrt{2} + 2\sqrt{3})$ ft or approximately 6.3 ft.

75. The overlap is $(2 - 2\sqrt{3}/3)$ m or approximately 0.85 m.

77. The perimeter is $36\sqrt{3}$ in. or approximately 62.4 in.

79. The area is $\frac{4}{3}(11 - 2\sqrt{3})$ cm².

Exercise Set 1.4, page 41

1. 0.2588 (0.25881905) 3. 0.3584 (0.35836795) 5. 0.7813 (0.78128563)
7. 1.3673 (1.3673275) 9. 0.1944 (0.19438031) 11. 0.8511 (0.85111667)
13. 0.9100 (0.90996127) 15. 0.9952 (0.99517046) 17. 0.9364 (0.93643840)
19. 1.1243 (1.1243328) 21. $89 \sin 26° \approx 39.105032$ 23. $55 \sec 41° \approx 72.875714$
25. $189 \tan 51° \approx 233.39556$ 27. $15.7 \sin 24° \approx 6.3857653$
29. $9.65 \sec 61° \approx 19.904721$ 31. $234 \tan 15° \approx 62.700111$
33. $20.25 \tan 21.5° \approx 7.9766871$ 35. $3.875 \sec 39°20' \approx 5.0098802$
37. $.034 \csc 69.5° \approx 0.0362987$ 39. $1247 \sin 2.45° \approx 53.306178$
41. The missile is 492 m high (491.80181). 43. The height is 70 ft (69.952271).
45. The altitude is 107 ft (106.51617). 47. The radius is 2.31 cm (2.3071736).
49. The bridge is 1053 ft high (1053.2265). 51. I'LLSEE (33577.1)

Exercise Set 1.5, page 49

1. cos 53° 3. sin 68° 5. csc 75° 7. sin 36° 9. cos 34° 11. 62° 13. 39°
15. 57° 17. 17° 19. 49° 21. 1.0921 (1.0921053) 23. 0.5121 (0.51209297)
25. 0.5918 (0.59179105) 27. 0.9954 (0.99542404) 29. 0.6401 (0.64010970)
31. 45°46′ (45.765768°) 33. 77°14′ (77.232225°) 35. 48°25′ (48.417350°)
37. 41°3′ (41.049348°) 39. 23°52′ (23.866064°)
41. The post is 10.7 ft high (10.711206). 43. The angle is 35.2° (35.150136°).
45. The shadow is 5 ft 10 in. (5.8057222). 47. The ship is 940 ft offshore (939.63144).
49. The Transamerica Pyramid is 853 ft tall (853.07503). 51. 45°35′ (45.587628°)
53. 827 (827.35845) 55. 29°45′ (29.756561°) 57. 14°40′ 59. 922
61. The Verrazano Narrows bridge is 4257 ft long (4257.1073).
63. The MLC Building is 786 ft tall (786.03998).
65. "LET NO ONE IGNORANT OF GEOMETRY ENTER MY DOOR." (Plato)
67. $\frac{s/2}{r} = \sin\left(\frac{180}{n}\right)^{\circ}$

$$\frac{s}{2} = r \sin\left(\frac{180}{n}\right)^{\circ}$$

$$s = 2r \sin\left(\frac{180}{n}\right)^{\circ}$$

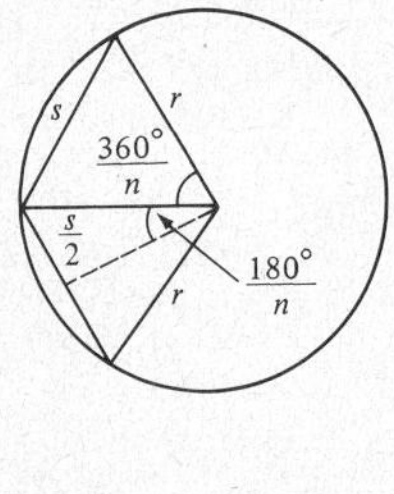

Chapter Review, page 52

1. coterminal 2. complementary 3. proportional 4. $1{:}1{:}\sqrt{2}$ 5. 2 (two)
6. $\sin P$ or $\cos R$ 7. horizontal 8. secant 9. (a) 74° (b) 45° (c) 135° (d) 45°
(e) 106° 10. (a) 3 (b) $3\sqrt{2}$ (c) $3\sqrt{3}$ (d) $3\sqrt{6}$ 11. (a) $\frac{s}{r}$ (b) $\frac{t}{r}$ (c) $\frac{s}{t}$ (d) $\frac{s}{r}$ (e) $\frac{r}{s}$

12. (a) $\sin P$ or $\cos H$ (b) $\sec P$ or $\csc H$ (c) $\tan H$ or $\cot P$ 13. (a) $\dfrac{40}{9}$ (b) $\dfrac{37}{12}$ (c) $\dfrac{2}{\sqrt{13}}$

14. (a) 30° (b) $9\sqrt{3}$ (c) 9 (d) $9\sqrt{2}$ (e) $\frac{9}{2}\sqrt{6}$ 15. (a) no solution (b) 31°28′ (c) 1.1810 (1.1809851) (d) 28°46′ (28.766651°) (e) 32°14′ (32.231431°)

16. (a) 13.6235 (13.623472) (b) 14.4689 (14.468939) (c) 166.8544 (166.85439) 17. (a) 30° (b) 48°46′ (48.763580°) (c) 35°6′ (35.102937°)

18. You were back 300 ft (300.01019). (Note that 49.4° was extraneous.)

19. The angle of elevation is 69° (68.962489°). 20. The rise is 11 ft vertically.

CHAPTER TWO. The Trigonometric Functions

Exercise Set 2.1, page 61

	1.	3.	5.	7.	9.	11.	13.	15.	17.
sin	$\frac{12}{13}$	$\frac{4}{5}$	$-\frac{2}{3}$	$\frac{\sqrt{3}}{3}$	$-\frac{35}{37}$	$-\frac{1}{2}$	$-\frac{2}{\sqrt{5}}$	$-\frac{9}{41}$	1
cos	$\frac{5}{13}$	$-\frac{3}{5}$	$-\frac{\sqrt{5}}{3}$	$-\frac{\sqrt{6}}{3}$	$-\frac{12}{37}$	$\frac{\sqrt{3}}{2}$	$\frac{1}{\sqrt{5}}$	$-\frac{40}{41}$	0
tan	$\frac{12}{5}$	$-\frac{4}{3}$	$\frac{2}{\sqrt{5}}$	$-\frac{1}{\sqrt{2}}$	$\frac{35}{12}$	$-\frac{1}{\sqrt{3}}$	-2	$\frac{9}{40}$	∞
cot	$\frac{5}{12}$	$-\frac{3}{4}$	$\frac{\sqrt{5}}{2}$	$-\sqrt{2}$	$\frac{12}{35}$	$-\sqrt{3}$	$-\frac{1}{2}$	$\frac{40}{9}$	0
sec	$\frac{13}{5}$	$-\frac{5}{3}$	$-\frac{3}{\sqrt{5}}$	$-\frac{3}{\sqrt{6}}$	$-\frac{37}{12}$	$\frac{2}{\sqrt{3}}$	$\sqrt{5}$	$-\frac{41}{40}$	∞
csc	$\frac{13}{12}$	$\frac{5}{4}$	$-\frac{3}{2}$	$\frac{3}{\sqrt{3}}$	$-\frac{37}{35}$	-2	$-\frac{\sqrt{5}}{2}$	$-\frac{41}{9}$	1

19. $\sin\gamma = \dfrac{q}{\sqrt{p^2+q^2}}$, $\cos\gamma = \dfrac{p}{\sqrt{p^2+q^2}}$, $\tan\gamma = \dfrac{q}{p}$, $\cot\gamma = \dfrac{p}{q}$, $\sec\gamma = \dfrac{\sqrt{p^2+q^2}}{p}$, $\csc\gamma = \dfrac{\sqrt{p^2+q^2}}{q}$ 21. QI 23. QIII 25. QIV 27. QIV 29. QIII

	31.	33.	35.	37.	39.	41.	43.	45.	47.	49.
sin	$\frac{12}{13}$	$-\frac{8}{17}$	$-\frac{4}{5}$	$-\frac{60}{61}$	$-\frac{84}{85}$	$\frac{15}{113}$	1	$\frac{1}{\sqrt{2}}$	-1	-1
cos	$-\frac{5}{13}$	$-\frac{15}{17}$	$\frac{3}{5}$	$-\frac{11}{61}$	$-\frac{13}{85}$	$\frac{112}{113}$	0	$-\frac{1}{\sqrt{2}}$	0	0
tan	$-\frac{12}{5}$	$\frac{8}{15}$	$-\frac{4}{3}$	$\frac{60}{11}$	$\frac{84}{13}$	$\frac{15}{112}$	∞	-1	∞	∞

Exercise Set 2.2, page 71

1. $\frac{\pi}{6}$ 3. $-\frac{\pi}{3}$ 5. $-\frac{5\pi}{36}$ 7. $-\frac{7\pi}{18}$ 9. $\frac{11\pi}{18}$ 11. $\frac{10\pi}{9}$ 13. $\frac{14\pi}{9}$ 15. 4π

17. $\frac{50\pi}{9}$ 19. $\frac{\pi^2}{180}$ 21. 45° 23. −120° 25. 150° 27. −225° 29. 36°

31. 540° 33. 810° 35. $\left(\frac{120}{\pi}\right)^\circ$ 37. $\left(\frac{540}{\pi}\right)^\circ$ 39. $(180\pi)^\circ$ 41. 40° 43. 305°

45. $\frac{5\pi}{3}$ 47. 140° 49. π 51. 85° 53. $\frac{\pi}{4}$ 55. 49° 57. 59° 59. $\frac{\pi}{3}$ 61. $\frac{1}{2}$

63. $\sqrt{3}$ 65. −2 67. $\frac{1}{2}$ 69. $\frac{1}{\sqrt{3}}$ 71. ∞ 73. $-\frac{1}{\sqrt{2}}$ 75. 0 77. −2 79. $-\frac{1}{2}$

81. sin 58° 83. tan 24° 85. $-\sec\frac{3\pi}{7}$ 87. $-\sin\frac{\pi}{7}$ 89. −tan 58° 91. −sec 49°

93. −sin 29° 95. tan 19° 97. $\sec\frac{\pi}{2}$ 99. $\sin(\pi - 2)$ 101. 45 103. $\frac{1600}{3}$

105. $\frac{5\pi}{32}$ 107. 2400 109. 16,000 111. 45 113. 100 115. 700 117. 73.2375

119. 1200

Exercise Set 2.3, page 81

1. Identity 3. Contradiction 5. Identity 7. Identity 9. Contradiction

11. $\csc\alpha$ 13. 1 15. $\sec\gamma$ 17. $\sin\delta$ 19. $4\sin\lambda\cos\lambda$

Only sine, cosine, and tangent are shown for 21 and 23.

	21.	23.
sin	$\pm\sqrt{1-\cos^2\alpha}$	$\pm\sqrt{\frac{1}{1+\cot^2\gamma}}$
cos	—	$\pm\sqrt{\frac{\cot^2\gamma}{1+\cot^2\gamma}}$
tan	$\frac{\pm\sqrt{1-\cos^2\alpha}}{\cos\alpha}$	$\frac{1}{\cot\gamma}$

25. $2\cos\alpha$ 27. $49\sec^2\gamma$ 29. $\sin\lambda$

31. $\tan\alpha = \sin\alpha\sec\alpha$
$$= \sin\alpha\frac{1}{\cos\alpha}$$
$$= \frac{\sin\alpha}{\cos\alpha}$$
$$\tan\alpha = \tan\alpha \quad \checkmark$$

33. $\cos\beta\tan\beta\csc\beta = 1$
$$\cos\beta\frac{\sin\beta}{\cos\beta}\frac{1}{\sin\beta} =$$
$$1 = 1 \quad \checkmark$$

35. $\tan\gamma(\tan\gamma + \cot\gamma) = \sec^2\gamma$

$$\tan^2\gamma + 1 =$$
$$\sec^2\gamma = \sec^2\gamma \quad \checkmark$$

37. $\tan\delta + \cot\delta = \sec\delta\csc\delta$

$$\frac{\sin\delta}{\cos\delta} + \frac{\cos\delta}{\sin\delta} =$$
$$\frac{\sin^2\delta + \cos^2\delta}{\cos\delta\sin\delta} =$$
$$\frac{1}{\cos\delta\sin\delta} =$$
$$\frac{1}{\cos\delta}\frac{1}{\sin\delta} =$$
$$\sec\delta\csc\delta = \sec\delta\csc\delta \quad \checkmark$$

39. $\csc\lambda - \sin\lambda = \cot\lambda\cos\lambda$

$$\frac{1}{\sin\lambda} - \sin\lambda =$$
$$\frac{1-\sin^2\lambda}{\sin\lambda} =$$
$$\frac{\cos^2\lambda}{\sin\lambda} =$$
$$\frac{\cos\lambda}{\sin\lambda}\cos\lambda =$$
$$\cot\lambda\cos\lambda = \cot\lambda\cos\lambda \quad \checkmark$$

41. $\dfrac{\tan x}{\sin^2 x + \cos^2 x} = \dfrac{1}{\cot x}$

$$\frac{\tan x}{1} =$$
$$\frac{1}{\cot x} = \frac{1}{\cot x} \quad \checkmark$$

43. $(\sin y + \cos y)^2 = 1 + 2\sin y\cos y$

$$\sin^2\gamma + 2\sin y\cos y + \cos^2 y =$$
$$\sin^2 y + \cos^2 y + 2\sin y\cos y =$$
$$1 + 2\sin y\cos y = 1 + 2\sin y\cos y \quad \checkmark$$

45. $\sec^2 z + \csc^2 z = \sec^2 z\csc^2 z$

$$\frac{1}{\cos^2 z} + \frac{1}{\sin^2 z} =$$
$$\frac{\sin^2 z + \cos^2 z}{\cos^2 z\sin^2 z} =$$
$$\frac{1}{\cos^2 z\sin^2 z} =$$
$$\frac{1}{\cos^2 z}\frac{1}{\sin^2 z} =$$
$$\sec^2 z\csc^2 z = \sec^2 z\csc^2 z \quad \checkmark$$

47. $\dfrac{\cos^2\theta - \sin^2\theta}{\sin\theta\cos\theta} = \cot\theta - \tan\theta$

$$\frac{\cos^2\theta}{\sin\theta\cos\theta} - \frac{\sin^2\theta}{\sin\theta\cos\theta} =$$
$$\frac{\cos\theta}{\sin\theta} - \frac{\sin\theta}{\cos\theta} =$$
$$\cot\theta - \tan\theta = \cot\theta - \tan\theta \quad \checkmark$$

49. $\dfrac{1}{1-\sin\sigma} + \dfrac{1}{1+\sin\sigma} = 2\sec^2\sigma$

$$\frac{1+\sin\sigma+1-\sin\sigma}{1-\sin^2\sigma} =$$
$$\frac{2}{\cos^2\sigma} =$$
$$2\sec^2\sigma = 2\sec^2\sigma \quad \checkmark$$

51. If $\lambda = \pi$, $\cos\lambda = -(1-\sin^2\lambda)^{1/2}$.

53. $\log(\cos\sigma) + \log(\tan\sigma) + \log(\csc\sigma) = 0$

$$\log(\cos\sigma\tan\sigma\csc\sigma) = 0$$
$$\log(1) = 0$$
$$0 = 0 \quad \checkmark$$

Exercise Set 2.4, page 95

1. $A = 0.1564$ 3. $C = -0.6745$ 5. $E = 203°11'$ 7. $G = 125°23'$
9. $I = -1.1824$ 11. $K = -0.9287$ 13. $M = 34°46'$ 15. $\phi = -1.3968$
17. $Q = 203°27'$ 19. $S = 45°0'$ 21. $U = 14.0350$ 23. $W = -81.3504$
25. $\alpha = 0°55'$ 27. $Y = -0.0894$ 29. $\gamma = 35°13'$ 31. $D = 1.3101$ 33. $D = 1.7348$
35. $\theta_m = 14°43'$ 37. $\theta_a = 30°45'$ 39. $\theta_m = 6°26'$ 41. 1121 cm^3
43. 4094 ft^3 or 152 yd^3 45. 210 lb 47. 12 ft^3 1088 lb 49. $V = \theta r^2 h$
51. $2.2397988 \approx 2.24$ 53. $20.559473 \approx 20.6$
55. $\theta = (30)(2\pi)t = 60\pi t$
$d = r + a - r\cos 60\pi t - \sqrt{a^2 - r^2\sin^2 60\pi t}$
57. 8.00
59. $a = 4r$
$d = r + 4r - r\cos\theta - \sqrt{16r^2 - r^2\sin^2\theta}$
$d = 5r - r\cos\theta - r\sqrt{16 - \sin^2\theta}$
61. 978.727 cm/s^2 63. 979.513 cm/s^2 65. 979.660 cm/s^2 67. 980.291 cm/s^2
69. 980.691 cm/s^2
71. "MATHEMATICS IS . . . THE STUDY OF SELF-EVIDENT THINGS." (Felix Klein)

Chapter Review, page 100

1. k/d 2. tangent, cotangent 3. $\cos\theta$ 4. radian 5. $\theta + 2\pi k$ or $180° - \theta$
6. $\pi - \theta$ 7. identity 8. Pythagorean 9. ratio 10. reciprocal 11. (a) $\sin\theta = \frac{8}{17}$, $\cos\theta = \frac{15}{7}$, $\tan\theta = \frac{8}{15}$ (b) $\sin\theta = \frac{2}{3}$, $\cos\theta = -\frac{\sqrt{5}}{3}$, $\tan\theta = -\frac{2}{\sqrt{5}}$ (c) $\sin\theta = -\frac{63}{65}$, $\cos\theta = \frac{16}{65}$, $\tan\theta = -\frac{63}{16}$ (d) $\sin\theta = -\frac{9}{\sqrt{130}}$, $\cos\theta = -\frac{7}{\sqrt{130}}$, $\tan\theta = \frac{9}{7}$ (e) $\sin\theta = -\frac{21}{29}$, $\cos\theta = \frac{20}{29}$, $\tan\theta = -\frac{21}{20}$ 12. (a) -1 (b) $-\frac{15}{17}$ (c) 0 (d) $\frac{4}{\sqrt{13}}$ (e) 0 13. (a) $\frac{3\pi}{4}$ (b) $-60°$ (c) $300°$ (d) $-\frac{145\pi}{18}$ (e) $\frac{\pi^2}{360}$ 14. (a) $\sin 38°$ (b) $-\cos\frac{\pi}{3}$ (c) $\tan 20°$

15. $\cos\theta = \pm\sqrt{1 - \sin^2\theta}\left[\text{or } \sin\left(\frac{\pi}{2} - \theta\right)\right]$, $\tan\theta = \frac{\sin\theta}{\pm\sqrt{1 - \sin^2\theta}}$, $\cot\theta = \frac{\pm\sqrt{1 - \sin^2\theta}}{\sin\theta}$,
$\sec\theta = \frac{1}{\pm\sqrt{1 - \sin^2\theta}}$, $\csc\theta = \frac{1}{\sin\theta}$

16. (a) $\dfrac{\sin\theta}{\csc\theta} = \sin\theta\dfrac{1}{\csc\theta}$

$= \sin\theta\sin\theta$

$= \sin^2\theta$ √

(b) $\cot^2\theta(1+\tan^2\theta) = \cot^2\theta(\sec^2\theta)$

$= \dfrac{\cos^2\theta}{\sin^2\theta}\dfrac{1}{\cos^2\theta}$

$= \dfrac{1}{\sin^2\theta}$

$= \csc^2\theta$ √

(c) $\dfrac{1+\cot^2\theta}{1+\tan^2\theta} = \dfrac{\csc^2\theta}{\sec^2\theta}$

$= \dfrac{1/\sin^2\theta}{1/\cos^2\theta}$

$= \dfrac{\cos^2\theta}{\sin^2\theta}$

$= \cot^2\theta$ √

17. (a) $A \approx 0.9969$ (b) $\beta \approx 59°24'$ (c) $C \approx 1.4690$ (d) $\delta \approx 218°42'$ (e) $E \approx -1.1999$ 18. $D \approx 1.4587$ 19. $\theta_m \approx 11.54°$ or $11°33'$ 20. (a) 139 cm^3 (b) $200\pi - 150\sqrt{3} \approx 368.51091$ (c) 27.80 m^3

CHAPTER THREE. Graphs of the Trigonometric Functions

Exercise Set 3.1, page 112

1.

x	cos x	x	cos x
0	1	π	−1
$\frac{\pi}{6}$	0.87	$\frac{7\pi}{6}$	−0.87
$\frac{\pi}{4}$	0.71	$\frac{5\pi}{4}$	−0.71
$\frac{\pi}{3}$	0.5	$\frac{4\pi}{3}$	−0.5
$\frac{\pi}{2}$	0	$\frac{3\pi}{2}$	0
$\frac{2\pi}{3}$	−0.5	$\frac{5\pi}{3}$	0.5
$\frac{3\pi}{4}$	−0.71	$\frac{7\pi}{4}$	0.71
$\frac{5\pi}{6}$	−0.87	$\frac{11\pi}{6}$	0.87
π	−1	2π	1

3.

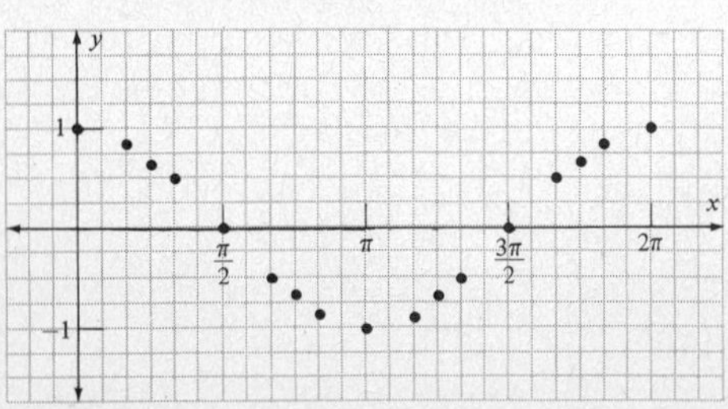

5.

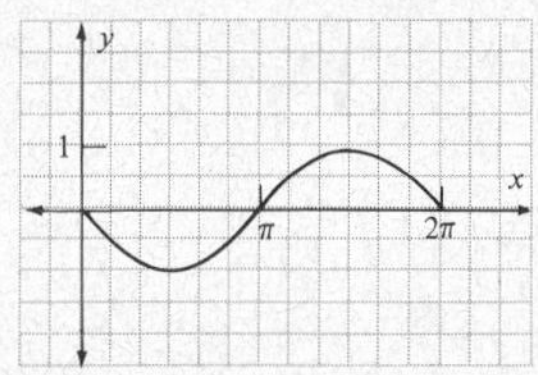

7.

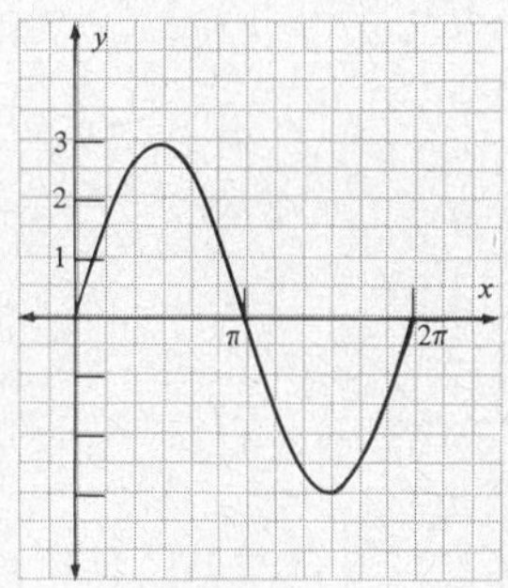

9.

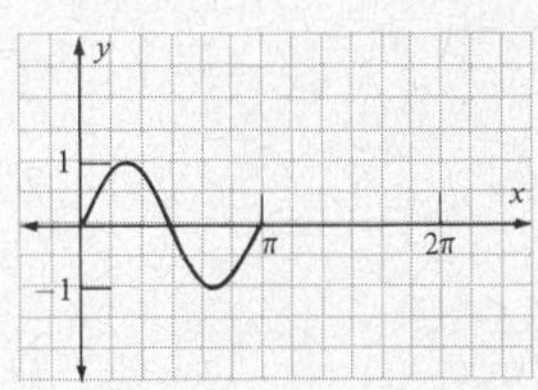

11.

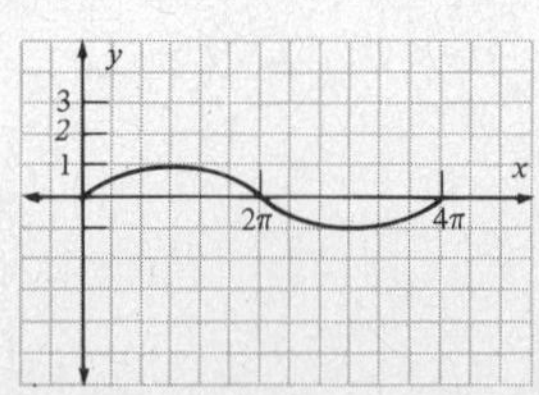

13. 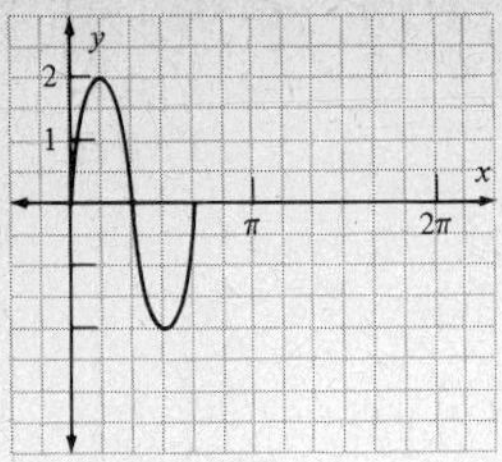

15. 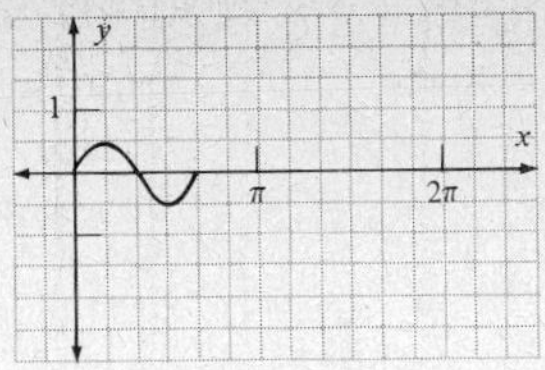

17. 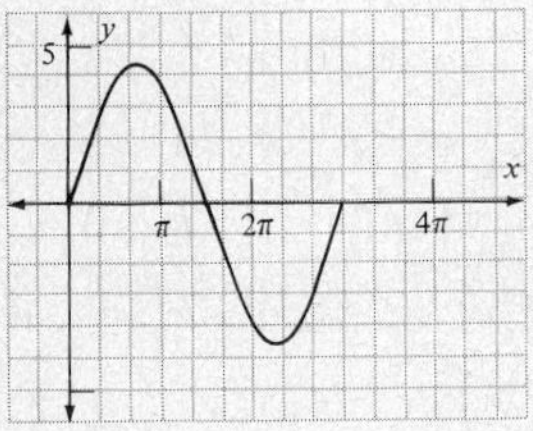

19. 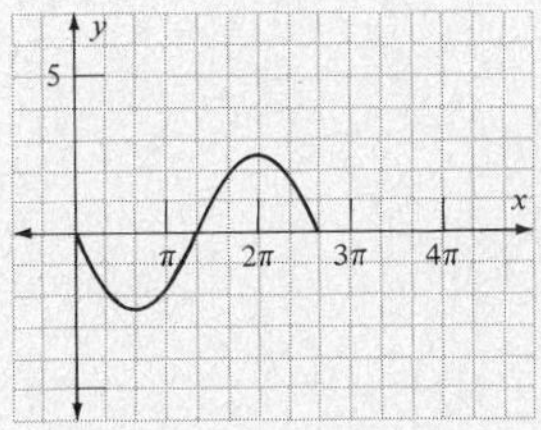

21. $(\pi, 2)$, $a = 3$, $b = 4$, $p = \dfrac{\pi}{2}$ 23. $\left(-\dfrac{\pi}{2}, 3\right)$, $a = 2$, $b = \dfrac{1}{2}$, $p = 4\pi$

25. $\left(-\dfrac{\pi}{3}, -\dfrac{3}{2}\right)$, $|a| = \dfrac{1}{2}$, $b = 2$, $p = \pi$ 27. $\left(-\dfrac{4\pi}{3}, \dfrac{1}{3}\right)$, $a = 3$, $b = \dfrac{2}{3}$, $p = 3\pi$

29. $\left(\dfrac{5\pi}{6}, -1\right)$, $a = \dfrac{5}{2}$, $b = \dfrac{3}{4}$, $p = \dfrac{8\pi}{3}$

31.

32.

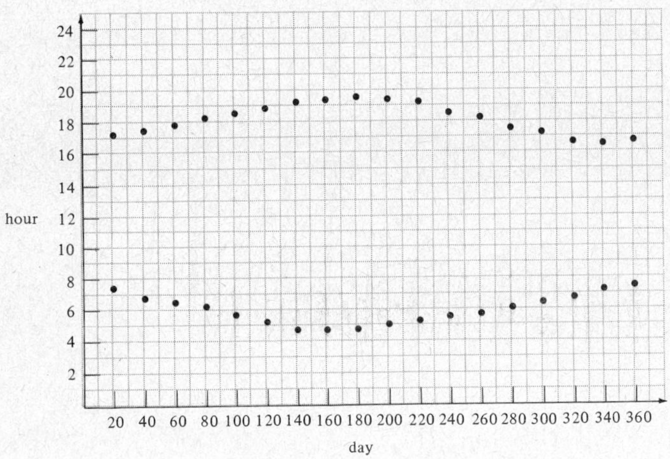

Equations will vary, but for axes as shown they should be approximately
(31) $y - 18.05 = 1.39 \sin 2\pi(x - 70)$

33. 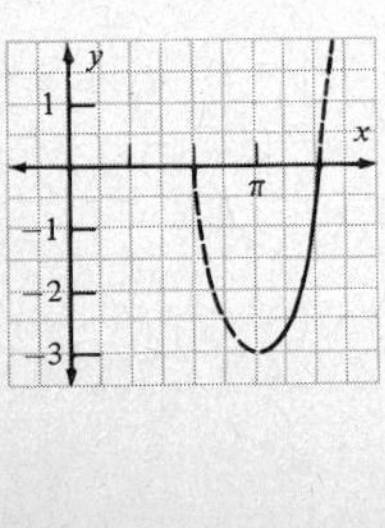

35. 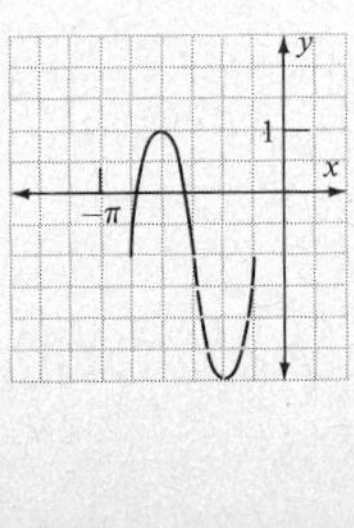

37. 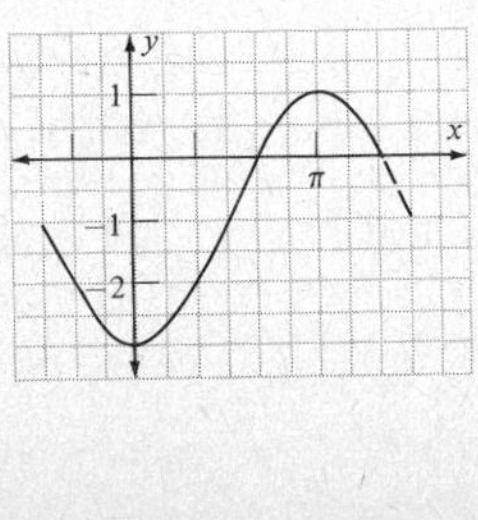

39. 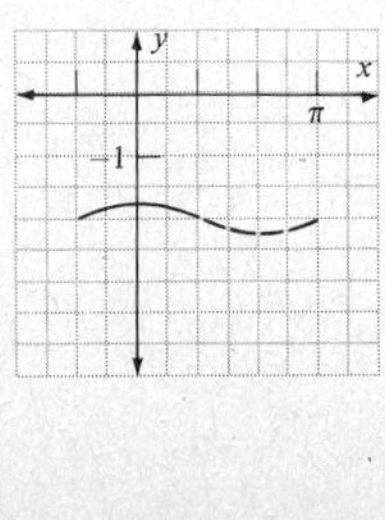

Exercise Set 3.2, page 124

1. 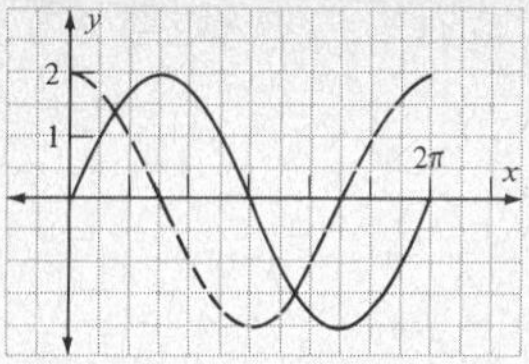

3. 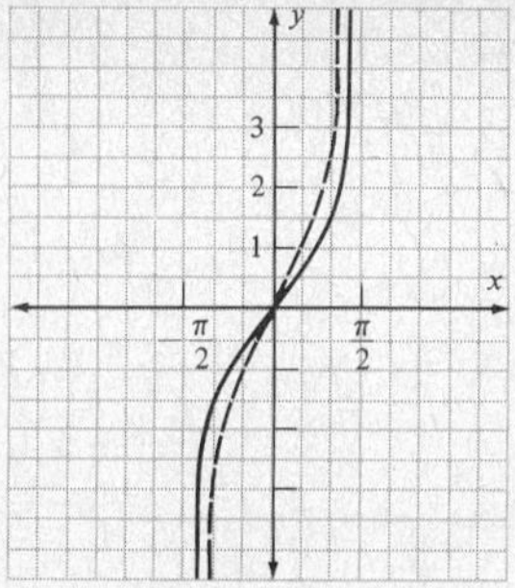

5. 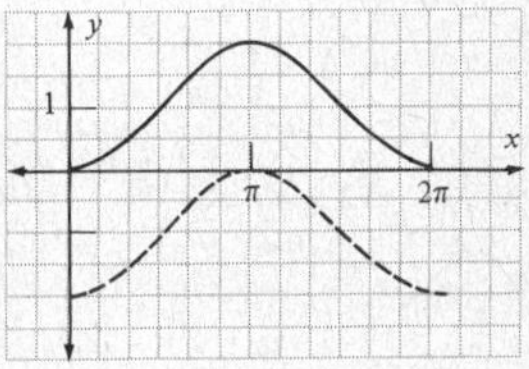

7. 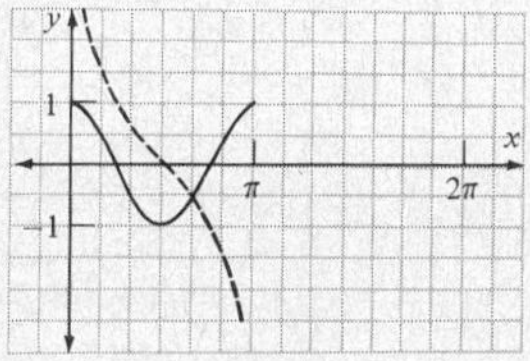

9. 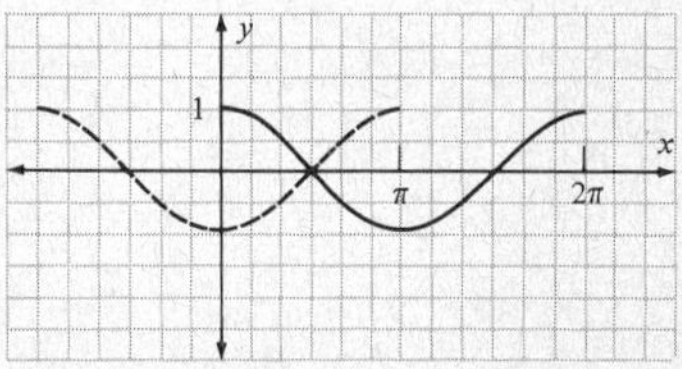

11. $y - 2 = 3 \sin 4\left(x - \frac{\pi}{2}\right); \left(\frac{\pi}{2}, 2\right), 3, 4, \frac{\pi}{2}$ 13. $y + \frac{3}{2} = 2 \cos\left(x - \frac{\pi}{6}\right); \left(\frac{\pi}{6}, -\frac{3}{2}\right), 2, 1, 2\pi$

15. $y - 2 = \frac{2}{3} \tan 2\left(x - \frac{\pi}{6}\right); \left(\frac{\pi}{6}, 2\right), \frac{2}{3}, 2, \frac{\pi}{2}$ 17. $y - \frac{5}{6} = -\frac{1}{2} \sin^2\left(x + \frac{4\pi}{3}\right);$ $\left(-\frac{4\pi}{3}, \frac{5}{6}\right), -\frac{1}{2}, 2, \pi$ 19. $y - \frac{9}{8} = -\frac{1}{6} \tan 6\left(x - \frac{3\pi}{4}\right); \left(\frac{3\pi}{4}, \frac{9}{8}\right), -\frac{1}{6}, 6, \frac{\pi}{6}$

21. 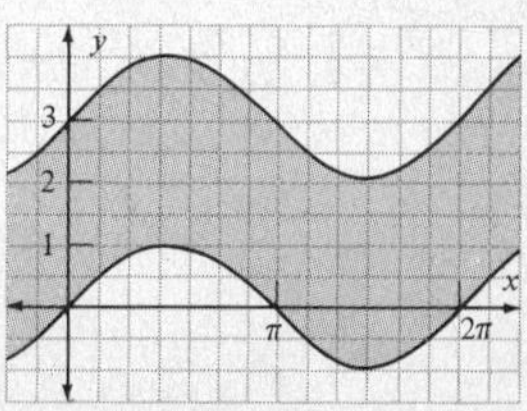

23. 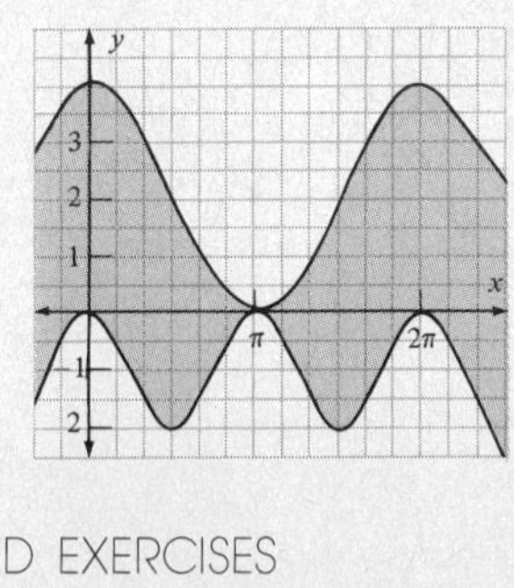

25. 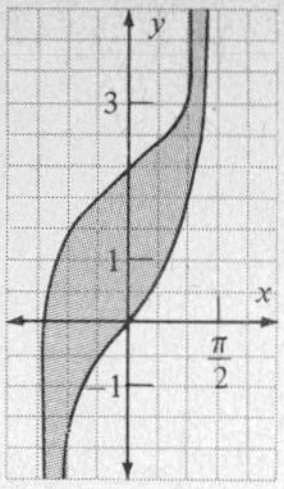

27. 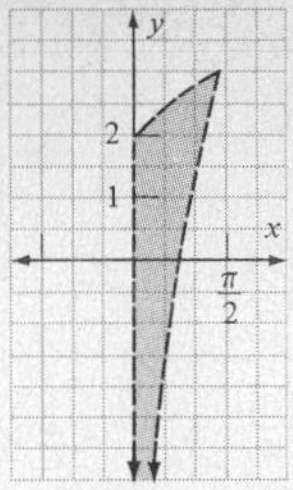

29. 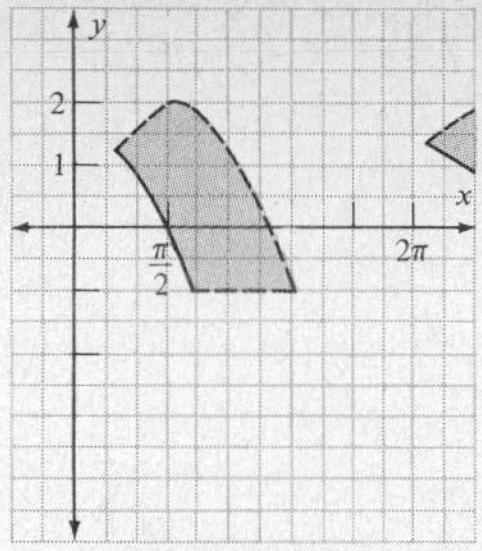

31. $\sqrt{k/m}$ 33. 10 cm 35. $2\sqrt{10}$ s ≈ 6.32 s 37. $2\pi\sqrt{l/g}$ 39. $\dfrac{2\pi\sqrt{3}}{3}$

41. 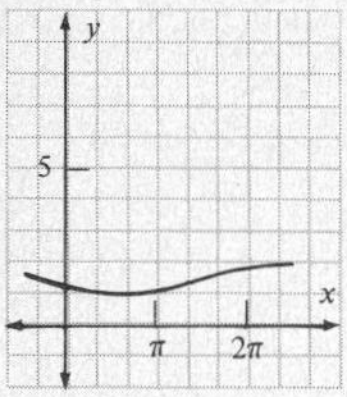

43. 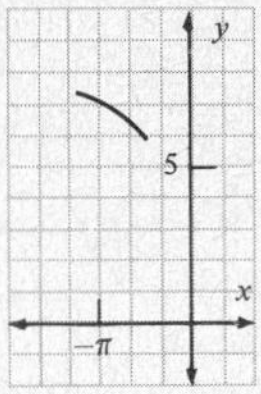

45. 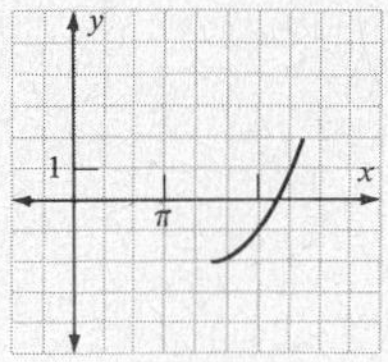

47. 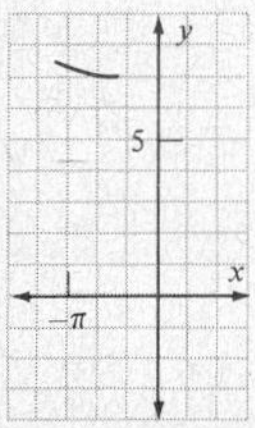

49. 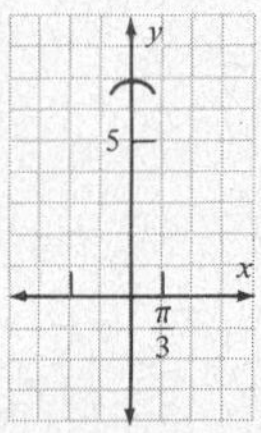

Exercise Set 3.3, page 132

1. 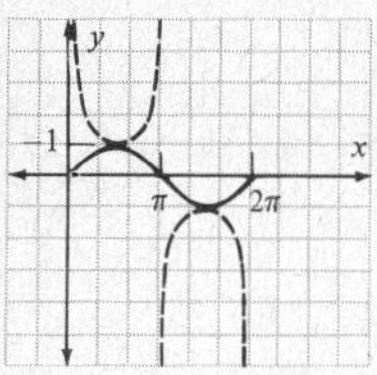

3. 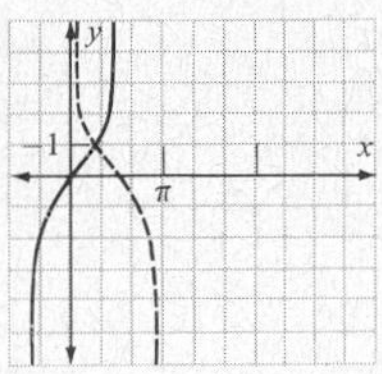

5. 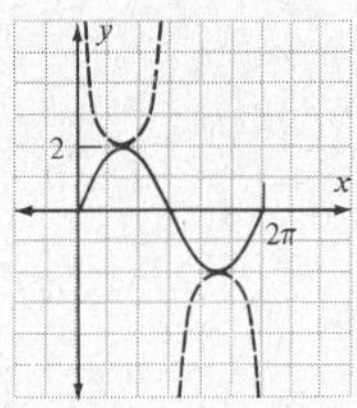

7. 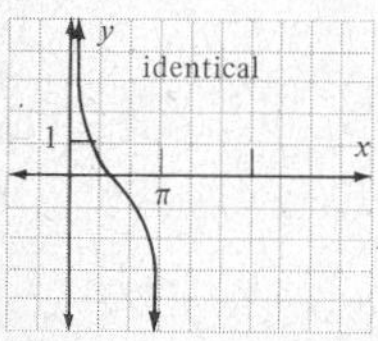

9. 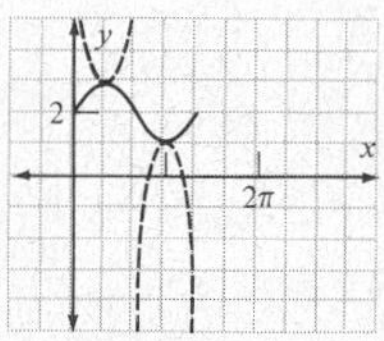

11. $y = \dfrac{1}{3} \csc \dfrac{2}{3}\left(x - \dfrac{\pi}{2}\right)$; $\left(\dfrac{\pi}{2}, 0\right), \dfrac{1}{3}, \dfrac{2}{3}, 3\pi$ 13. $y = \dfrac{2}{3} \sec \dfrac{3}{2}(x + \pi)$; $(-\pi, 0), \dfrac{2}{3}, \dfrac{3}{2}, \dfrac{4\pi}{3}$

15. $y - 4 = -\tan\frac{9}{4}\left(x - \frac{\pi}{9}\right)$; $\left(\frac{\pi}{9}, 4\right)$, $-1, \frac{9}{4}, \frac{4\pi}{9}$ 17. $y + \frac{20}{3} = -\frac{1}{3}\sin\frac{3}{4}(x + \pi)$; $\left(-\pi, -\frac{20}{3}\right)$, $-\frac{1}{3}, \frac{3}{4}, \frac{8\pi}{3}$ 19. $y - 2 = \sec\left(x + \frac{9\pi}{8}\right)$; $\left(-\frac{9\pi}{8}, 2\right)$, 1, 1, 2π.

21. 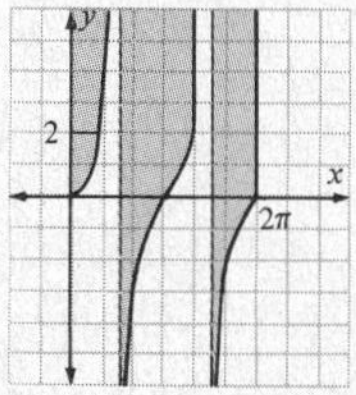

23. 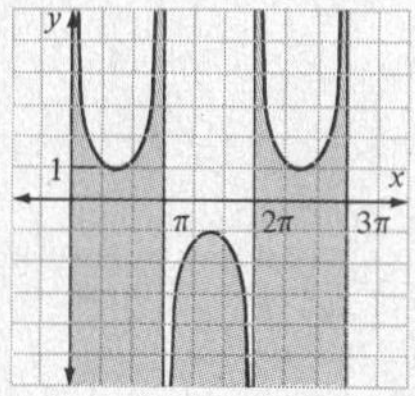

25. 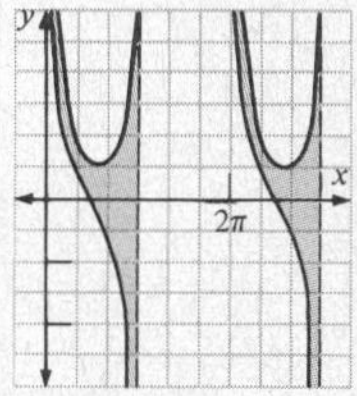

27. 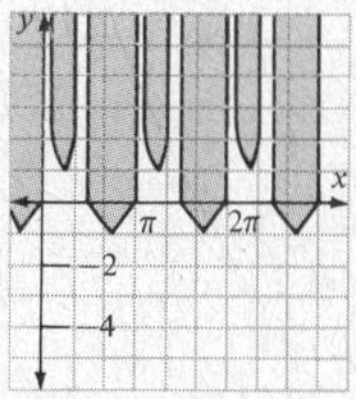

29. 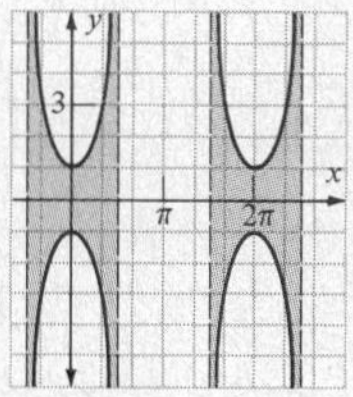

31. 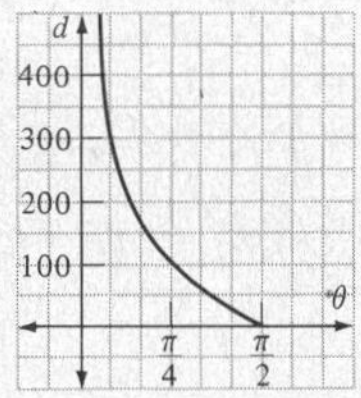

33. $\pi/\sqrt{MC}$ 35. $-\sqrt{MC}/2$ 37. T is the period, the length of one cycle of a wave.
39. 8

41. 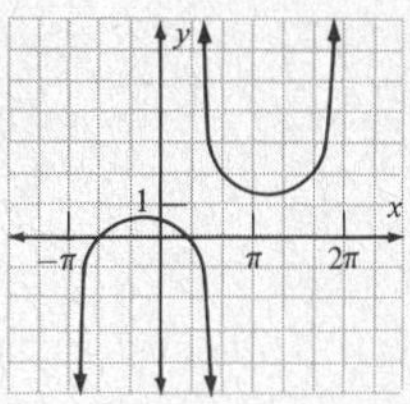

43. 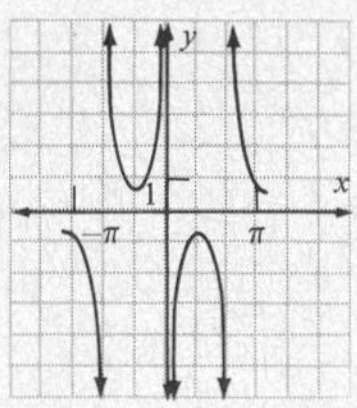

45. 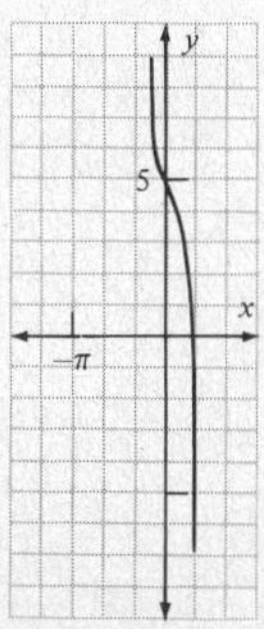

47. 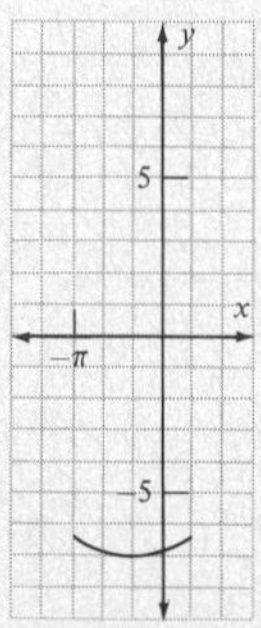

49. 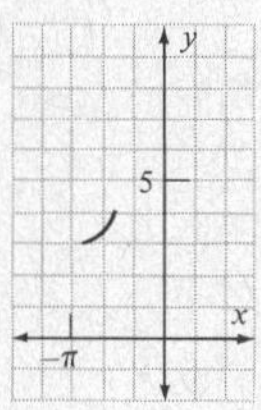

Exercise Set 3.4, page 142

1. 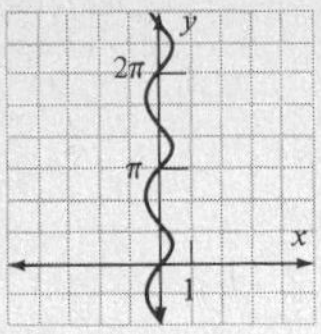

3. 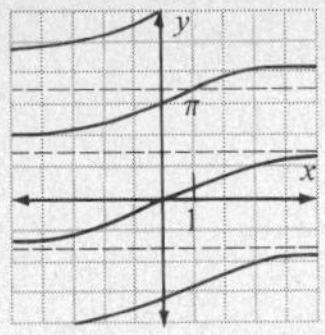

5. 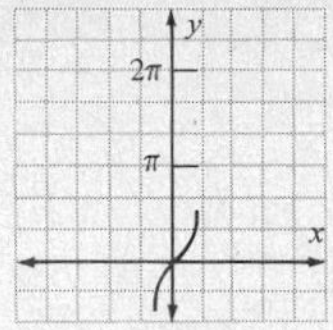

7. 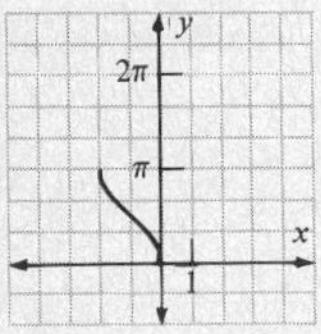

9. 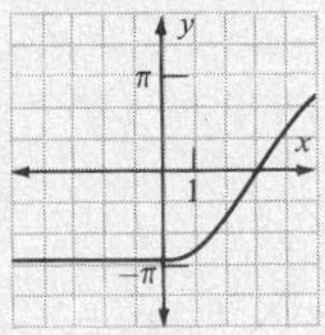

11. $\dfrac{\pi}{2} + \pi k$ 13. πk 15. πk 17. $\dfrac{\pi}{4}$ 19. $\dfrac{\pi}{4}$ 21. $\dfrac{3\pi}{4}$ 23. $\dfrac{\pi}{3}$ 25. $\dfrac{\pi}{3}$ 27. $-\dfrac{\pi}{6}$

29. $\dfrac{\pi}{6}$ 31. $A = \dfrac{1}{2}$ 33. $E = -1$ 35. $I = \dfrac{4}{5}$ 37. $M = -\dfrac{5}{12}$ 39. $P = \dfrac{2\sqrt{6}}{5}$

41. $S = 0$ 43. $U = -\dfrac{\sqrt{2}}{2}$ 45. $\dfrac{\sqrt{1 - x^2}}{x}$ 47. x 49. $2x$ 51. $\theta \approx 11.31° \approx 11°19'$

53. If $M = R_1 + R_2$, then the circles are tangent and $\theta = \pi/2$. Substitute and verify.

55. $\theta = \text{Csc}^{-1} \dfrac{V}{\pi R^2 H}$ or $\text{Sin}^{-1} \dfrac{\pi R^2 H}{V}$ 57. $\theta = \text{Sin}^{-1} \dfrac{1}{m}$ 59. $\theta \approx 19.77° \approx 19°46'$

61. $\theta = 2 \tan^{-1} \frac{1}{18} \approx 6.36° \approx 6°22'$

Exercise Set 3.5, page 154

1. 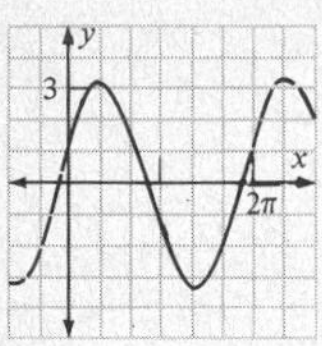

3. 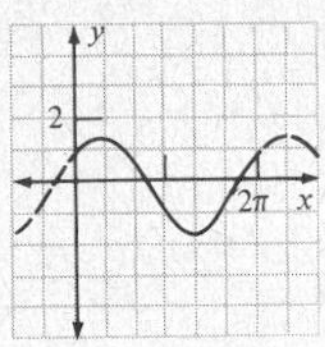

5. 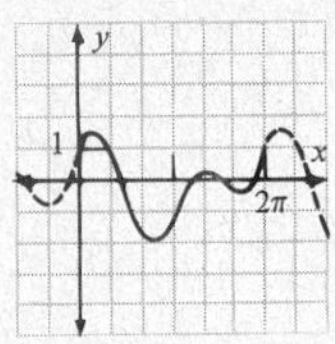

7. 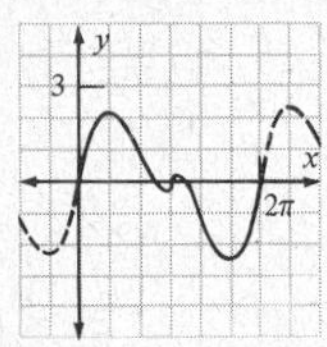

9. 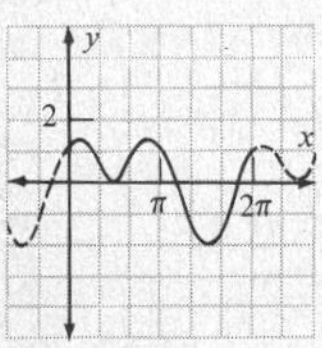

11. 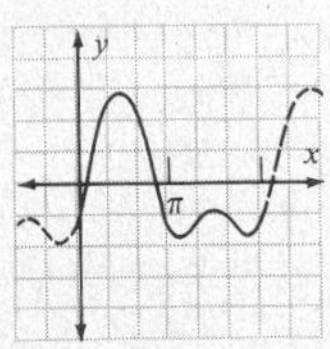

13. 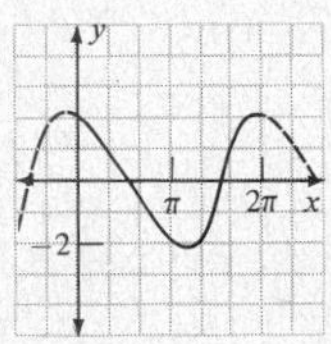

15. 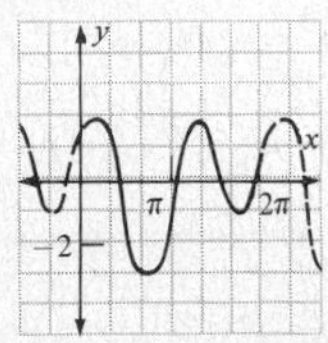

17. 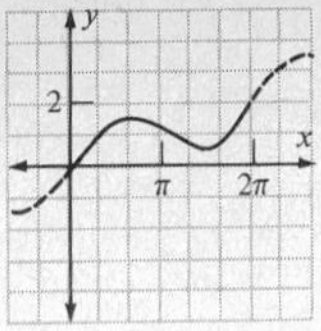

19. 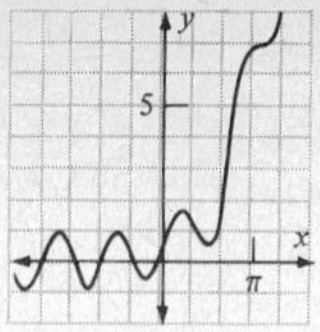

21. 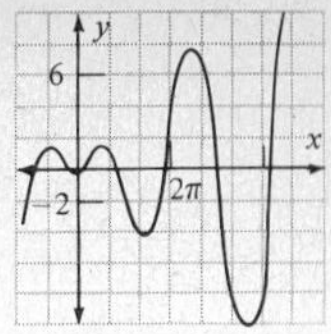

23. 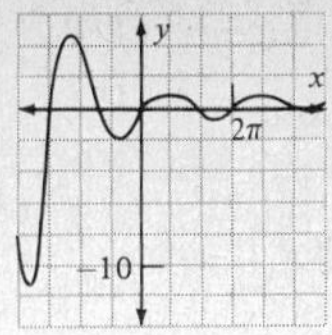

25. 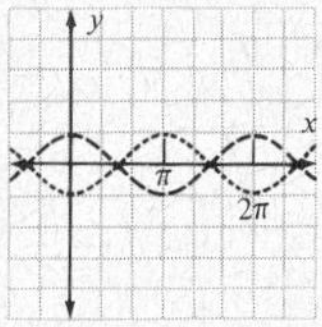

27. 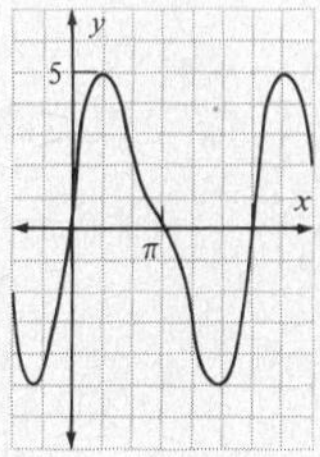

29. 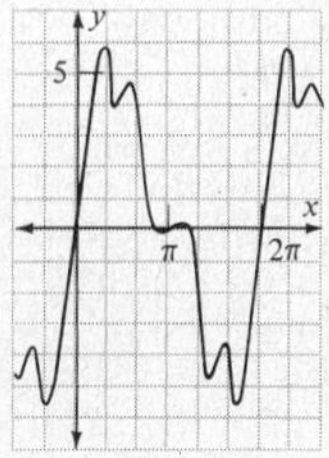

31. 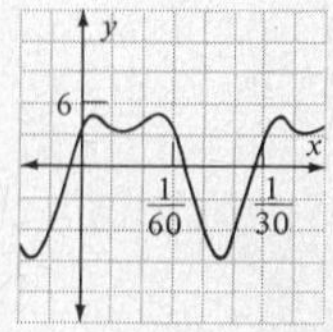

33. Answers will vary. One such equation is

$$y = \left| \left[\frac{x}{\pi} \right] \sin 2x \right|$$

where [ ] is the greatest integer function.

Chapter Review, page 157

1. amplitude, frequency 2. period 3. h, k 4. π/b 5. asymptote
6. $y - k > a \cos b(x - h)$ 7. $2\pi/b$ 8. $x = \sin y$, $-\pi/2 \le y \le \pi/2$ 9. angle, tangent
10. adding, ordinates 11. $y = \frac{2}{5} \sin 3\left(x - \frac{2x}{3}\right)$; $\left(\frac{2\pi}{3}, 0\right)$, $a = \frac{2}{5}$, $p = \frac{2\pi}{3}$
12. $y - 2 = \frac{3}{2} \tan 2(x + \pi)$; $(-\pi, 2)$, $a = \frac{3}{2}$, $p = \frac{\pi}{2}$ 13. $y + 2 = 3 \sec \frac{3}{2}\left(x - \frac{4\pi}{9}\right)$; $\left(\frac{4\pi}{9}, -2\right)$, $a = 3$, $p = \frac{4\pi}{3}$

14. (a) 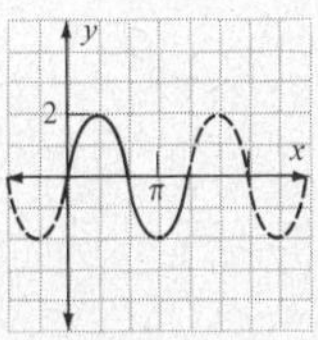

(b) 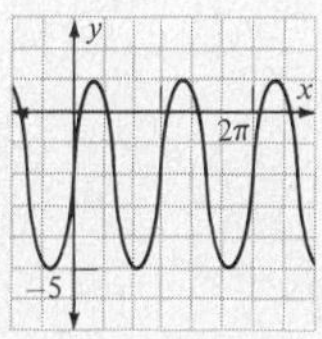

15. (a) 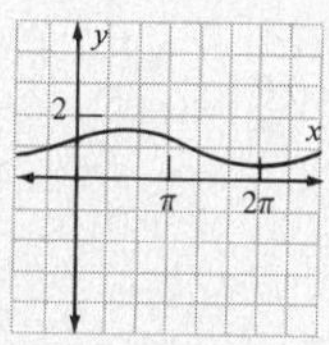

(b) 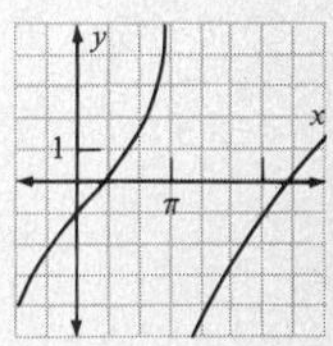

16. (a) 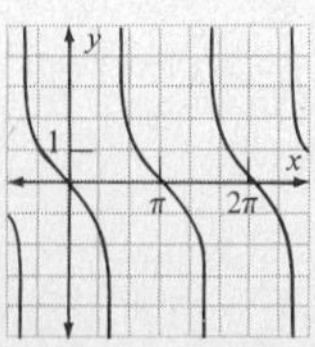

(b) 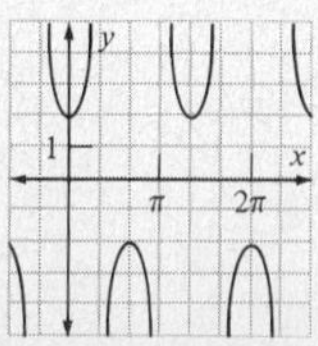

17. (a) 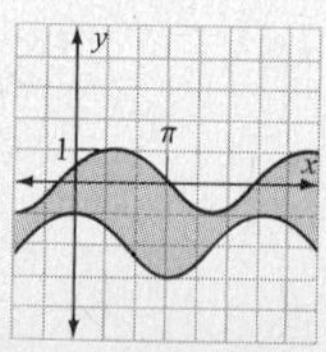

(b) 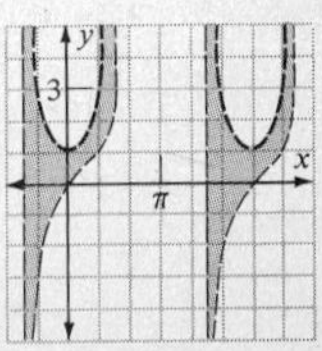

18. (a) 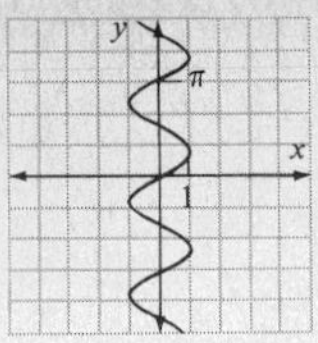

(b) 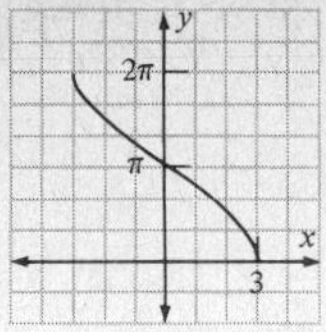

19. (a) 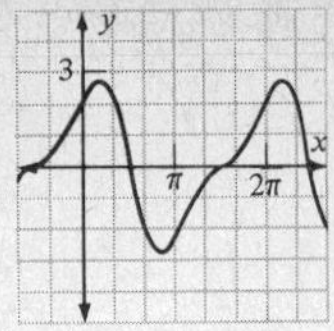

(b) 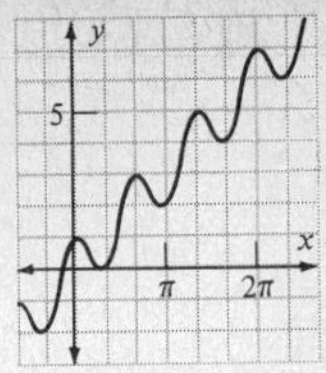

20. $\frac{\sqrt{2}}{2}$ 21. $\frac{17}{15}$ 22. 1 23. $\sqrt{\frac{x}{1-x}}$ $\alpha = \cos^{-1}\frac{hrpg}{2T}$ 25. $\alpha \approx 35.018° \approx 35°1'$

CHAPTER FOUR. Identities and Equations

Exercise Set 4.1, page 166

1. $$\begin{aligned}\cos\theta + \tan\theta\sin\theta &= \cos\theta + \frac{\sin\theta}{\cos\theta}\sin\theta\\ &= \frac{\cos^2\theta + \sin^2\theta}{\cos\theta}\\ &= \frac{1}{\cos\theta}\\ &= \sec\theta \quad \checkmark\end{aligned}$$

3. $$\begin{aligned}1 - 2\sin^2\lambda &= 1 - 2(1 - \cos^2\lambda)\\ &= 1 - 2 + 2\cos^2\lambda\\ &= 2\cos^2\lambda - 1 \quad \checkmark\end{aligned}$$

5. $$\begin{aligned}\cot x + \tan x &= \cot x(1 + \tan^2 x)\\ &= \cot x \sec^2 x \quad \checkmark\end{aligned}$$

7. $$\begin{aligned}(1 - \cos\alpha)(\csc\alpha + \cot\alpha) &= (1 - \cos\alpha)\left(\frac{1}{\sin\alpha} + \frac{\cos\alpha}{\sin\alpha}\right)\\ &= (1 - \cos\alpha)\left(\frac{1 + \cos\alpha}{\sin\alpha}\right)\\ &= \frac{1 - \cos^2\alpha}{\sin\alpha}\\ &= \frac{\sin^2\alpha}{\sin\alpha}\\ &= \sin\alpha \quad \checkmark\end{aligned}$$

9. $$\begin{aligned}\cot^2\beta + \cos^2\beta + \sin^2\beta &= \cot^2\beta + 1\\ &= \csc^2\beta \quad \checkmark\end{aligned}$$

11. $$\begin{aligned}\frac{\tan A - \cot A}{\tan A + \cot A} &= \frac{(\sin A/\cos A) - (\cos A/\sin A)}{(\sin A/\cos A) + (\cos A/\sin A)}\\ &= \frac{(\sin^2 A - \cos^2 A)/(\cos A\sin A)}{(\sin^2 A + \cos^2 A)/(\cos A\sin A)}\\ &= \sin^2 A - \cos^2 A\\ &= \sin^2 A - (1 - \sin^2 A)\\ &= 2\sin^2 A - 1 \quad \checkmark\end{aligned}$$

13. $$\begin{aligned}\frac{\sin x}{\csc x - \cot x} &= \frac{\sin x(\sin x)}{\sin x(\csc x - \cot x)}\\ &= \frac{\sin^2 x}{1 - \cos x}\\ &= \frac{\sin^2 x(1 + \cos x)}{1 - \cos^2 x}\\ &= 1 + \cos x \quad \checkmark\end{aligned}$$

15. $$\begin{aligned}\frac{\cot^2\alpha - 1}{1 + \cot^2\alpha} &= \frac{(\cos^2\alpha/\sin^2\alpha) - 1}{\csc^2\alpha}\\ &= \frac{\cos^2\alpha - \sin^2\alpha}{\sin^2\alpha\csc^2\alpha}\\ &= \cos^2\alpha - (1 - \cos^2\alpha)\\ &= 2\cos^2\alpha - 1 \quad \checkmark\end{aligned}$$

17. $\dfrac{1 + \csc \beta}{\csc \beta - 1} = \dfrac{\sin \beta(\csc \beta + 1)}{\sin \beta(\csc \beta - 1)}$

$= \dfrac{\sin \beta \csc \beta + \sin \beta}{\sin \beta \csc \beta - \sin \beta}$

$= \dfrac{1 + \sin \beta}{1 - \sin \beta} \quad \surd$

19. $\dfrac{1 - \cos A}{\tan A \cos A} = \dfrac{(1 - \cos A)(1 + \cos A)}{(\sin A/\cos A)[\cos A(1 + \cos A)]}$

$= \dfrac{1 - \cos^2 A}{\sin A(1 + \cos A)}$

$= \dfrac{\sin^2 A}{\sin A(1 + \cos A)}$

$= \dfrac{\sin A}{1 + \cos A} \quad \surd$

21. $\dfrac{1}{1 + \sin \theta} + \dfrac{1}{1 - \sin \theta} = \dfrac{1 - \sin \theta + 1 + \sin \theta}{1 - \sin^2 \theta}$

$= \dfrac{2}{\cos^2 \theta}$

$= 2 \sec^2 \theta \quad \surd$

23. $\dfrac{\cos \lambda}{1 - \csc \lambda} + \dfrac{\cos \lambda}{1 + \csc \lambda} = \dfrac{\cos \lambda(1 + \csc \lambda + 1 - \csc \lambda)}{1 - \csc^2 \lambda}$

$= \dfrac{2 \cos \lambda}{-\cot^2 \lambda}$

$= -2 \cos \lambda \tan^2 \lambda$

$= -2 \cos \lambda \dfrac{\sin \lambda}{\cos \lambda} \tan \lambda$

$= -2 \sin \lambda \tan \lambda \quad \surd$

25. $\dfrac{1}{\cot x} + \dfrac{\cos x}{1 + \sin x} = \dfrac{\sin x}{\cos x} + \dfrac{\cos x}{1 + \sin x}$

$= \dfrac{\sin x + \sin^2 x + \cos^2 x}{\cos x(1 + \sin x)}$

$= \dfrac{\sin x + 1}{\cos x(1 + \sin x)}$

$= \dfrac{1}{\cos x}$

$= \sec x \quad \surd$

27. $\dfrac{\sec \alpha}{\cos \alpha} - \dfrac{\tan \alpha}{\cot \alpha} = \sec^2 \alpha - \tan^2 \alpha$

$= 1 + \tan^2 \alpha - \tan^2 \alpha$

$= 1 \quad \surd$

29. $\dfrac{\csc \beta}{\sin \beta} - \dfrac{\cot \beta}{\tan \beta} = \csc^2 \beta - \cot^2 \beta$

$= 1 + \cot^2 \beta - \cot^2 \beta$

$= 1 \quad \surd$

31. $\dfrac{\sin^4 \alpha - \cos^4 \alpha}{\sin^2 \alpha - \cos^2 \alpha} = \dfrac{(\sin^2 \alpha - \cos^2 \alpha)(\sin^2 \alpha + \cos^2 \alpha)}{\sin^2 \alpha - \cos^2 \alpha}$

$= 1 \quad \surd$

33. $\dfrac{\sin^3 \beta + \sin \beta \cos^2 \beta}{\cos \beta} = \dfrac{\sin \beta(\sin^2 \beta + \cos^2 \beta)}{\cos \beta}$

$= \dfrac{\sin \beta}{\cos \beta}$

$= \tan \beta \quad \surd$

35. $\cos^4 x - \sin^4 x = (\cos^2 x - \sin^2 x)(\cos^2 x + \sin^2 x)$
$= \cos^2 x - \sin^2 x \quad \checkmark$

37. $\dfrac{\tan^3 \theta - 1}{\tan \theta - 1} = \dfrac{(\tan \theta - 1)(\tan^2 \theta + \tan \theta + 1)}{\tan \theta - 1}$
$= \sec^2 \theta - 1 + \tan \theta + 1$
$= \sec^2 \theta + \tan \theta \quad \checkmark$

39. $\sin^6 \theta + \cos^6 \theta = (\sin^2 \theta + \cos^2 \theta)(\sin^4 \theta - \sin^2 \theta \cos^2 \theta + \cos^4 \theta)$
$= \sin^4 \theta + 2 \sin^2 \theta \cos^2 \theta + \cos^4 \theta - 3 \sin^2 \theta \cos^2 \theta$
$= (\sin^2 \theta + \cos^2 \theta)^2 - 3 \sin^2 \theta \cos^2 \theta$
$= 1 - 3 \sin^2 \theta \cos^2 \theta \quad \checkmark$

41. $\begin{vmatrix} \tan \theta & \sec \theta \\ \sec \theta & \tan \theta \end{vmatrix} = \begin{vmatrix} \cot \theta & \csc \theta \\ \csc \theta & \cot \theta \end{vmatrix}$
$\tan^2 \theta - \sec^2 \theta = \cot^2 \theta - \csc^2 \theta$
$1 = 1 \quad \checkmark$

43. $\log(\tan \alpha) + \log(\sec \alpha) = \log(\sin \alpha) - 2 \log(\cos \alpha)$

$$\log(\tan \alpha \sec \alpha) = \log\left(\frac{\sin \alpha}{\cos^2 \alpha}\right)$$

$$\log\left(\frac{\sin \alpha}{\cos \alpha} \frac{1}{\cos}\right) =$$

$$\log\left(\frac{\sin \alpha}{\cos^2 \alpha}\right) =$$

Exercise Set 4.2, page 176

1. $A = -\theta$ 3. $C = \dfrac{\pi}{2} - x$ 5. $B = -x$ 7. $A = 50°$ 9. $A = 3x$ 11. $\dfrac{\sqrt{6} + \sqrt{2}}{4}$

13. $\dfrac{\sqrt{6} - \sqrt{2}}{4}$ 15. $\dfrac{\sqrt{2} - \sqrt{6}}{4}$ 17. $\dfrac{4}{\sqrt{2} - \sqrt{6}}$ 19. $\dfrac{\sqrt{6} - \sqrt{2}}{\sqrt{6} + \sqrt{2}}$ or $2 - \sqrt{3}$

21. $\cos(\theta - 180°) = -\cos \theta$ 23. $\sin\left(x + \dfrac{\pi}{6}\right) = \dfrac{1}{2}(\sqrt{3} \sin x + \cos x)$

25. $\sin(\theta + 60°) = \dfrac{1}{2}(\sin x + \sqrt{3} \cos x)$ 27. $\tan(x - \pi) = \tan x$

29. $\cos\left(\theta + \dfrac{\pi}{4}\right) = \dfrac{\sqrt{2}}{2}(\cos \theta - \sin \theta)$ 31. $\dfrac{4}{5}$ 33. $-\dfrac{304}{425}$

35. $\cos(\alpha + \beta) + \cos(\alpha - \beta) = \cos \alpha \cos \beta - \sin \alpha \sin \beta + \cos \alpha \cos \beta + \sin \alpha \sin \beta$
$= 2 \cos \alpha \cos \beta \quad \checkmark$

37. $\sin(\alpha + \beta) \cos \beta - \cos(\alpha + \beta) \sin \beta = \sin \alpha \cos^2 \beta + \cos \alpha \sin \beta \cos \beta$
$- \cos \alpha \cos \beta \sin \beta + \sin \alpha \sin^2 \beta$
$= \sin \alpha(\cos^2 \beta + \sin^2 \beta)$
$= \sin \alpha \quad \checkmark$

39. $\sin(A + B) \sin(A - B) = \sin^2 A \cos^2 B - \cos^2 A \sin^2 B$
$= \sin^2 A(1 - \sin^2 B) - (1 - \sin^2 A) \sin^2 B$
$= \sin^2 A - \sin^2 A \sin^2 B - \sin^2 B + \sin^2 A \sin^2 B$
$= \sin^2 A - \sin^2 B \quad \checkmark$

41. $\tan(\gamma - \delta) - \tan(\delta - \gamma) = \tan(\gamma - \delta) + \tan(\gamma - \delta)$
$$= 2\tan(\gamma - \delta)$$
$$= \frac{2(\tan\gamma - \tan\delta)}{1 + \tan\gamma\tan\delta} \quad \checkmark$$

43. $\dfrac{\cos(x + h) - \cos x}{h} = \dfrac{1}{h}(\cos x\cos h - \sin x\sin h - \cos x)$
$$= \frac{1}{h}[\cos x(\cos h - 1) - \sin x\sin h]$$
$$= \cos x\left(\frac{\cos h - 1}{h}\right) - \sin x\left(\frac{\sin h}{h}\right) \quad \checkmark$$

45–50. Also sketch. 45. $y = \sqrt{2}\cos\left(x + \dfrac{\pi}{4}\right)$ or $\sqrt{2}\cos\left(x - \dfrac{3\pi}{4}\right)$

47. $y = 2\cos 3\left(x - \dfrac{\pi}{18}\right)$ 49. $y = 3\sqrt{2}\cos 4\left(x - \dfrac{\pi}{16}\right)$

51. $\tan\theta = \tan(\beta - \alpha) = \dfrac{\tan\beta - \tan\alpha}{1 + \tan\alpha\tan\beta}$ 53. $\dfrac{\pi}{4}$ 55. $\dfrac{5\pi}{6}$
$$= \frac{m_2 - m_1}{1 + m_1m_2}$$

57. $\sin(\pi - x) = \sin\pi\cos x - \cos\pi\sin x$
$$= (0)\cos x - (-1)\sin x$$
$$= \sin x \quad \checkmark$$

59. $\tan(x + y) = \dfrac{\sin(x + y)}{\cos(x + y)} = \dfrac{\sin x\cos y + \cos x\sin y}{\cos x\cos y - \sin x\sin y}$
$$= \frac{[(\sin x\cos y)/(\cos x\cos y)] + [(\cos x\sin y)/(\cos x\cos y)]}{[(\cos x\cos y)/(\cos x\cos y)] - [(\sin x\sin y)/(\cos x\cos y)]}$$
$$= \frac{\tan x + \tan y}{1 - \tan x\tan y} \quad \checkmark$$

61. $\begin{bmatrix} \cos\alpha & \sin\alpha \\ -\sin\alpha & \cos\alpha \end{bmatrix}\begin{bmatrix} \cos\beta & \sin\beta \\ -\sin\beta & \cos\beta \end{bmatrix} = \begin{bmatrix} \cos\alpha\cos\beta - \sin\alpha\sin\beta & \cos\alpha\sin\beta + \sin\alpha\cos\beta \\ -\sin\alpha\cos\beta - \cos\alpha\sin\beta & -\sin\alpha\sin\beta + \cos\alpha\cos\beta \end{bmatrix}$
$$= \begin{bmatrix} \cos(\alpha + \beta) & \sin(\alpha + \beta) \\ -\sin(\alpha + \beta) & \cos(\alpha + \beta) \end{bmatrix}$$

Exercise Set 4.3, page 184

1. $\sin 6A$ 3. $\cos 4x$ 5. $2\tan 2\theta$ 7. $\dfrac{1}{2}\sin\alpha$ 9. $\cos 2\alpha$ 11. $\sqrt{3}$ 13. 0

15. $\dfrac{\sqrt{2} + \sqrt{6}}{4}$ 17. $\frac{120}{169}, \frac{119}{169}, \frac{120}{119}$ 19. $-\frac{120}{169}, -\frac{119}{169}, \frac{120}{119}$ 21. $-\frac{240}{289}, \frac{161}{289}, -\frac{240}{161}$

23. $\dfrac{13}{\sqrt{170}}, -\dfrac{1}{\sqrt{170}}, -13$ 25. $\dfrac{3}{\sqrt{10}}, \dfrac{-1}{\sqrt{10}}, -3$ 27. $4\cos^3\alpha - 3\cos\alpha$

29. $8\cos^4\theta - 8\cos^2\theta + 1$ 31. $(32\sin^5\theta - 32\sin^3\theta + 6\sin\theta)\sin\left(x + \dfrac{\pi}{2}\right)$

33. $1 + \cos 2\alpha = 1 + (2\cos^2\alpha - 1)$
$= 2\cos^2\alpha \quad \checkmark$

35. $\cos 2\beta + 2\sin^2\beta = 1 - 2\sin^2\beta + 2\sin^2\beta$
$= 1 \quad \checkmark$

37. $\tan\frac{x}{2} = \frac{1 - \cos x}{\sin x}$
$= \frac{1}{\sin x} - \frac{\cos x}{\sin x}$
$= \csc x - \cot x \quad \checkmark$

39. $\sin\theta - \cos\theta\tan\frac{\theta}{2} = \sin\theta - \cos\theta\frac{(1 - \cos\theta)}{\sin\theta}$
$= \frac{\sin^2\theta - \cos\theta + \cos^2\theta}{\sin\theta}$
$= \frac{1 - \cos\theta}{\sin\theta}$
$= \tan\frac{\theta}{2} \quad \checkmark$

41. $\cot 2\lambda = \frac{1}{\tan 2\lambda}$
$= \frac{1 - \tan^2\lambda}{2\tan\lambda}\left(\frac{1/\tan^2\lambda}{1/\tan^2\lambda}\right)$
$= \frac{\cot^2\lambda - 1}{2\cot\lambda} \quad \checkmark$

43. $\cot A\sin 2A = \frac{\cos A}{\sin A}(2\sin A\cos A)$
$= 2\cos^2 A$
$= 1 + 2\cos^2 A - 1$
$= 1 + \cos 2A \quad \checkmark$

45. $2\cot 2B = 2\left(\frac{\cot^2 B - 1}{2\cot B}\right)$
$= \cot B - \frac{1}{\cot B}$
$= \cot B - \tan B \quad \checkmark$

47. $\tan\gamma + \tan\frac{1}{2}\gamma = \frac{\sin\gamma}{\cos\gamma} + \frac{1 - \cos\gamma}{\sin\gamma}$
$= \frac{\sin^2\gamma + \cos\gamma - \cos^2\gamma}{\cos\gamma\sin\gamma}$
$= \frac{\cos\gamma}{\cos\gamma\sin\gamma} - \frac{\cos^2\gamma - \sin^2\gamma}{\cos\gamma\sin\gamma}$
$= \frac{1}{\sin\gamma} - \frac{\cos 2\gamma}{\frac{1}{2}\sin 2\gamma}$
$= \csc\gamma - 2\cot 2\gamma \quad \checkmark$

49. $2(\sin 2\delta\cos\delta - \cos 2\delta\sin\delta)^2 = 2\sin^2\delta$
$= 1 + 2\sin^2\delta - 1$
$= 1 - (1 - 2\sin^2\delta)$
$= 1 - \cos 2\delta \quad \checkmark$

51. $\frac{1 - \cos 2x}{\sin x} = \frac{1 - 2\cos^2 x + 1}{2\sin x\cos x}$
$= \frac{1 - \cos^2 x}{\sin x\cos x}$
$= \frac{\sin^2 x}{\sin x\cos x}$
$= \frac{\sin x}{\cos x}$
$= \tan x \quad \checkmark$

53. $\frac{\sin 2\alpha}{\sin\alpha} - \frac{\cos 2\alpha}{\cos\alpha} = \frac{2\sin\alpha\cos\alpha}{\sin\alpha} - \frac{2\cos^2\alpha - 1}{\cos\alpha}$
$= 2\cos\alpha - 2\cos\alpha + \sec\alpha$
$= \sec\alpha \quad \checkmark$

55. $\dfrac{\cos 2\beta - \sin 2\beta}{\sin \beta \cos \beta} = \dfrac{\cos^2 \beta - \sin^2 \beta - 2 \sin \beta \cos \beta}{\sin \beta \cos \beta}$

$= \dfrac{\cos \beta}{\sin \beta} - \dfrac{\sin \beta}{\cos \beta} - 2$

$= \cot \beta - \tan \beta - 2 \quad \surd$

57. $\dfrac{\sec 2\theta - 1}{\sec 2\theta + 1} = \dfrac{(1/\cos 2\theta) - 1}{(1/\cos 2\theta) + 1} = \dfrac{1 - \cos 2\theta}{1 + \cos 2\theta}$

$= \dfrac{1 - (1 - 2 \sin^2 \theta}{1 + (2 \cos^2 \theta - 1)}$

$= \dfrac{2 \sin^2 \theta}{2 \cos^2 \theta}$

$= \tan^2 \theta \quad \surd$

59. $\dfrac{\sin 5x}{\sin x} - \dfrac{\cos 5x}{\cos x} = \dfrac{\sin 5x \cos x - \cos 5x \sin x}{\sin x \cos x}$

$= \dfrac{\sin 4x}{\frac{1}{2} \sin 2x}$

$= \dfrac{2 \sin 2x \cos 2x}{\frac{1}{2} \sin 2x}$

$= 4 \cos 2x \quad \surd$

61. $\sin \theta > \sin \theta \cos \theta$, $0 < \cos \theta < 1$, θ acute

$\sin \theta \cos \theta = \frac{1}{2} \sin 2\theta$

Therefore, $\sin \theta > \frac{1}{2} \sin 2\theta$, θ acute

63. $\begin{bmatrix} \cos A & \sin A \\ \sin A & \cos A \end{bmatrix} \begin{bmatrix} \sin A & \cos A \\ \cos A & -\sin A \end{bmatrix} = \begin{bmatrix} \sin A \cos A + \sin A \cos A & \cos^2 A - \sin^2 A \\ \sin^2 A + \cos^2 A & \sin A \cos A - \sin A \cos A \end{bmatrix}$

$= \begin{bmatrix} 2 \sin A \cos A & \cos^2 A - \sin^2 A \\ 1 & 0 \end{bmatrix}$

$= \begin{bmatrix} \sin 2A & \cos 2A \\ 1 & 0 \end{bmatrix}$

Exercise Set 4.4, page 192

1. $270° + 360°K$ 3. $0° + 180°K$ 5. $30° + 180°K$, $150° + 180°K$ 7. $45° + 90°K$

9. $90° + 180°K$, $45° + 90°K$ 11. $\{0°, 45°, 180°, 225°\}$ 13. $\{180°\}$

15. $\{45°, 108.435°, 225°, 288.435°\}$ 17. $\{54.736°, 125.264°, 234.736°, 305.264°\}$

19. None 21. $\left\{\frac{\pi}{8}, \frac{3\pi}{8}, \frac{5\pi}{8}, \frac{7\pi}{8}, \frac{9\pi}{8}, \frac{11\pi}{8}, \frac{13\pi}{8}, \frac{15\pi}{8}\right\}$

23, $\left\{0, \frac{\pi}{5}, \frac{2\pi}{5}, \frac{3\pi}{5}, \frac{4\pi}{5}, \ldots, \frac{9\pi}{5}\right\}$

25. $\left\{\frac{3\pi}{4}, \frac{7\pi}{4}\right\}$ 27. $\left\{\frac{5\pi}{6}, \frac{7\pi}{6}\right\}$ 29. $\left\{\frac{\pi}{12}, \frac{5\pi}{12}, \frac{7\pi}{12}, \frac{11\pi}{12}, \frac{13\pi}{12}, \frac{17\pi}{12}, \frac{19\pi}{12}, \frac{23\pi}{12}\right\}$

31. $\left\{\frac{\pi}{4}, \frac{\pi}{3}, \frac{3\pi}{4}, \frac{5\pi}{4}, \frac{4\pi}{3}, \frac{7\pi}{4}\right\}$ 33. $\left\{\frac{\pi}{20}, \frac{3\pi}{20}, \frac{5\pi}{20}, \frac{7\pi}{20}, \frac{9\pi}{20}, \ldots, \frac{39\pi}{20}\right\}$

35. $\left\{\frac{\pi}{3}, \frac{2\pi}{3}, \frac{5\pi}{4}, \frac{7\pi}{4}\right\}$ 37. $\left\{\frac{\pi}{6}, \frac{5\pi}{6}\right\}$ 39. $\left\{0, \frac{2\pi}{7}, \frac{4\pi}{7}, \frac{6\pi}{7}, \frac{8\pi}{7}, \frac{10\pi}{7}, \frac{12\pi}{7}\right\}$

41. {0°, 138.590°, 221.410°}

43. {71.565°, 161.565°, 251.565°, 341.565°} 45. {63.435°, 116.565°, 243.435°, 296.565°}

47. {45°, 71.565°, 225°, 251.565°} 49. {43.622°, 136.378°, 196.852°, 343.148°}

51. 2.584 cu ft 53. $V = \frac{\sqrt{2}\, R^3}{24}(10 - 3\pi)$ 55. $V = R^3 \tan\phi(\sin\theta - \frac{\sin^3\theta}{3} - \theta\cos\theta)$

Chapter Review, page 195

1. $\tan\theta$, $1/\cot\theta$ 2. 1 3. sine, cosine 4. $\sin(x + y)$ 5. $\cos(x - y)$ or $\cos(y - x)$

6. $A\cos B(x - h)$ 7. $\cos 2\theta$ 8. $\pm\sqrt{(1 - \cos x)/2}$ 9. $\tan 2A$

10. Pythagorean identities (or extraneous roots) 11. $\sin x$ 12. $\cos 6y$ 13. 84/85

14. $\sqrt{26}/26$, 120/169

15. $$\frac{1 - \sin^2\theta}{1 + \cot^2\theta} = \frac{\cos^2\theta}{\csc^2\theta} = \sin^2\theta\cos^2\theta \quad \checkmark$$

16. $$\frac{\sin(\alpha + \beta)}{\sin(\alpha - \beta)} = \frac{\sin\alpha\cos\beta + \cos\alpha\sin\beta(1/\cos\alpha\cos\beta)}{\sin\alpha\cos\beta - \cos\alpha\sin\beta(1/\cos\alpha\cos\beta)} = \frac{\tan\alpha + \tan\beta}{\tan\alpha - \tan\beta} \quad \checkmark$$

17. $$\begin{aligned}\sqrt{2}\cos\left(\theta - \frac{3\pi}{4}\right) &= \sqrt{2}\left(\cos\theta\cos\frac{3\pi}{4} + \sin\theta\sin\frac{3\pi}{4}\right)\\ &= \sqrt{2}\left(-\frac{1}{\sqrt{2}}\cos\theta + \frac{1}{\sqrt{2}}\sin\theta\right)\\ &= -\cos\theta + \sin\theta\\ &= \sin\theta - \cos\theta \quad \checkmark\end{aligned}$$

18. $$\begin{aligned}\sin 3A &= \sin(A + 2A)\\ &= \sin A\cos 2A + \cos A\sin 2A\\ &= \sin A - 2\sin^2 A + 2\sin A\cos^2 A\\ &= \sin A - 2\sin^2 A + 2\sin A - 2\sin^3 A\\ &= 3\sin A - 4\sin^3 A \quad \checkmark\end{aligned}$$

19. $$\begin{aligned}\sin 2B(1 + \cot^2 B) &= 2\sin B\cos B(\csc^2 B)\\ &= 2\sin B\cos B\left(\frac{1}{\sin^2 B}\right)\\ &= \frac{2\cos B}{\sin B}\\ &= 2\cot B \quad \checkmark\end{aligned}$$

20. (a) $\frac{\pi}{2} + k\pi, \frac{2\pi}{3} + 2k\pi, \frac{4\pi}{3} + 2k\pi$ (b) $71.163° + 360°K$, $288.837° + 360°K$

CHAPTER FIVE. Oblique Triangles

Exercise Set 5.1, page 203

1. 973 cm² (972.70042) 3. 5129 km² (5129.0840) 5. 35 (35.461882)
7. 9.56 (equilateral triangle) 9. 2626 (2625.8304) 11. 22.4 (22.377331)
13. 0.89 (0.88809221) 15. 168.9 (168.92241) 17. 1.165 (1.1651660)
19. 771 (770.82500) 21. 91.1 23. 0.51 25. 6943.1 27. 0.610 29. 275250
31. 25500 ft³ 33. 1072 yd 35. 22 cm (22.114969); 18 cm (17.637007)
37. 5.4 km (5.4300116) 39. 339 ft (339.12126)

Exercise Set 5.2, page 211

1. Two 3. Zero 5. One 7. 39° (38.682187) or 141° (141.31781)
9. 52° (52.181088) or 128° (127.81891) 11. $A \approx 28°$ (28.047400) 13. $c \approx 117$
15. $E \approx 79°$ 17. $G \approx 129°$ 19. $k \approx 44$ 21. $m \approx 78$ 23. $n \approx 74$ 25. $Q \approx 59°$
27. $s \approx 69$ 29. $U \approx 75°$ 31. 100.0 cm (99.997904) 33. 2.8 mi (2.8271104)
35. S85°31′E 37. 590 ft (589.55038)
39. "IT IS THE MAN NOT THE METHOD THAT SOLVES THE PROBLEM" (H. Maschke)

Exercise Set 5.3, page 218

1. 40 (39.563959) 3. 60° 5. 95 (95.241323) 7. 58° (58.411864) 9. 4 or 19
11. $a \approx 58.5$ 13. $C \approx 88.6°$ 15. $E \approx 62.8°$ 17. $G \approx 40.4°$ 19. $I \approx 49.2°$
21. $m \approx 56.4$ 23. $R \approx 10.8°$ 25. $T \approx 27.3°$ 27. $W \approx 29.3°$ 29. $x \approx 15.3$
31. 624 mi (624.12433) 33. 346 ft (346.25885) 35. 660 mi (661.21592)
37. 1668 yd (1667.5420) or 1929 yd 39. 898 yd (897.58795)
41. "ALGEBRA IS BUT WRITTEN GEOMETRY AND GEOMETRY IS BUT FIGURED ALGEBRA." (Sophie Germain)

Exercise Set 5.4, page 228

1. 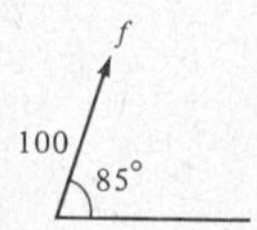

3. 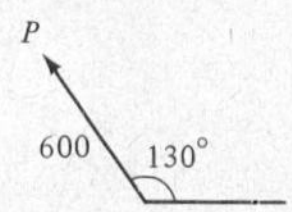

5. 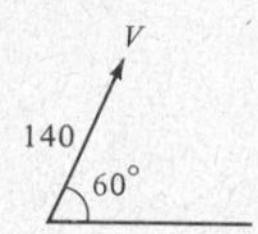

7. 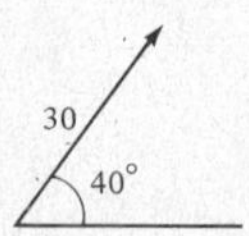

9. 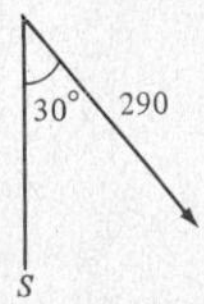

11. 270 mi N22°W 13. 192 mi/h at 261° 15. 230 lb; 40.6°
17. Horizontal, 227 kg; vertical, 446 kg 19. 170 lb 21. 88° (87.674862)
23. 33 mi/h (33.416067), 224 mi/h (224.15672)

25. 3 mi/h (3.3416166), 9 mi/h (9.0625337)

27. 22.4 mi/h (22.360679), N10°W (10.304846) 29. 484 mi (484.38637), S14°43′W

31. 28 lb (28.315594)

33. 22 lb "wasted," because only 73 lb (72.774222) horizontal force required

35. 26 tons (28.977774 − 3) 37. 15° (14.712458) 39. 2075 lb (2075.3168)

41.

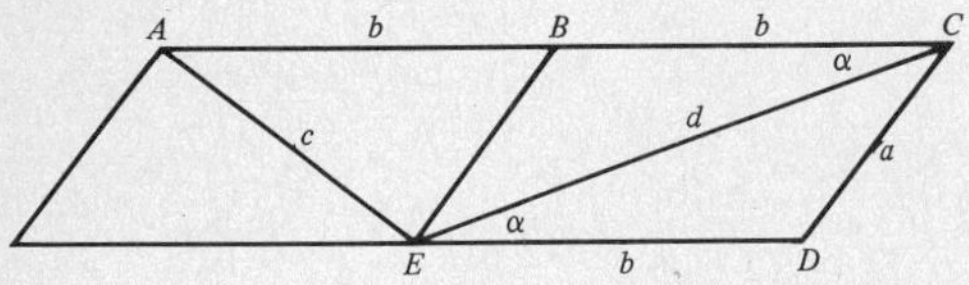

$$\begin{aligned}
\triangle ACE: \quad & c^2 = 4b^2 + d^2 - 4bd\cos\alpha \\
\triangle DEC: \quad & a^2 = b^2 + d^2 - 2bd\cos\alpha \\
& 2a^2 = 2b^2 + 2d^2 - 4bd\cos\alpha \\
\hline
& c^2 - 2a^2 = 2b^2 - d^2 \\
& c^2 + d^2 = 2a^2 + 2b^2 = 2(a^2 + b^2)
\end{aligned}$$

Chapter Review, page 231

1. oblique 2. one-half; product 3. proportional 4. ambiguous 5. no
6. right 7. one or two 8. cosines 9. clockwise; north 10. resultant
11. (a) 5805 (5804.5031) (b) 276.2 (276.22148) 12. (a) 45 (45.441620)
(b) 33.63 (33.633797) (c) No such triangle (d) 59° (e) 66° (f) 49.6° (g) No triangle
13. 44.8° 14. 70.6° 15. 64.0 16. 3.94 acres 17. 41 in., 72 in. 18. 142 mi
19. 16 mi (15.509843) 20. 11 mi (11.474031)

CHAPTER SIX. Complex Numbers and Polar Coordinates

Exercise Set 6.1, page 239

1. -1 3. i 5. 1 7. $-i$ 9. i 11. $2i$ 13. $11i$ 15. $3i\sqrt{3}$ 17. $4i\sqrt{3}$
19. $2i\sqrt{30}$ 21. $5 + i$ 23. $-1 + 5i$ 25. $10 + 0i$ or 10 27. $7 + i$ 29 $4 - 7i$
31. $16 - 11i$ 33. $4 + 3i$ 35. $8 + 6i$ 37. $8 + 38i$ 39. $2 - 4i$ 41. $\frac{1}{2} - \frac{1}{2}i$
43. $-\frac{1}{2} + \frac{1}{2}i$ 45. $0 - i$ or $-i$ 47. $\frac{1}{2} + \frac{3}{2}i$ 49. $-\frac{3}{10} - \frac{11}{10}i$ 51. $x = 2, y = -3$
53. $x = 3, y = 5$ 55. $x = -2, y = 5$ 57. $x = 21, y = 1$ 59. $x = 4, y = 17$
61. $G = -4$ 63. $K = 5i$ 65. $M = 2i$ 67. $P = 4i$ 69. $Y = 3i$ 71. $S = -2i$
73. $A = -i$ 75. $E = -1$ 77. "THE IMAGINARY IS ONE OF THE MASTER KEYS OF PHYSICAL SCIENCE." (Thomas Hill)
79. $$\begin{aligned}(a + bi)(c + di) &= ac + bdi^2 + bci + adi \\ &= (ac - bd) + (bc + ad)i\end{aligned}$$
which is a complex number because $ac - bd$ and $bc + ad$ are real.

Exercise Set 6.2, page 246

1. 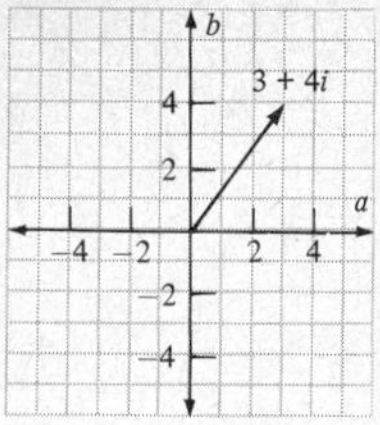

3. 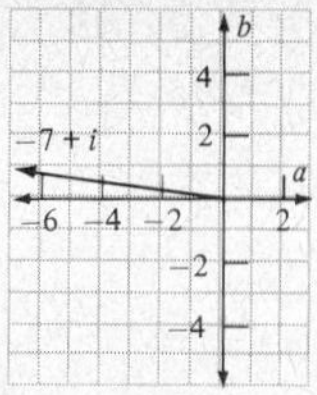

5. 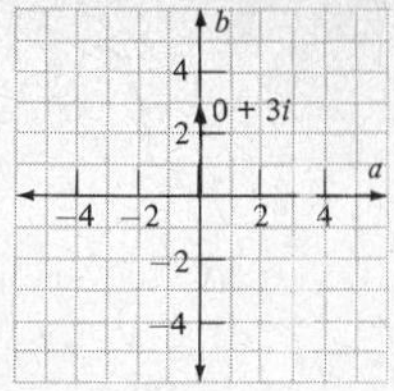

7. 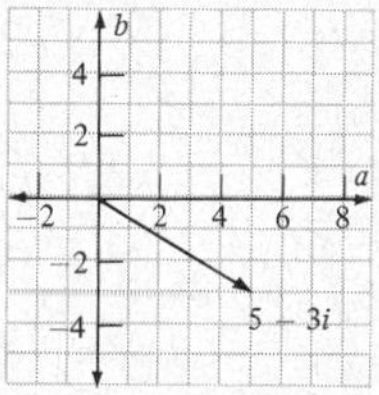

9. 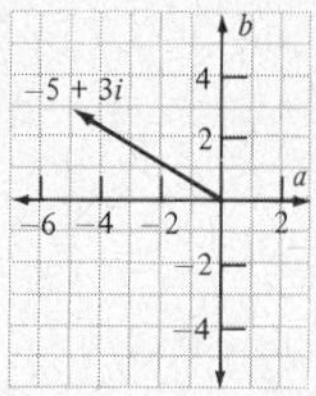

11. $\sqrt{2}$ cis 45° or $\sqrt{2}$ cis $\dfrac{\pi}{4}$ 13. 2 cis 0° or 2 cis 0   15. 2 cis 330° or 2 cis $\dfrac{11\pi}{6}$

17. 4 cis 330° or 4 cis $\dfrac{11\pi}{6}$ 19. 6 cis 240° or 6 cis $\dfrac{4\pi}{3}$ 21. $\dfrac{3}{2} + \dfrac{3\sqrt{3}}{2}i$ 23. $\sqrt{2} + \sqrt{2}i$

25. $-1 + i$ 27. $-2 + 2\sqrt{3}i$ 29. $\dfrac{\sqrt{15}}{2} - \dfrac{\sqrt{5}}{2}i$ 31. 5 cis 53°8′   33. $0.62 + 1.90i$

35. 17 cis 118°4′   37. $-0.86 - 2.50i$ 39. 2.24 cis 309°14′

Exercise Set 6.3, page 252

1. 6 cis 45°   3. 35 cis 70°   5. $3\sqrt{2}$ cis 167°   7. 2 cis 70°   9. $\dfrac{\sqrt{3}}{3}$ cis 87°

11. 64 cis π 13. 8 cis $\dfrac{\pi}{2}$ 15. 32 cis 160°   17. cis 120°   19. 32 cis 90°

21. 3 cis 22.5°
3 cis 202.5°

23. 3 cis $\dfrac{\pi}{9}$
3 cis $\dfrac{7\pi}{9}$
3 cis $\dfrac{13\pi}{9}$

25. $2^{1/10}$ cis 9°
$2^{1/10}$ cis 81°
$2^{1/10}$ cis 153°
$2^{1/10}$ cis 225°
$2^{1/10}$ cis 297°

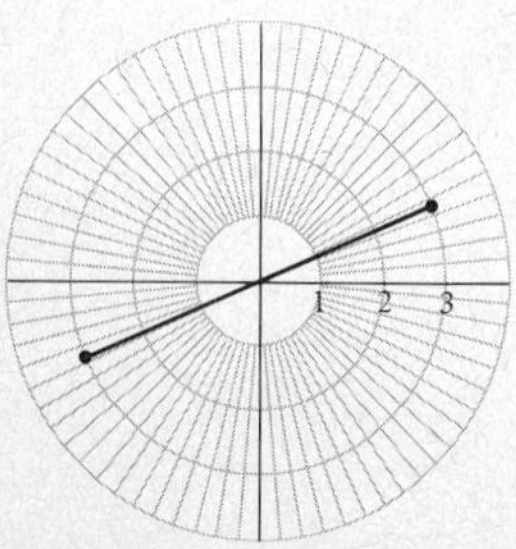

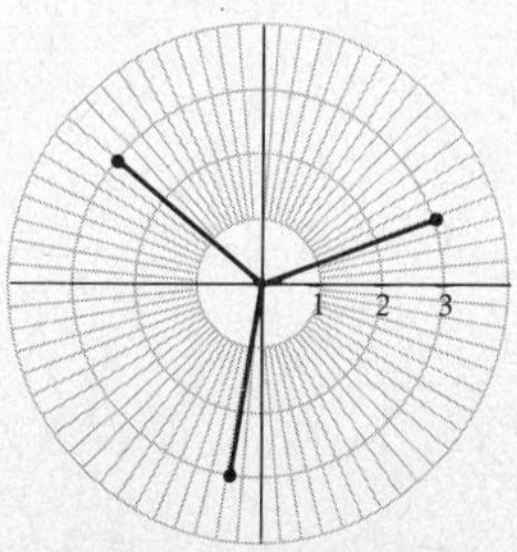

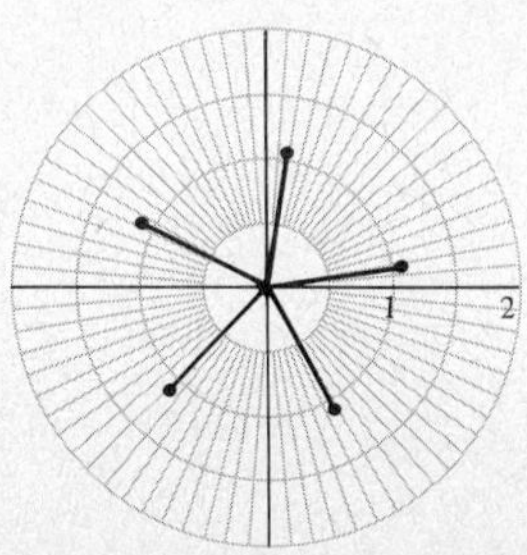

27. $1 \text{ cis } 90° = i$

$1 \text{ cis } 210° = -\frac{\sqrt{3}}{2} - \frac{1}{2}i$

$1 \text{ cis } 330° = \frac{\sqrt{3}}{2} - \frac{1}{2}i$

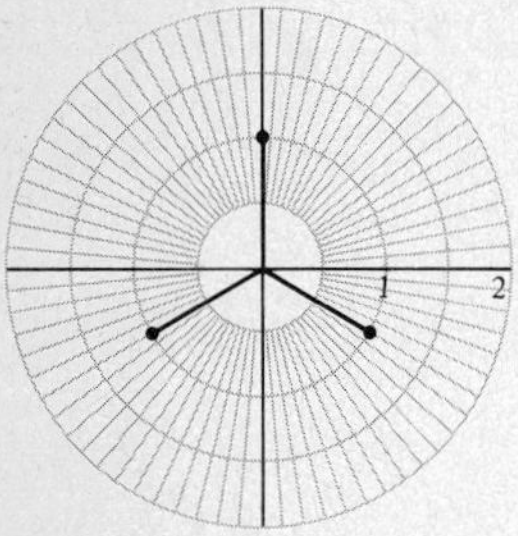

29. $1 \text{ cis } 0° = 1$

$1 \text{ cis } 90° = i$

$1 \text{ cis } 180° = -1$

$1 \text{ cis } 270° = -i$

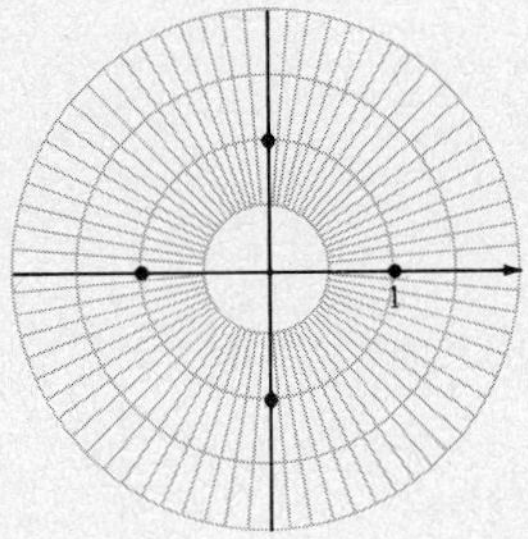

31. $1 \text{ cis } 30° = \frac{\sqrt{3}}{2} + \frac{1}{2}i$

$1 \text{ cis } 150° = -\frac{\sqrt{3}}{2} + \frac{1}{2}i$

$1 \text{ cis } 270° = 0 - i$

33. $2 \text{ cis } 110°$

$2 \text{ cis } 230°$

$2 \text{ cis } 350°$

35. $1 \text{ cis } 60° = \frac{1}{2} + \frac{\sqrt{3}}{2}i$

$1 \text{ cis } 180° = -1$

$1 \text{ cis } 300° = \frac{1}{2} - \frac{\sqrt{3}}{2}i$

37. $1 \text{ cis } 45° = \frac{1}{\sqrt{2}} + \frac{1}{\sqrt{2}}i$

$1 \text{ cis } 225° = -\frac{1}{\sqrt{2}} - \frac{1}{\sqrt{2}}i$

39. $3 \text{ cis } 7°$

$3 \text{ cis } 79°$

$3 \text{ cis } 151°$

$3 \text{ cis } 223°$

$3 \text{ cis } 295°$

41.

$$x^3 = 1$$

$$x^3 - 1 = 0$$

$$(x - 1)(x^2 + x + 1) = 0$$

$$x - 1 = 0 \quad x^2 + x + 1 = 0$$

$$x = 1 \qquad x = \frac{-1 \pm \sqrt{1 - 4}}{2}$$

$$x = \frac{-1}{2} \pm \frac{\sqrt{3}}{2}i$$

Exercise Set 6.4, page 260

1. $(2, -310°)$
   $(-2, -130°)$
   $(-2, 230°)$

3. $(1.5, -200°)$
   $(-1.5, 340°)$
   $(-1.5, -20°)$

5. $(3, 145°)$
   $(3, -215°)$
   $(-3, -35°)$

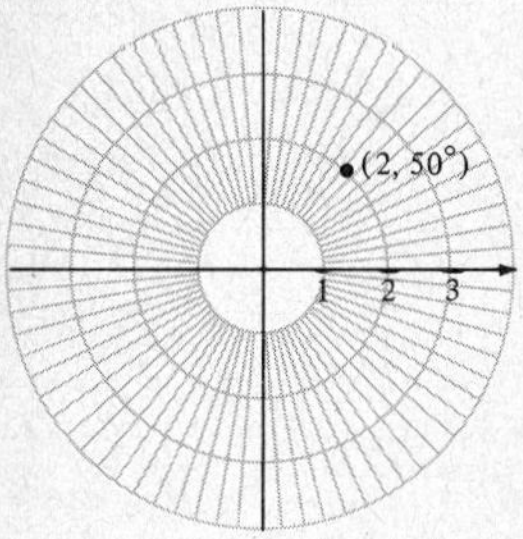

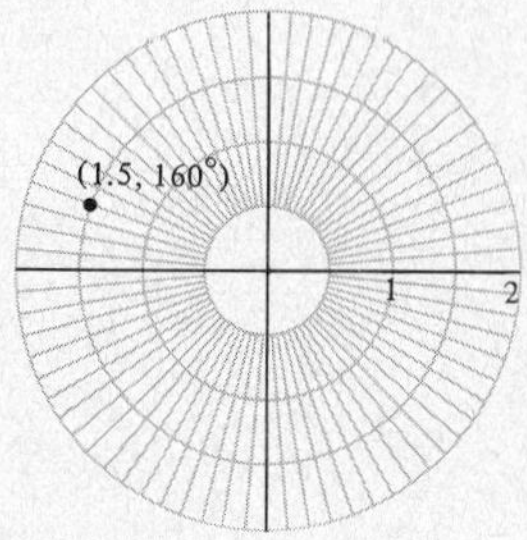

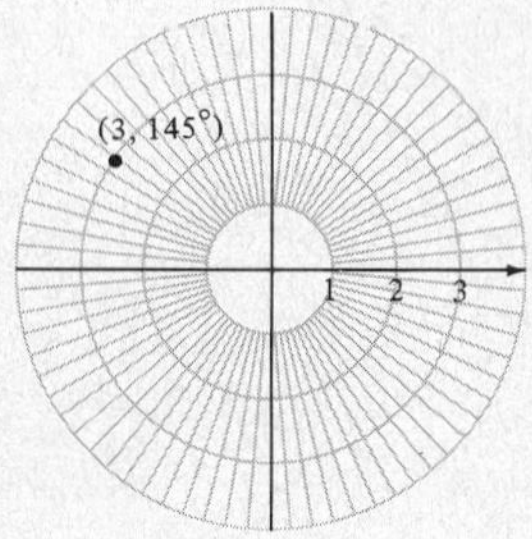

7. $\left(-\sqrt{3}, \dfrac{7\pi}{4}\right)$
   $\left(\sqrt{3}, \dfrac{5\pi}{4}\right)$
   $\left(\sqrt{3}, -\dfrac{5\pi}{4}\right)$

9. $\left(3, \dfrac{4\pi}{5}\right)$
   $\left(3, -\dfrac{6\pi}{5}\right)$
   $\left(-3, -\dfrac{\pi}{5}\right)$

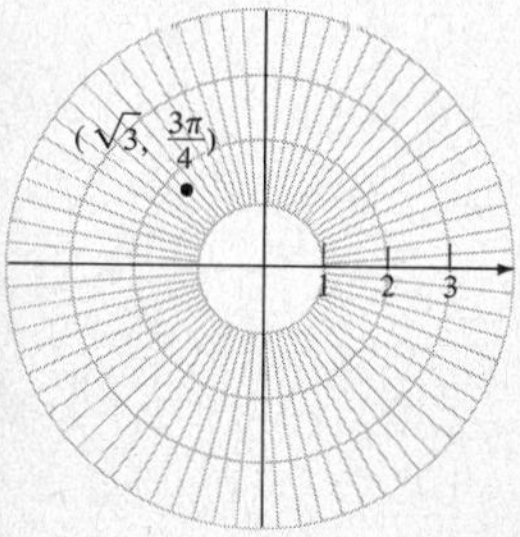

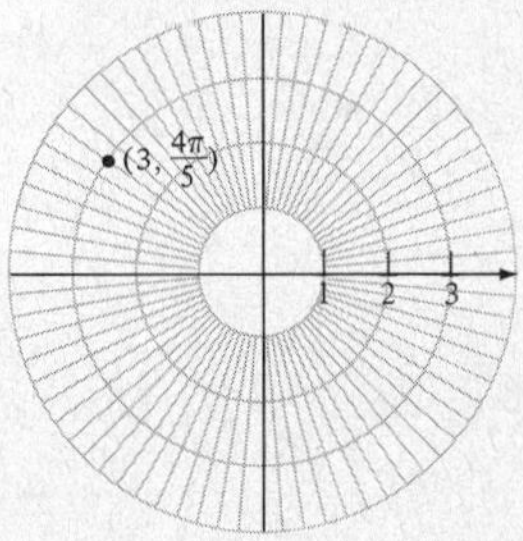

11. $(2, 2\sqrt{3})$ 13. $(-3\sqrt{3}, -3)$ 15. $(1, -1)$ 17. $(0, 0)$ 19. $(-1, -\sqrt{3})$

21. $(2, 0°)$ 23. $(3\sqrt{2}, 45°)$ 25. $(1, 120°)$ 27–30. Given to the nearest minute.

27. $(5, 126°52')$ 29. $(29, 226°24')$ 31. $y = 2$ 33. $x^2 + y^2 = 16$

35. $(x^2 + y^2)^2 = 4(x^2 - y^2)$ 37. $x^2 + y^2 = (y + 9)^2$ 39. $y - x = 2$ 41. $r \cos \theta = -2$

43. $r^2 = 25$ 45. $r \sin 2\theta = 2 \sin \theta + 2 \cos \theta$ 47. $r^2 = \cos^2 2\theta$

49. $9r^2 \cos^2 \theta + r^2 \sin^2 \theta = 9$

51.
$$\begin{aligned} r &= 2a \cos \theta \\ r^2 &= 2ar \cos \theta \\ x^2 + y^2 &= 2ax \\ x^2 - 2ax + y^2 &= 0 \\ x^2 - 2ax + a^2 + y^2 &= a^2 \\ (x - a)^2 + y^2 &= a^2 \end{aligned}$$

Circle center $(a, 0)$
Radius a

Exercise Set 6.5, page 271

1.

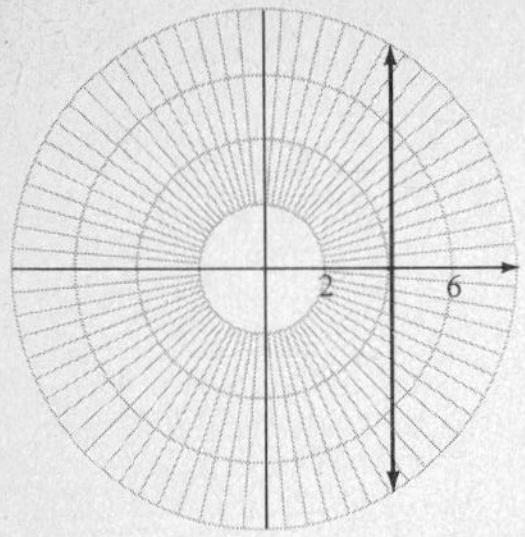

3.

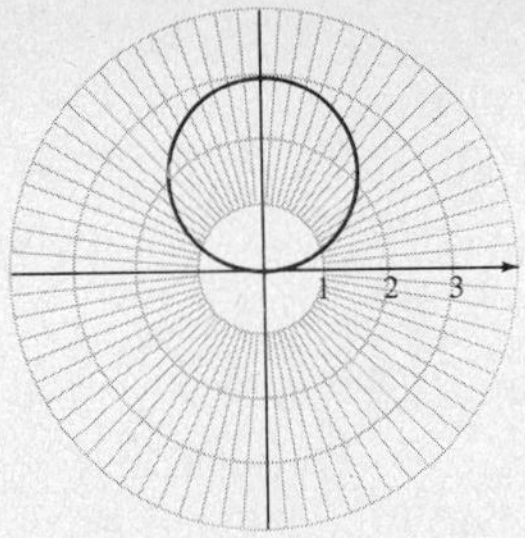

5.

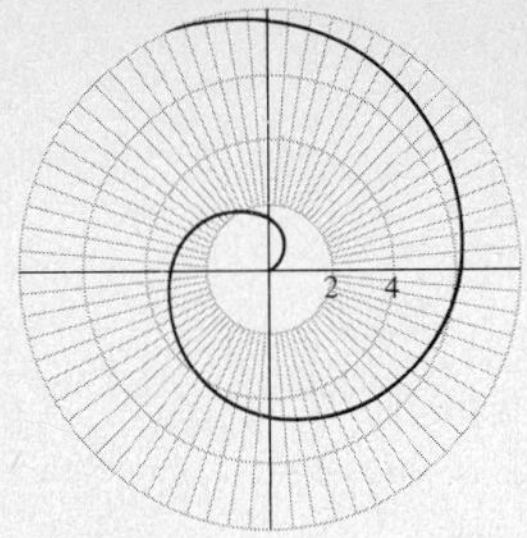

7.

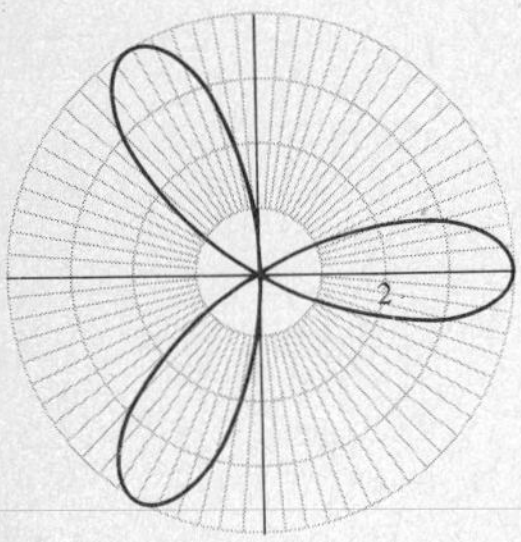

9.

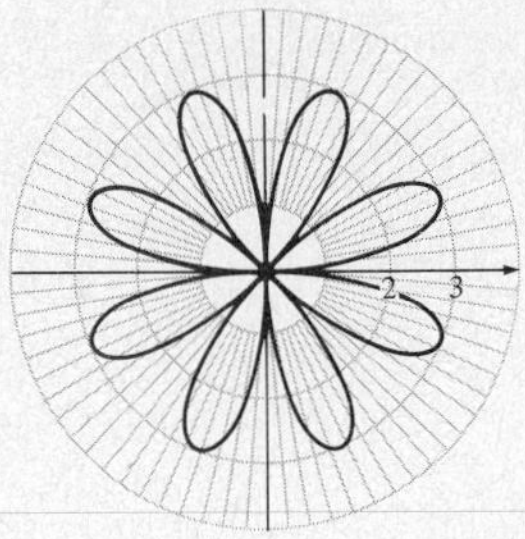

11.

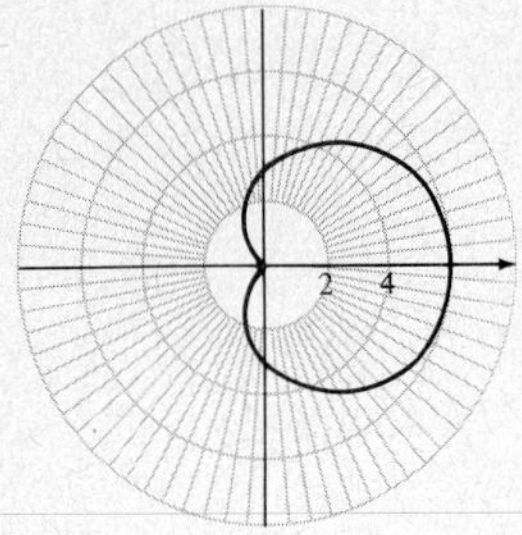

13.

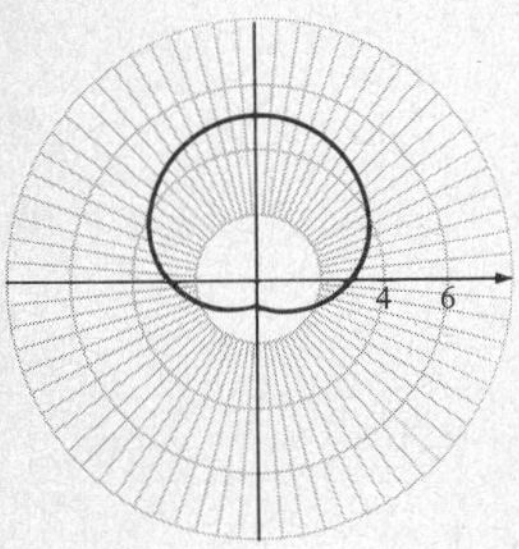

15.

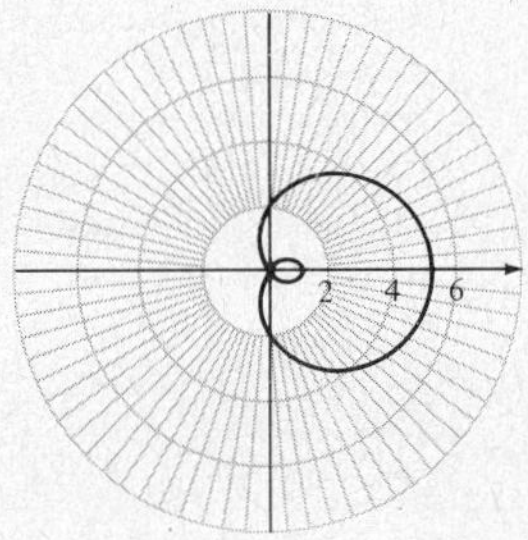

17.

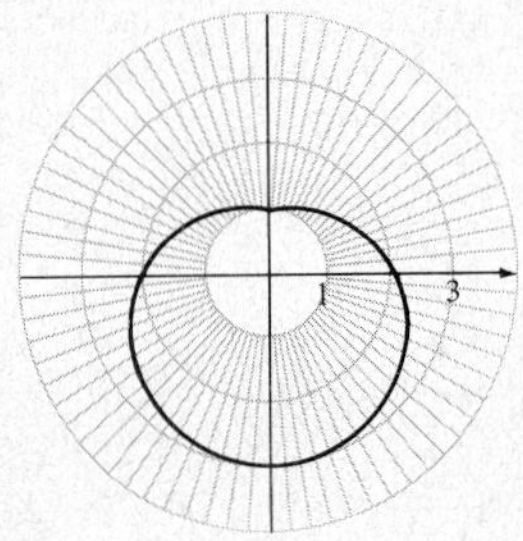

19.

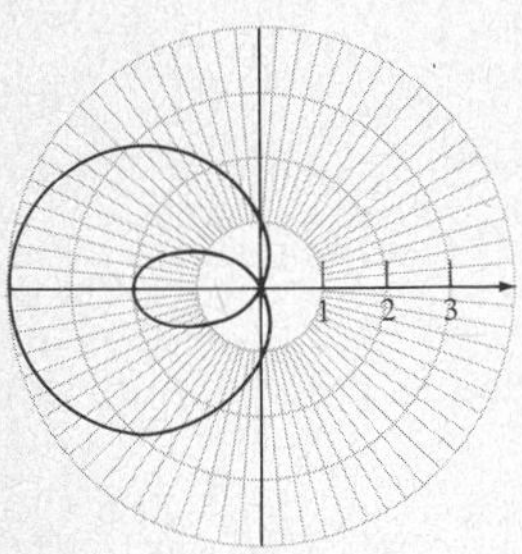

21.

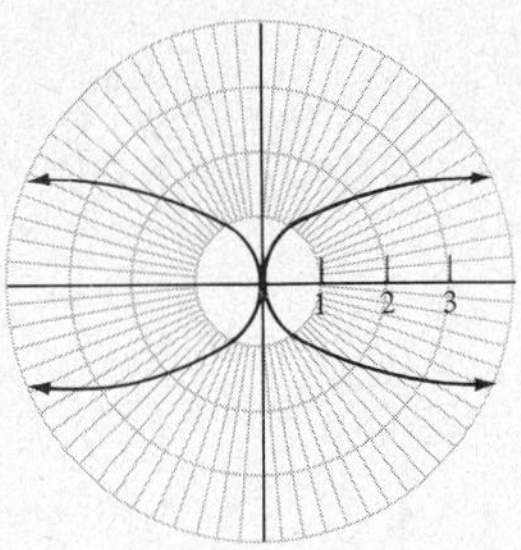

23.

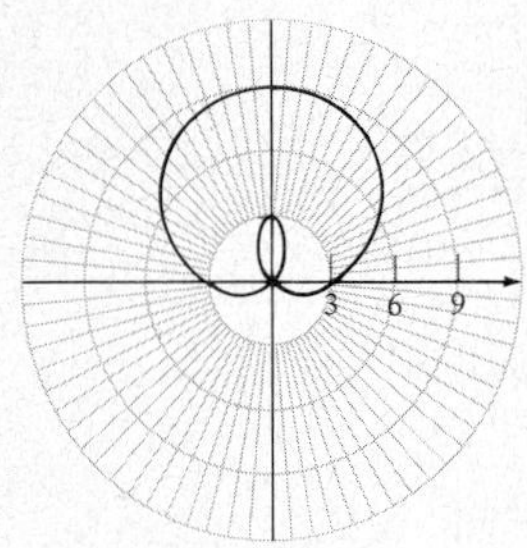

25.

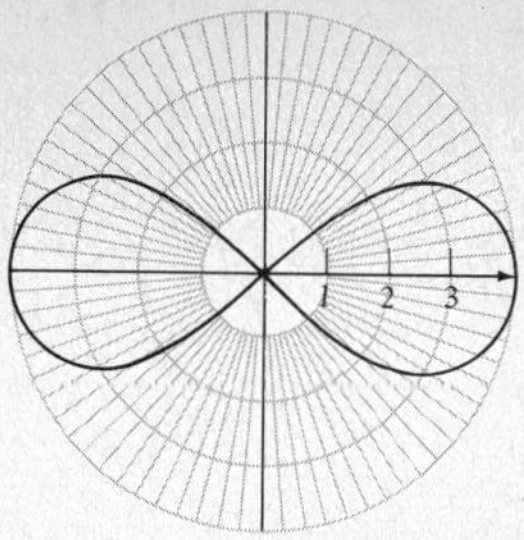

27.

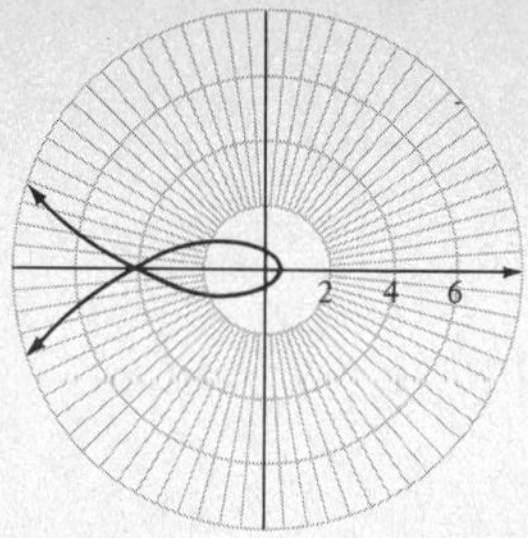

29.

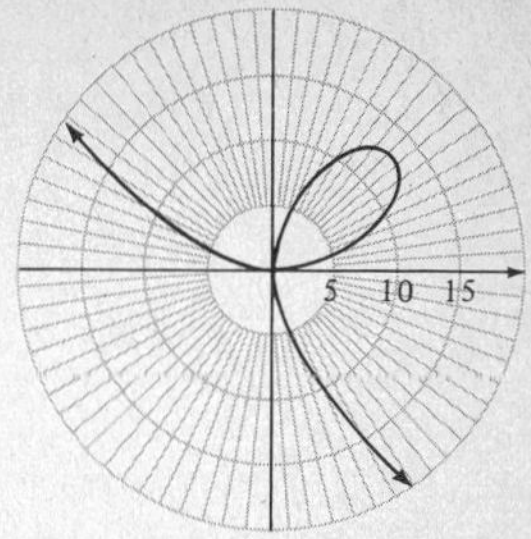

31.

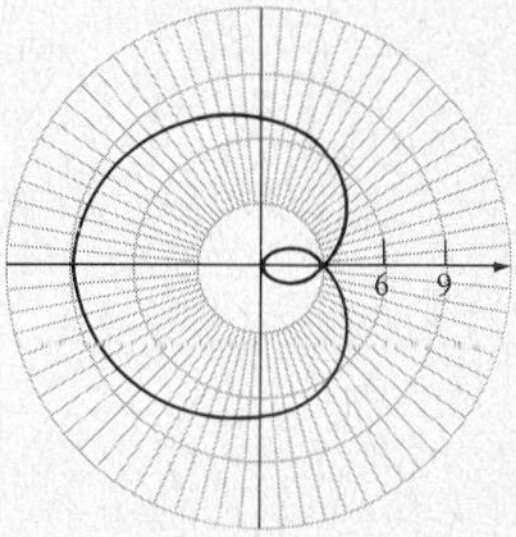

33. $r = \sin\theta \qquad \sin\theta = \cos\theta$

$r = \cos\theta \qquad \dfrac{\sin\theta}{\cos\theta} = 1$

$\tan\theta = 1, \ \theta = \dfrac{\pi}{4} + \pi k$

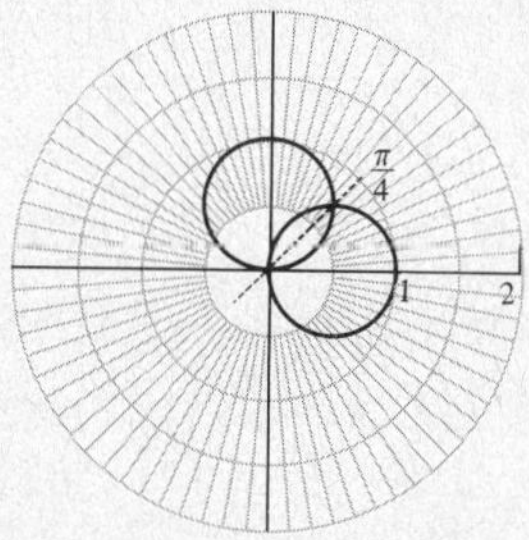

Both (0, 0) and $\left(\dfrac{\sqrt{2}}{2}, \dfrac{\pi}{4}\right)$ appear to be solutions on the graph. Actually, (0, 0) is a point on $r = \sin\theta$ and $\left(0, \dfrac{\pi}{2}\right)$ is a point on $r = \cos\theta$. So while they both pass through the pole, they do so at different "times."

35. The points (r, θ) and $(r, \theta + \pi)$ are symmetric with respect to the pole. Hence $r = f(\theta)$ is symmetric with respect to the pole if $r = f(\theta)$ and $r = f(\theta + \pi)$ are equivalent.

37.

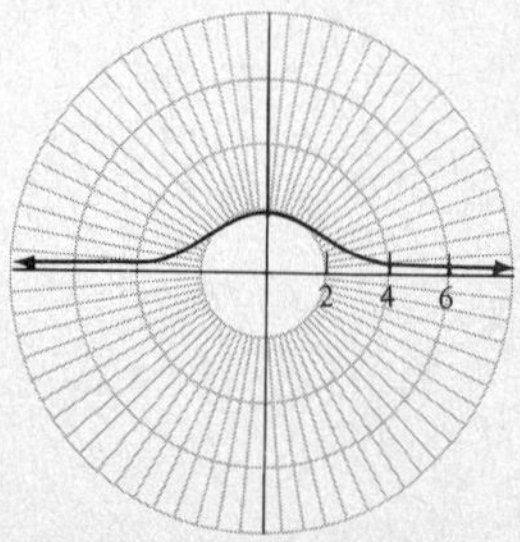

Chapter Review, page 273

1. complex part 2. 1 3. $a = c$, $b = d$ 4. conjugate
5. polar (or trigonometric), $r \text{ cis } \theta$ 6. $\sqrt{a^2 + b^2}$, $\arctan b/a$ 7. $(A/B) \text{ cis}(\alpha - \beta)$
8. $r^n \text{ cis } n\theta$ 9. polar axis 10. same 11. (a) $3i\sqrt{3}$ (b) i (c) $2 + i$ (d) $1 - i$
(e) $3 + i$ (f) $-1 - 5i$ (g) $8 - i$ (h) $11 + 3i$ (i) $\frac{1}{5} + \frac{2}{5}i$ (j) $-\frac{4}{13} - \frac{7}{13}i$ 12. (a) $2\sqrt{2}$ cis 315°
(b) 2 cis 120° 13. (a) $-\frac{3\sqrt{2}}{2} - \frac{3\sqrt{2}}{2}i$ (b) $-\frac{\sqrt{3}}{2} - \frac{3}{2}i$

14. (a) $6 \text{ cis } \frac{5\pi}{6}$ (b) $\frac{\sqrt{3}}{3}$ cis 145° 15. (a) $8\sqrt{2}$ cis 315° (b) $4\sqrt{2}$ cis 115°

16. (a) $\sqrt{5}$ cis 15°
$\sqrt{5}$ cis 105°
$\sqrt{5}$ cis 195°
$\sqrt{5}$ cis 285°

(b) 2 cis 90° = $2i$
2 cis 210° = $-\frac{\sqrt{3}}{2} - \frac{1}{2}i$
2 cis 330° = $\frac{\sqrt{3}}{2} - \frac{1}{2}i$

(c) 1 cis 22.5°
1 cis 112.5°
1 cis 202.5°
1 cis 292.5°

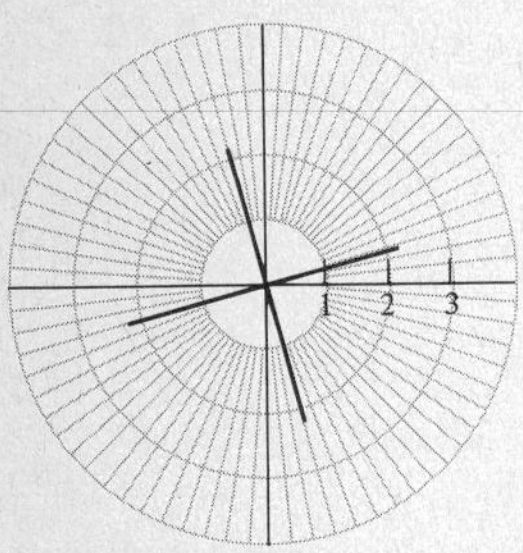

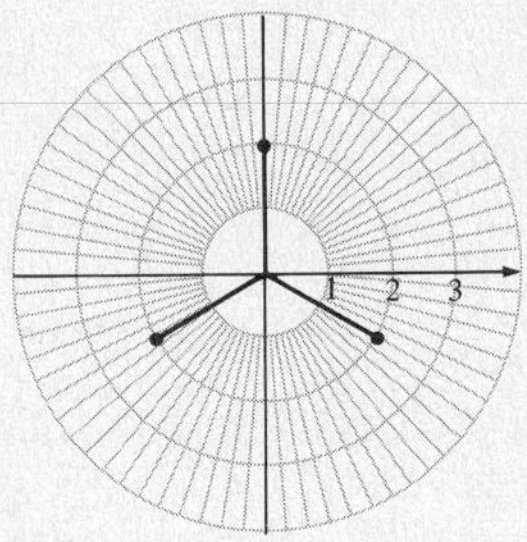

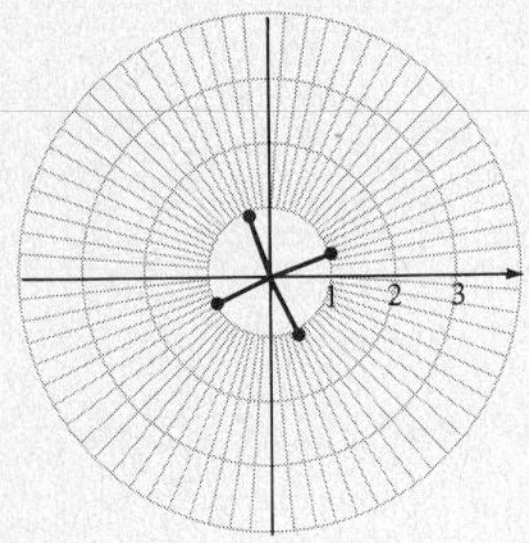

17. (a) (2, 330°) (−2, 150°), (−2, −210°): rect. $(\sqrt{3}, -1)$ (b) $(\sqrt{3}, -135°)$,
$(-\sqrt{3}, 45°)$, $(-\sqrt{3}, -315°)$: rect. $\left(-\frac{\sqrt{6}}{2}, -\frac{\sqrt{6}}{2}\right)$

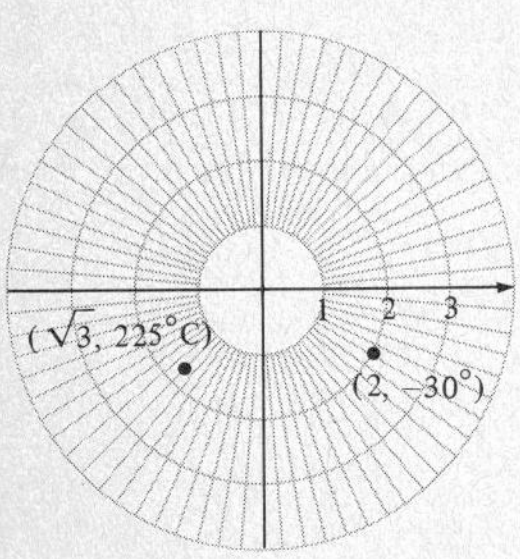

18. (a) $\left(2, \frac{\pi}{4}\right)$ (b) $\left(2, \frac{5\pi}{6}\right)$ 19. (a) $x = 3$ (b) $x^2 + y^2 = y - x$

20. (a)

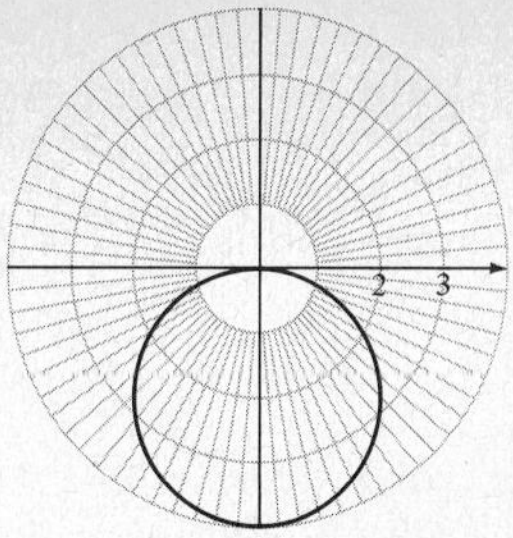

(b)

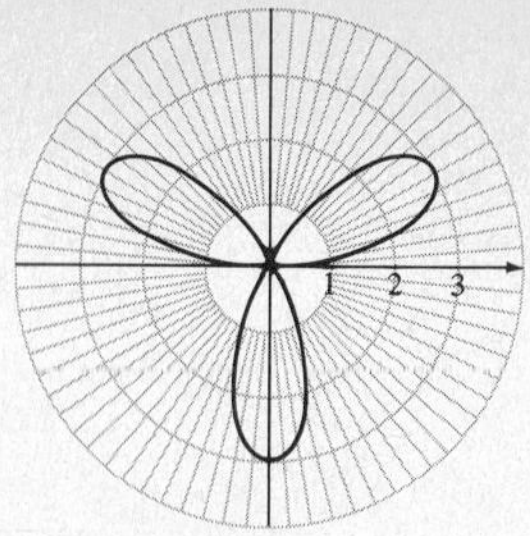

(c)

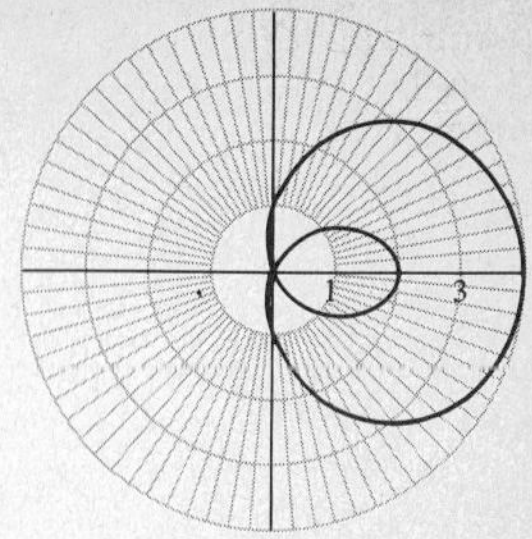

(d)

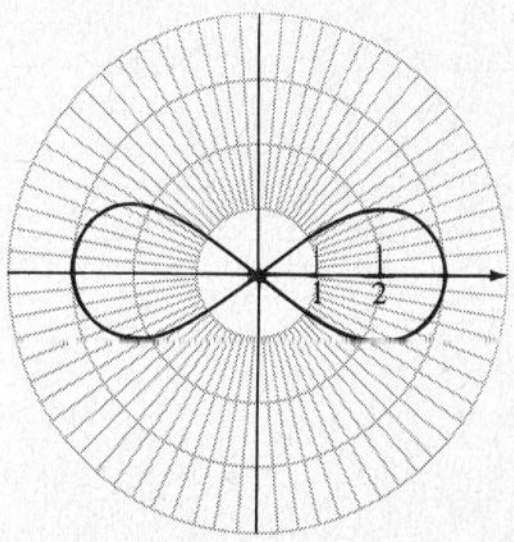

(e)

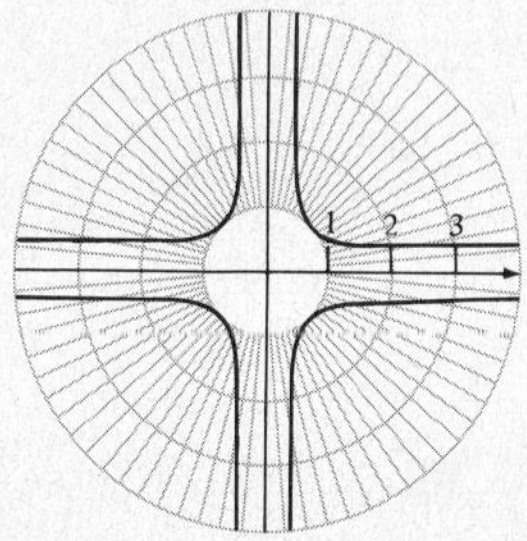

(f)

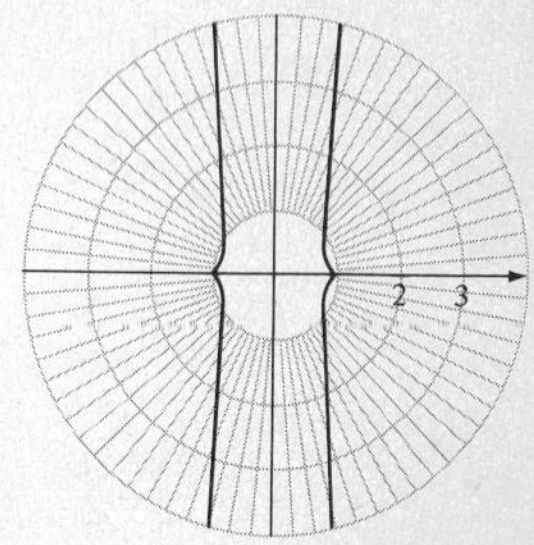

APPENDIX A. Functions and Graphs

Exercise Set A.1, page 281

1-2.

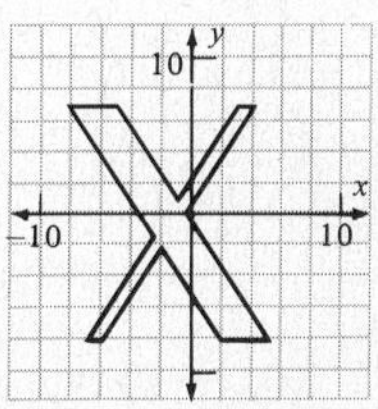

3.

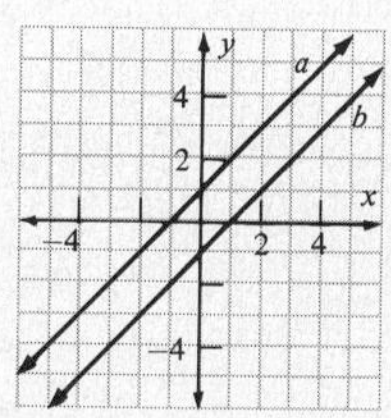

5.

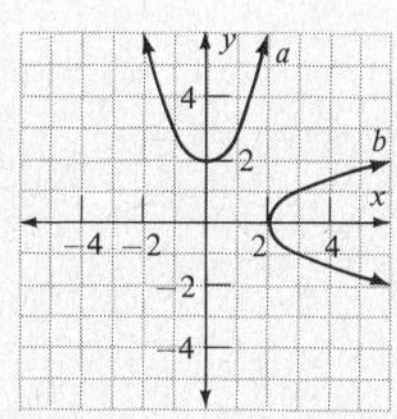

7.

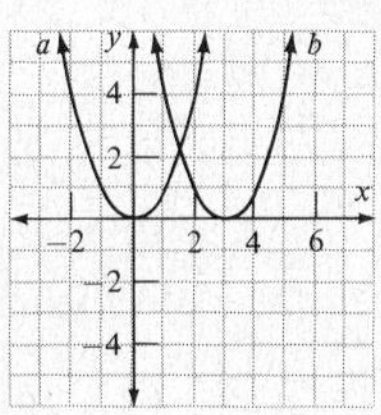

9.

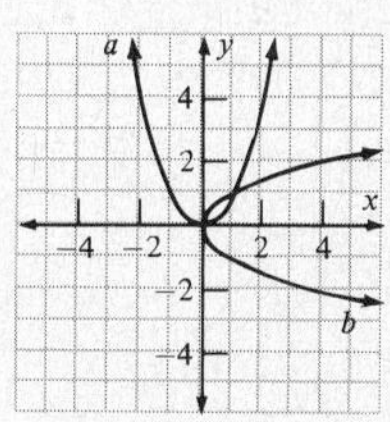

11.

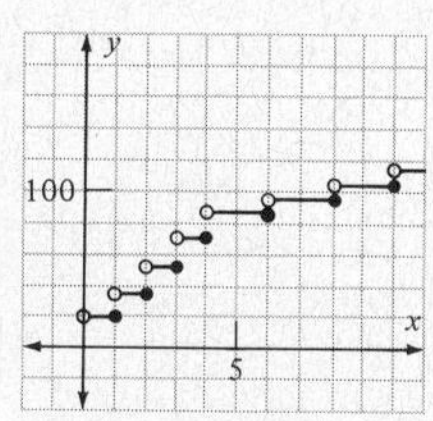

13.

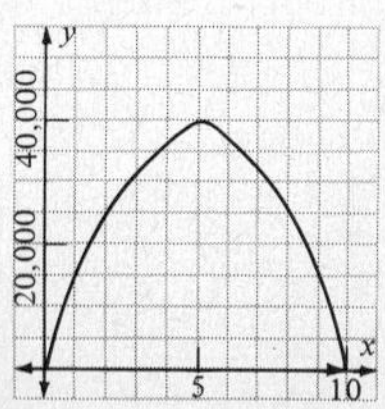

15.

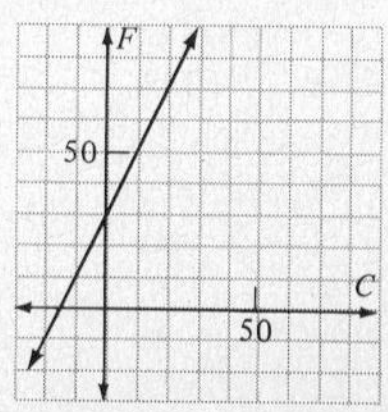

17.

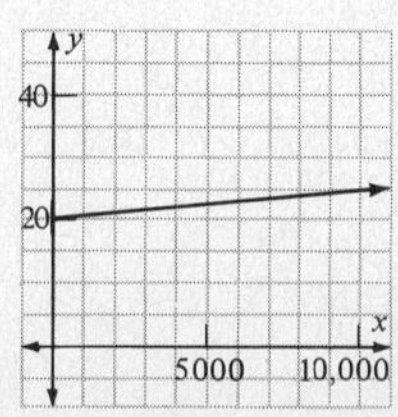

Exercise Set A.2, page 286

1. -1 3. -5 5. -15 7. -7 9. -1 11. (a) 0.153 (b) 0.076 (c) 0.057 (d) 0.55 13. (a) 13.4 (b) -0.39 (c) 1.02 (d) 0.025 15. (a) 0.119 (b) The change in beef prices from 1960 to 1975. 17. (a) 122, 127.5, 136 (b) 135 19. (a) 26.925, 62.625, 142.8, 254.925, 399 (b) 113 ft

Exercise Set A.3, page 294

1. (a) Function, inverse is a function (b) Function, inverse is not a function (c) Function, inverse is not a function (d) Not a function 3. (a) $y = \dfrac{x-1}{3}$ (b) Does not exist

5. (a)

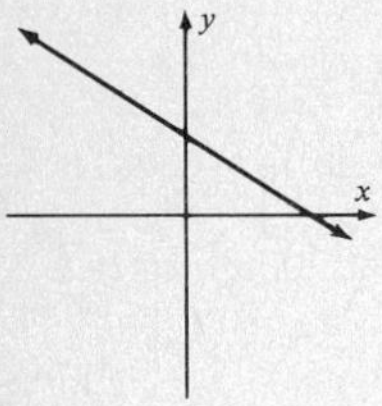

(b)

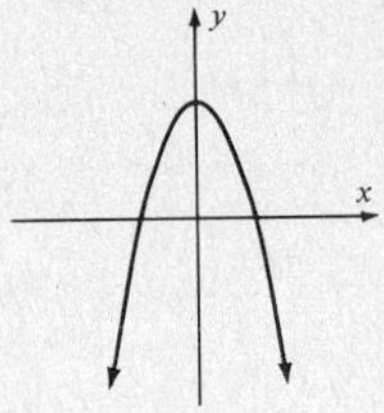

7. (a)

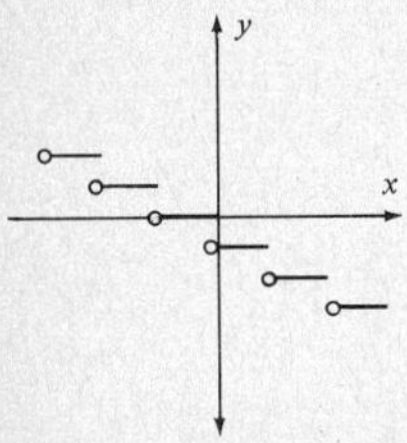

(b)

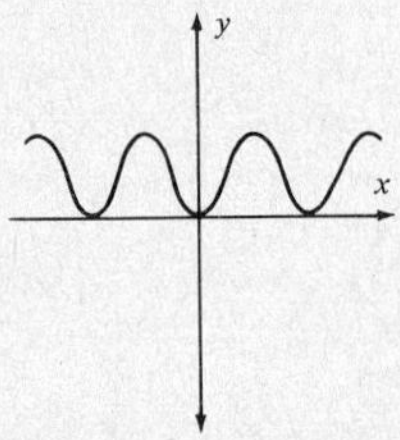

9. (a) $y = x^2 + 8x + 9$ (b) $y = -(x-2)^2$

11. (a)

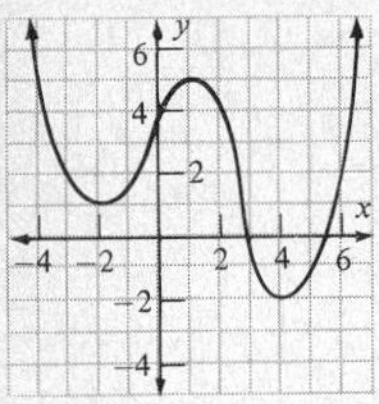

(b)

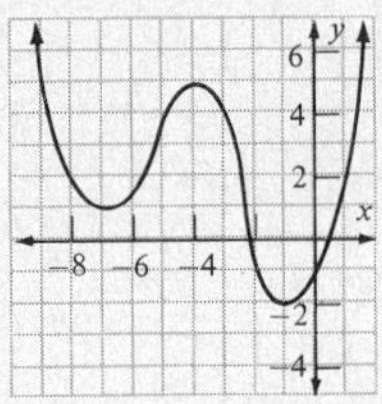

13. (a)

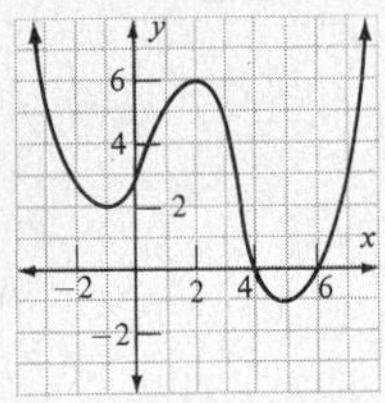

(b)

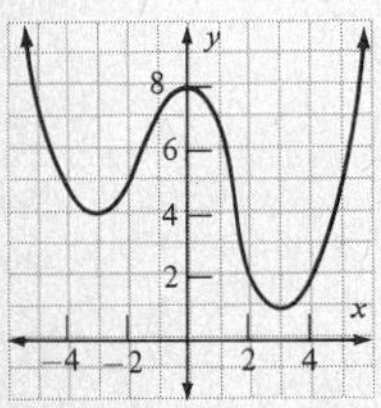

15. (a)

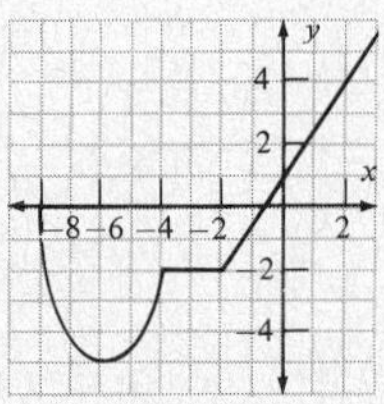

(b)

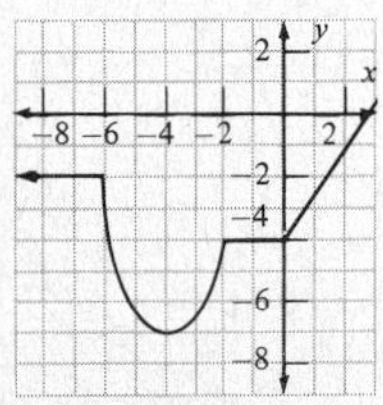

17. $f^{-1}(x) = \dfrac{65 - x}{0.0055}$ 19. $p^{-1}(a) = 2(a - 110)$

Exercise Set A.4, page 303

1. 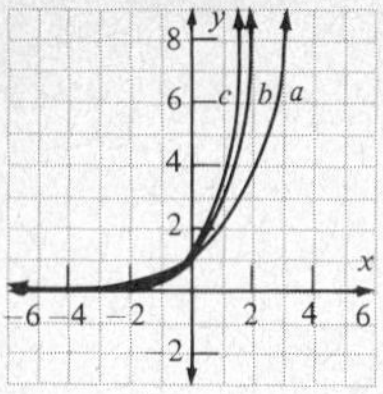

3. 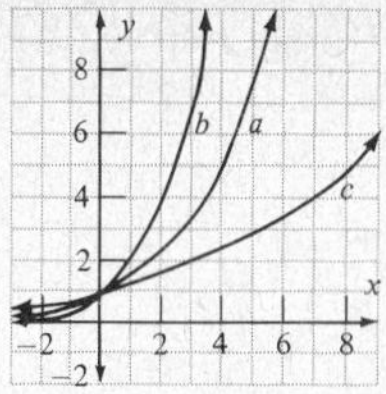

5. 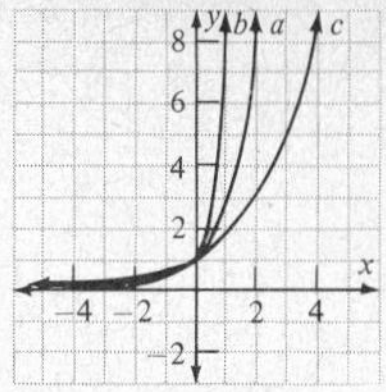

7. 9.95 mg remains (9.9478752) 9. \$2742.24 balance (2742.2353)

11. 1,400,000 in 1975 (1,398,259.3) 13. 6.5 lb/in.^{2} (6.5492301)

15. 372 times (371.57669) 17. 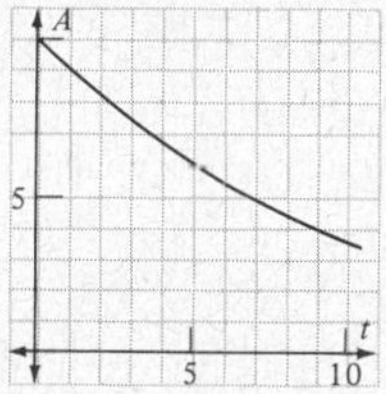

19. 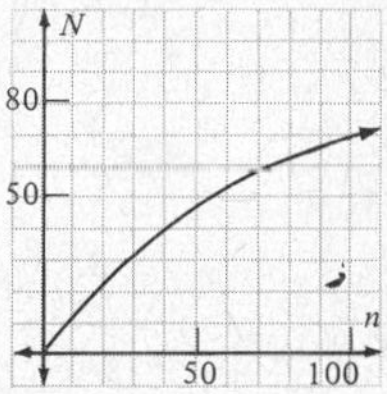

Exercise Set A.5, page 311

1. (a) $3 = \log_5 125$ (b) $5 = \log_3 243$ 3. (a) $2 = \log 100$ (b) $-2 = \log 0.01$

7. (a) $\frac{3}{2} = \log_{16} 64$ (b) $\frac{4}{3} = \log_{27} 81$ 7. (a) $5^4 = 625$ (b) $4^5 = 1024$ 9. (a) $10^3 = 1000$ (b) $10^{-3} = 0.001$ 11. (a) $81^{3/4} = 27$ (b) $8^{-2/3} = \frac{1}{4}$ 13. (a) 0.67394200 (b) 1.5518088

15. (a) 0.0047931240 (b) 7.5257499 17. 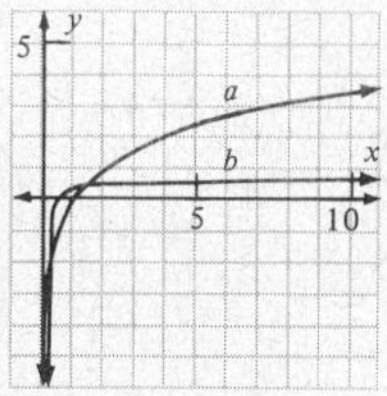

19. 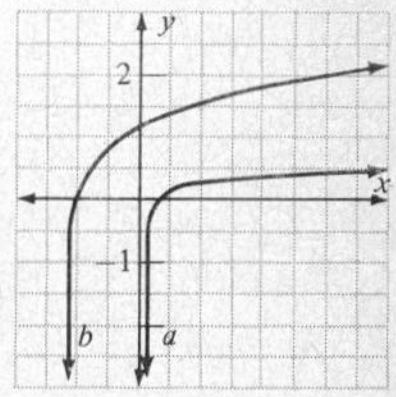

21. (a) $a = \dfrac{1}{0.21} \ln \dfrac{14.7}{P}$ (b) $p = \log \dfrac{3.21}{H}$ 23. (a) $t = 10 \ln \dfrac{A_0}{A}$ (b) $k = \dfrac{1}{t} \ln \dfrac{S_0}{S}$

25. 44.4 years (44.432512) 27. 1.75 percent (.017466807)

29. (a) 3.98×10^{-13} ($3.9810717 \times 10^{-13}$) (b) 1.12×10^{-12} ($1.1220185 \times 10^{-12}$) (c) 2.00×10^{-7} (1.9952623×10^{-7}) (d) 3.98×10^{-8} (3.9810717×10^{-8}) (e) 6.31×10^{-6} (6.3095735×10^{-6})

Exercise Set A.6, page 320

1. $\log_b 2 + \log_b \pi + \log_b r$ 3. $\log 4 - \log 3 + \log \pi + 3 \log r$ 5. $\frac{1}{3}(\ln c - 2 \ln s)$

7. $2 \log_b s - \log_b r - \frac{1}{2} \log_b t$ 9. $\ln S_0 - kt$ 11. $\log_b A^2B$

13. $\log_b A_0(\frac{1}{2})^{-t/h}$ or $\log_b A_0 \cdot 2^{t/h}$ 15. $\log \dfrac{M\sqrt{3}}{N^3}$ 17. $\ln \dfrac{\sqrt{2e}}{\pi}$ 19. $\log 10P$

21. $x = 3$ 23. $y = \frac{4}{3}$ 25. $b = 4$ 27. $n = -4$ 29. $x = \frac{5}{3}$ 31. $x = 4$

33. $N = \frac{1}{49}$ 35. $b = 4$ 37. $x = 3$ 39. $x = \frac{9}{8}$ 41. $N = 80(1 - e^{-0.018n})$

43. $P = 14.7e^{-0.21a}$ 45. $x = 30$ 47. $x = 1$, (-10 extraneous)

49. $\dfrac{25 - x}{10^2} = \dfrac{1}{x}$: $x = 5, 20$ 51. $N = 2.04 \log e \approx 0.88596074$

53. $x = \dfrac{\ln 6.52}{\ln 3.78} = \dfrac{\log 6.52}{\log 3.78} \approx 1.4099726$ 55. $n = \dfrac{\log \pi}{\log 0.59} = \dfrac{\ln \pi}{\ln 0.59} \approx -2.1695581$

57. $y = \dfrac{\log 11}{\log 2} = \dfrac{\ln 11}{\ln 2} \approx 3.4594316$ 59. $L = \dfrac{\log 1454}{\log 3.12} = \dfrac{\ln 1454}{\ln 3.12} \approx 6.3999494$

index

83 84 85 9 8 7 6 5 4 3 2 1